# NOT ON QUEEN VICTORIA'S BIRTHDAY

# NOT ON QUEEN VICTORIA'S BIRTHDAY

## the story of the Rio Tinto mines

DAVID AVERY

COLLINS
*St James's Place, London, 1974*

William Collins Sons & Co Ltd
London · Glasgow · Sydney · Auckland
Toronto · Johannesburg

First published 1974

ISBN 0 00 211334 1

Set in Dante
Printed in Great Britain by
Westerham Press, Westerham, Kent
Designed by Melvyn Gill Design Associates

# *Contents*

# PREFACE

The writing of this book has been a revelation in friendship. In the course of my research I have received generous assistance from many people in Britain and Spain, most of them previously unknown to me, whose enthusiasm and kindness have been overwhelming. It must not be assumed, of course, that they necessarily agree with the opinions or interpretations contained in this history. My views and my prejudices are mine alone.

My thanks are due first of all to the board of the Rio Tinto-Zinc Corporation (RTZ) and the *consejo* of the Compañía Española de Minas de Río Tinto (now Unión Explosivos Río Tinto, S.A.) for providing the facilities which have enabled me to study the records of the former Rio Tinto Company in both countries. Sir Val Duncan, OBE, chairman and chief executive of RTZ and the Right Honourable the Lord Byers, OBE, until recently an executive director of RTZ (both of whom were directors of the old Rio Tinto Company), and the Excmo D. Antonio de Torres Espinosa, managing director of the Spanish company, in particular have all taken a considerable interest in the work, provided me with much encouragement and arranged help and hospitality for me in Madrid, Rio Tinto, Huelva and Seville. In addition, Mr Gerald Coke, CBE, formerly chairman of the Rio Tinto Company and now a director of RTZ, Mr Roy Wright, CBE, deputy chairman and deputy chief-executive of RTZ, Sir Mark Turner, deputy chairman of RTZ, and Mr Hugh Saunders, OBE, TD, formerly deputy managing director of RTZ, have all found time to read and comment on various portions of this work. D. J-M. Naharro, a director and *secretario-general* of Unión Explosivos Río Tinto, made the facilities of his company freely available to me when in Spain, and Mr R. J. Altham, director of both the British and the Spanish companies, acted as my link between the two concerns.

Mr P. H. Truscott, formerly secretary of RTZ and a member of the *consejo* of the former Compañía Española de Minas de Río Tinto, has assisted me at every stage of my enquiry. I have constantly profited from his knowledge of Rio Tinto and from the stimulus of my conversations with him. Mr Leonard U. Salkield, OBE, formerly head of the smelting department at the Rio Tinto mines and metallurgical consultant to Río Tinto-Patiño S.A., has been most generous in placing at my disposal the product of his own considerable researches into the history of the mines, in acting as my instructor and guide during my exploration of the Rio Tinto area and in advising me on numerous technical points. Both have read and reread every word of the manuscript, providing me with detailed comments on every page and drawing my attention to source materials in Spain and in England which otherwise I might have overlooked. It is impossible to emphasize sufficiently the debt which I owe them.

Dr Antonio Blanco and Dr J-M. Luzón of Seville university have kindly helped me concerning the pre-historic and Roman archaeology of Rio Tinto. Mr Robert Wilton Brown, former head of Rio Tinto Company's labour office at the mines, told me much of industrial relations there and ferreted out all sorts of material; the late Ilsmo D. José Sánchez-Mora, former head of the company's legal department in Huelva, who had such an intimate bond with the old British company through his own and through his father's many years of service with it, gave me a complete week of his time just two months before his death.

I am indebted, too, to a number of other former employees and widows of former employees of the old company who have had the patience to answer my questions: in England, Mr Max Baker and Mr W. Liddell, until recently directors of RTZ Services; Mr Eddie Bourne, now a consultant of RTZ, and Mr Phil Newland, formerly of the sales department; Mr E. Brown assistant secretary of RTZ; Mrs Dorothy Herald (formerly Miss Coughtrey) Lord Byers's secretary; and Mrs B. Gough, Mrs E. Howell and Mrs N. Hill (who also made available for my inspection a number of family documents); in Spain, Mr Joe Jowers, until recently *secretario-general* at the mines of the former Compañía Española de Minas de Rió Tinto, who also (with Mrs Jowers) has provided me with the most congenial hospitality on many occasions; the late D. Juan Wilkins Rincón, formerly of the mines' accounts department; Mr Eddie Rowe, formerly of the heaps department; D. Francisco Martín-Pernil, until recently the company's agent in Seville for almost five decades; and D. José Diaz, formerly head of the Rió Tinto-Patiño company's sulphuric acid plant.

I am also grateful to D. Arcadio Calvo recently manager at the mines and D. Carlos Marti the present manager who ensured that my visits there were pleasant as well as useful; to Mrs Erica Adam and her son Mr John Adam for their kindness in entertaining me at their lovely home of San Antonio de Clavinque near Seville and for showing me mementos of the late Walter Browning; to my great friend D. Antonio Sánchez-Gijón y Martinez in whose home in Madrid I have worked so agreeably on a number of occasions; to D. Oscar Weinberg and D. Enrique Lucena of Madrid, who took me to various places I needed to see and acted as my interpreters when my Spanish proved insufficient; to Señorita Dolores Vazquez of the mines' public relations section who did the same for me at Rio Tinto; to the Revd Fr Gil Varon, SJ, former head of the Rio Tinto technical school, for information about his early days at Rio Tinto; to Mr S. R. Brady of Queensland, Australia, for details of Huelva at the beginning of this century; to D. Benito Acebal, formerly of the Madrid office, who helped me in my search for documents; to Mr John Hunt, *secretario-general* at the mines of Río Tinto-Patiño S.A., and to Mrs Maria-Teresa Hunt, for their hospitality; to D. Francisco Gracia, formerly assistant secretary of Unión Explosivos Río Tinto and now general manager of Editorial Labor, S.A. of Barcelona, who for the sake of friendship and scholarship has unstintingly applied himself to smoothing many paths on my behalf in Spain; and to Dr Theo Vega, formerly head of the Rio Tinto medical service who as a medical historian was able to give me considerable assistance with that specialist subject, and took care of my health when I was at the mines.

I have received much help and advice in my search for source materials and background information from the officials of the London Library, the National Register of Archives and the Business Archives Council in London, and the Biblioteca Naciónal and the Museo Arqueológico Nacional in Madrid.

D. Santiago de Arechavaleta y Uriguen of the RTZ language training unit helped me to overcome many problems where the Spanish source materials used archaic words; and Miss Frances Vale, the head of the RTZ public relations department, and Mr B. Brown formerly of the publications section, and his successor Mr M. Williams, have devoted much time and trouble arranging for the design and publication of this book and seeing it into print. It has been a great pleasure to be able to work in close co-operation with Mr Melvyn Gill, the designer of the book, and with Mr Laurence Cotterell who acted as literary adviser.

The manuscript was read for style by two great friends, Mr Laurie Wellsted and Mr David Pam. As a result of their advice the book is more readable than it would otherwise have been.

More is owed than I am able to say to the members of the RTZ history section. They have been involved in every stage of this book's development and assisted in innumerable ways. Miss Pennie Mason helped me with much research during a long period when I lay in hospital. Mrs Margaret Bailey, too, has assisted on certain aspects of the research and is responsible for making the index. She has also, during my spells of ill health, including a second period of hospitalization, carried the burdens of administration and acted as intermediary between me and many of those concerned with the production of the book. Miss Jane Denison (now Mrs Heukalon), Mrs Jacki Logan, Miss Barbara Reed, Miss Kerry Hudleston, Miss Ruth Courtland and Señorita Marisa Bonaño y Morueta typed the manuscript in its various drafts, while Miss Hudleston also constructed the glossary and Señorita Bonaño helped with some of the translations from the Spanish.

Mr E. J. Cocks and Mrs Charlotte Willson-Pepper found time from their own researches into various aspects of the mining history of Australia and the United States to give me their opinion on certain points. Mr Charles Harvey, of the University of Bristol Department of Economic History, during his research into the Rio Tinto Company archives, made many valuable comments and drew my attention to certain aspects of the company's activities in the early part of the twentieth century which led me to reconsider and amend some sections of my manuscript; while Dr L. Pratt of the University of Alberta generously made available to me the results of some of his research into Anglo-Spanish relations between the two world wars.

Finally, a few words about this book itself. It is not an 'official' history and does not try to represent the views of any institution or of any person save the author. Although certain directors and employees of the Rio Tinto-Zinc Corporation have been amongst those who have given me considerable help, their assistance was concerned entirely with providing me with free and complete access to source materials and advising on the technical aspects of printing, design and illustration. The only occasions when my treatment of the subject has been mentioned was when Lord Byers made plain that the company's attitude was one of complete acceptance of the fact that the author alone could decide what went into the book, and when Sir Val Duncan expressed the hope that it would 'tell the full story, warts and all'.

In fact to tell the complete story in every detail would be impossible: for each chapter it would have been easy to write a small book. My problem, one which faces all historians, was how to organize all the material; how to keep a sense of proportion in deciding what to include and what not. Inevitably, those familiar with Rio Tinto and its history will be able to point to parts of the text and complain that this or that detail has been omitted. But my own concern, of necessity, has been not that this book should be all-inclusive but that it should give a representative and significant picture of the subject.

The book could of course have been treated in a variety of ways: for example, as an economic history or as a history of mining engineering and technology – though such a book could not have been written by me. It is none of these subjects, although these matters have had to be referred to in places. My own inclination is for things political and social; my understanding of statistics and machines less than my comprehension of people. And so I have written about the men and women of Rio Tinto: their dealings with each other, their attitudes and beliefs, and their attempts to preserve or to transform (largely the latter) the society in which they lived.

If this book has a principal theme, then it is the relationship between the small British colony at Rio Tinto and the mass of Spanish workers and their families surrounding the British enclave. The attempts of both communities to understand each other and to preserve their separate identities, the fears and apprehensions which both sides experienced, the occasional eruptions into violence, is the stuff of which this story is largely made. I personally find the people of the two communities largely admirable, but there was foolishness (and often blindness) on both parts. The lack of comprehension of one for the other, which dominated so much of the history of the British and Spaniards at Rio Tinto, is analyzed at length in chapter 12, 'Not on Queen Victoria's birthday'. Hence I have given that title to the book as a whole. That the mystification and misunderstanding was largely overcome by the year 1954, when Rio Tinto passed from British control, was a fortunate outcome. (It is a lucky historian who is able to give his book a happy ending.)

This book has been written largely from research in the archives of the former Rio Tinto Company, and I am fortunate that the source materials have been so abundantly available and so freely accessible. In my writing I have formed my own interpretations of facts, and I am aware that the wide diversity of views that are held in connexion with many aspects of Spanish history may mean that not all my conclusions will be acceptable to every reader. However, there can be no doubt that it is time for an appraisal to be made of the history of Rio Tinto. And at the very least, this book should help draw attention to the rich collections of documents to be found in the keeping of business houses and awaiting the attention of historians. There is such a profusion of manuscript material in their minutes, their letter-books, their files and their ledgers, and of printed information in their prospectuses and annual reports, from which our conception and portrayal of the past may be deepened and enriched.

I

# *Prologue*

At the *gare d'*Austerlitz in Paris, on a February morning in 1873, a consignment of metal-bound wooden chests was stowed on board a train under the watchful eye of a quiet Scotsman. So began a carefully planned and secret journey in the course of which the strong-boxes were taken, at times by train and at other times in bullock-drawn carts, by indirect routes through France and across the Pyrenees into Spain. Their destination was Madrid; their taciturn custodian, William Macfarlane, an employee of the London banking-house of Matheson and Company; their contents, £422,680 in gold specie.

The object of the journey was to take the treasure undetected through the lines of the warring factions engaged in civil conflict, and to deliver it into the hands of the financially hard-pressed government of the new Spanish republic in return for a deed of sale for the celebrated copper mines of Rio Tinto. The gold coins in the boxes represented the first instalment on the sum of £3,500,000 for which the state had contracted to sell one of the country's major assets to a consortium of European businessmen.

The start of that year had seen the government of King Amadeo of Spain in a state of near collapse and the country in complete turmoil. The despatch of a large part of the army to Cuba to suppress a colonial uprising had given the enemies of the régime the chance to rise in revolt themselves. The government found itself unable to exercise control over large areas of Spain and in many of the provinces law and order gave way to the rule of the gun or of the mob. To many observers it seemed that the rebellions of the three factions most hostile to the king (the socialists, the anarchists and the supporters of the Carlist claimant to the throne) signalled the start of the country's disintegration.

There seemed little to be hoped for from the king's ministers, divided as they were by jealousy and party rivalry and preoccupied by their continual squabbles with Amadeo himself. By 11 February the king was too weary to continue and on that day abdicated his precarious throne. Most of his ministers were reluctant to surrender the burdens and the pleasures of high office, and decided to ignore the provisions of the constitution which required the summoning of a constituent assembly at such a time. Instead they proclaimed the first

Spanish republic with Estanislao Figueras as president.

Only three days later, on 14 February, Figueras issued a decree declaring the intention of his government to sell the royal mines of Rio Tinto in the province of Huelva to the group of international businessmen, headed by the London banker Hugh Matheson, which had offered 92,756,592 *pesetas*, equivalent to £3,500,000 sterling, for the freehold of the property. At the same time the finance minister, José Echegaray, issued the draft of a law to authorize the sale and to give the new owners the right to build and to operate the railway which they would need if they were ever to work the mines effectively.

For the republican government the sale was a matter of utmost urgency, for it desperately needed money to suppress its opponents and to re-establish control over the countryside. The decision to sell, however, had been taken long before. As early as June 1869 it had been resolved to dispose of the mines which had been a constant drain on government revenue over many years due to their lack of profitability and to civil service mismanagement. However, the reputation of the Rio Tinto mines as one of Europe's most onerous and expensive burdens proved a serious obstacle to selling them, and the feeble state of the government led most onlookers to the opinion that any purchaser would be foolhardy in the extreme.

The collapse of the monarchy while Matheson was still engaged in his negotiations only confirmed in the minds of most European investors the dangers of investment in Spain; the announcement of the purchase by Matheson and Company was condemned by most as an ill-considered gamble. To Hugh Matheson and his friends, however, it appeared neither foolish nor unnecessarily risky.

Matheson had been persuaded to organize a bid by Heinrich (otherwise Enrique) Doetsch and Wilhelm Sundheim, two Germans with mining and trading interests in the province of Huelva who knew the mines well. During a visit to London they were able to persuade Matheson that the mines had been a failure under the Spanish crown only because the government of the country had been unable to provide the necessary capital for developing them and had frustrated all attempts by its own men on the spot to make the mines' administration more flexible. Doetsch and Sundheim were convinced that a large injection of cash and the employment of skilled mining engineers, coupled with the construction of a railway to reduce transportation costs, would enable the mines to become highly profitable. Unable to organize and mount a bid themselves they were able to convince Matheson that he should take the lead in assembling the financial and technical resources as a prelude to the negotiations with the Spanish government.

Doetsch and Sundheim were also shrewd judges of Spanish politics. Despite the public disorder prevailing in Spain they were convinced that a strong man (or group of strong men) would emerge in the normal Spanish political tradition and put an end, for a few years at least, to the chaos. Their assessment of the situation was accepted by Matheson who, ignoring the fears for Spain's future expressed by so many of his contemporaries, went on to complete the negotiations.

Hugh Matheson was one of those rare men who see possibilities in situations where most people see only dangers. The instability of the Spanish government and its desperation for money was for him an opportunity to get the best possible terms including a reduction in price from £4,160,000, powers for the compulsory acquisition of land for the building of the essential railway to link the mines with the sea, and (what made his purchase unique in mining history) not only the mineral rights but the absolute freehold in perpetuity of all the land and buildings owned by the Spanish government within the Rio Tinto boundaries.

As soon as Matheson and his consortium acquired the ownership, they set in motion the procedures for forming a limited company under English law to operate the mines and to obtain public participation as a way of getting some of the cash necessary for transforming the run-down workings into the most profitable mining enterprise in Europe. On 29 March 1873 the Rio Tinto Company was registered in London and the syndicate assigned its rights and obligations to the new concern, taking £1 million in debentures and shares in return for the deposit it had paid of £422,000.

Matheson became chairman of the new company and until his death twenty-five years later directed its fortunes so successfully that he made the company as much a legend with the share-buying public as the mines themselves already were with mining men. The company he created achieved fame, too, in other ways for under its enthusiastic chairman it provided its Spanish workmen and their families with a range of social services without parallel in previous Spanish history: a medical service which won unstinted admiration; housing which reached high standards for the period; a system of elementary education which attracted some of the country's best teachers; and recreational facilities which represented a considerable improvement on many of the taverns and the brothels which they were designed to replace. (It was the staff of the Rio Tinto Company who introduced football to Spain and helped form the first Spanish football club in Huelva.) These are the things for which undoubtedly Matheson would have liked to have been remembered.

But in the event, the memory of Matheson's good intentions has been overlaid by the bitterness and rancour in which his company found itself enmeshed. For though the first chairman was benevolent he was also a despot, and the company he fashioned inherited from him both characteristics. Because it became the largest single employer of labour in Spain, and the most important contributor to the economy of Huelva province, his successors as chairman, together with the company's senior staff at the mines, exercised for many years political and economic powers which today seem almost unbelievable. At the time they led the company into frequent conflicts with the government and its own workpeople, winning for it unwanted newspaper headlines throughout Europe. It is a sad fact that most Spaniards who know something of the history of Rio Tinto seem to remember first of all the strikes, the violence and the bloodshed, and normally lay the blame for these at the door of the British company.

In 1954, eighty years after Matheson's purchase, the Rio Tinto Company sold the mines to the newly formed Compañía Española de Minas de Río Tinto, retaining only a one-third interest. Thus the effective ownership and control reverted again to Spaniards and the British company, though retaining close links with Spain, turned its eyes and energies to other parts of the world. Great changes have occurred at Rio Tinto even since 1954, but they are beyond the scope of this book and are best left for Spanish historians to describe.

To those who visit Rio Tinto, most things about it seem remarkable. It lies buried deep in the hills of the *sierra de* Aracena, some thirty miles inland from the gulf of Cadiz as the crow flies, and forty-two miles by modern roads from the ancient city of Seville. It is a labyrinth of strangely sculptured and brilliantly coloured interlocking valleys, where the hands and machines of men have torn the hillsides apart into gigantic fissures and carved them into a series of steeply descending angular terraces. For the most part the cliffs are coloured a dusty-pink which changes, with the decline of the strong sun and the advent of the evening light, into a rich mulberry-red. Embedded in the rock face and relieving the dominant pink with wedges of deep lavender and yellowy-white, speckled leaf-green and gold, are great outcrops of slate and porphyry. Competing with the hills for eminence are the high escarpments of black and purple slag and tall, levelled banks of leached pyrites, silver-grey.

Down the hillsides run translucent rivulets stained bright green by the metal in their waters; and in the valley bottoms lie lakes of blood-red,* whose runlets trickle away to form a slow, sinuous river: the *río tinto*. The name, inviting comparison as it does with the common wine

*The colour-change from green to red occurs as the iron salts in the water are oxidized by contact with the air.

of the country, the *vino tinto*, is apt. For the river Tinto, which derives its name from the staining of its water by the minerals in the earth, and now has given that name back to the mines whose ores are the cause of its discoloration, is the deepest of all the reds to be seen in this landscape, the purple-red of wine.

Spread along the valleys and over the slopes of the hills are the workshops and railway yards of the mines themselves, and the crowded miners' villages of white terraced cottages with scarlet pantile roofs. There are the shells of other buildings, too; some half-buried in slag, most falling into ruins and all blackened with age: among them the now smokeless chimney of an abandoned smelter; a broken stone tower set on a solitary hillside, recognizable as a Cornish pump-house (a strange architectural intruder into an Andalusian landscape); gaunt unroofed walls poised precariously at the edge of an immense quarry, remnants of an eighteenth-century town, half of which has been swallowed by the great pit.

Dating further back in time than these, and old even when Christopher Columbus sailed from the lower reaches of the same river Tinto on his epoch-making voyage of discovery to the new world, are reminders of other men who once toiled underground in these mountain valleys: the ruins of a warehouse, solitary upright columns and the mutilated gravestones of a necropolis; all fashioned when imperial Rome held sway in the province of Iberia and her emissaries ruled here in the city which they called Urium.

More ancient still, for they were already eight hundred or more years old when the Caesars were upstarts, are the foundations of the houses of the unknown men who built a town here on the hill called *cerro* Salomón, from which they sent their precious silver to embellish and adorn the cities of the prehistoric world. The people of the district will tell the interested visitor that these mines of Rio Tinto were those of the celebrated King Solomon who ruled the tribes of Israel in the tenth century B.C. and gave his name to *cerro* Salomón and the towns of Zalamea la Vieja (now known as Nerva) and Zalamea la Real.*

Less obvious, perhaps, but also there to be seen by those who have the eyes to notice such things, is the way in which the past has become part of the present at Rio Tinto. The observant stranger will discern in the oddly neat street-plan of the nineteenth and early twentieth-century villages the attempts of British engineers to impose their own ideas of orderliness in what was to them an alien and often disorderly land. He will note that the laden railway waggons are hauled over tracks laid on ballast platforms of dark-coloured slag the sharp-edged pieces of which, with their bubbled and whirled surfaces, are the refuse of prehistoric metal workers. He may observe how the modern

*See appendix I for the archaeology and folklore of Rio Tinto.

*View of the north lode open-cast pit from cerro Salomón.*

excavating machinery, tearing away the rock face in the open-cast pits, lays bare ancient galleries and shafts, long-collapsed but still providing the hazards of sudden rock falls. In the haunting music known as the *esquila*, unique to Rio Tinto, he will hear preserved the melodies of an ancient vanished culture, possibly middle-eastern in origin. In the club at Bella Vista, still known as *el club inglés* although the management of the mines was transferred to Spaniards in 1954, he will observe the Spanish executives reading the daily weather chart with its temperatures given in the Fahrenheit scale, surrounded by mementos of former cricket matches and English prints on the walls. And noticing such things he will conclude that Rio Tinto is an amalgam to which men and women from many parts of the world have contributed, a society which bears the marks of a number of nationalities and which is demonstrably unique.

The Rio Tinto mines are probably the best known in the world. They are also, with the exception of those in Cyprus, the oldest still being worked. Every mining man knows of them, and will tell you that their eight square miles are part of the most massive deposits in the world and have provided the largest single source of European copper in ancient and modern times. For the ore bodies at Rio Tinto contain the largest mass of cupriferous pyrite known to man together with some gold and silver, sulphur and iron. Herein lies the explanation of the red soil that dominates the landscape, for it results from oxidization of the metal-bearing rocks over many millions of years.

And herein, too, lies the explanation why so many people from widely different nations have travelled to Rio Tinto to live in a succession of mining communities, often enduring great hardship, attempting to extract the minerals from the depths; people whose labours have helped to sustain first the ancient Mediterranean culture, and then the Roman and later the British empires at the time of their power and glory. The British have impressed their mark on Rio Tinto as clearly as any of their predecessors; perhaps more so. The company which Hugh Matheson founded achieved successes unknown at Rio Tinto since the departure of the Romans in the fifth century; and the technical and financial achievements of its employees, attained in the face of many difficulties (the chief of which were political), are something of which they had every right to be proud. Black marks were earned, and deservedly so, in some respects, but the story of the British at Rio Tinto is worth the telling and most will find it worthy of admiration.

The difficulties which these men overcame, their successes and their failures, will become more apparent if placed in the context of the chain of events that led up to the sale of the mines to the Matheson consortium. Only by doing so is it possible to provide that historical perspective without which there can be no proper understanding of the nature of their mistakes and of their triumphs.

# Part One:

# AND ALL'S TO DO AGAIN

*The cerro Salomón at dawn.*

# 2 *The hill of Solomon*

## The rediscovery of the Rio Tinto mines, 1556

In the dry and dusty month of June 1556 a small band of horsemen rode through the Andalusian hills searching, on the orders of King Philip the second, for lost metal mines. Led by a guide from Zalamea la Real they were on their way to the small and remote hill-village of Zalamea la Vieja to investigate rumours of disused workings in that area.

In charge of the party was Francisco de Mendoza, a member of the council of finance, who had been instructed to inspect and report on a number of old Roman mines found in Seville province between 1551 and 1555, including the rich silver and lead mines of Guadalcanal, and then to look for others. The search was a direct result of the Spanish crown's shortage of money; for despite the vast inflow of wealth into Spain from the newly discovered American continent the Spanish government was in a state of constant financial crisis.

For many years King Charles the first of Spain had also been the Emperor Charles the fifth of the holy Roman empire and in that capacity had been involved in numerous intrigues and wars which had forced him time and again to raise loans at ruinous rates of interest from his German and Genoese bankers. Weary with the fighting, and the continual struggle to raise money to pay his mercenary armies and to meet the expenses of his court with its great swarm of bureaucrats and hangers-on, the emperor had decided to abdicate.

Early in 1556 his Spanish kingdom went to his son, Philip the second (who had received the Spanish Netherlands only three months earlier), a young man of twenty-eight and the husband of the ailing forty-one year old queen of England, Mary Tudor. The new king's government was facing bankruptcy (which indeed occurred the following year, 1557, when the crown formally defaulted on its debts) and the council of finance was prepared to seize any possibility of replenishing the royal coffers. This was what lay behind Mendoza's mission.

The members of the search party had already investigated a number of ancient shafts in the districts of Aracena and Valverde del Camino, and noted a few of them for further examination, before they arrived at Zalamea la Real, a small agricultural town in the foothills of the *sierra de* Aracena. There they were told of abandoned shafts and galleries in the hills to the north and, since there was no road they

*The town of Zalamea la Real: photo of 1910.*

could follow to get there, were provided with a local man to show them the way.

Some two miles from their destination they cleared a ridge and saw ahead of them a network of valleys in which great mounds of slag stood in dark contrast against the green shrub-covered hillsides. As they rode on they recognized with mounting excitement the signs of ancient Roman occupation: carved columns, the dressed stone of tumbled walls and immense drainage adits emerging from the hill-face. They had rediscovered some of the greatest mines of the ancient world, a complex of workings which the world knows today as the mines of Rio Tinto.

Mendoza, from his brief inspection of the surface, realized that he was looking at the evidence for ancient mining carried out on a greater scale than at any of the other places he had visited. But he was in a hurry to get away and resume his duties at the council table and, since it was obvious that a prolonged investigation would be necessary, he wrote a preliminary report for the king on his journey so far and stated that though he was about to return home he would leave two of his subordinates behind at the scene of the latest discovery to make a detailed inspection.[1] These were a Madrid priest called Diego Delgado and a gentleman from Castro Nuño, Pedro de Aguilar.

The two men obtained accommodation for themselves in the village of Zalamea la Vieja, which was already being called Rio Tinto (Delgado used both names in his report*) close to a high ridge the two peaks of which dominated even that hilly landscape. Delgado and Aguilar decided to concentrate their efforts on the hill close to the village at the south-eastern end of the ridge, now known as *cerro*

*It has since undergone yet another change of name and is now the town of Nerva.

Salomón although in Delgado's reports it was left unnamed, and to ignore the peak at the north-western end (now known as *cerro* Colorado and recently revealed also as the site of mining operations in antiquity). Their examination took the whole of July and the first two weeks of August.

Delgado and Aguilar began by questioning the villagers and employing one of them to take them to a great cave 'like a church' inside the hill which the man knew as 'Salitre'. By the flames of their torches they made an examination and judged it to be 80 yards in length and 60 yards in breadth. The height they could not estimate but merely noted as considerable. Several galleries ran from the cave even deeper into the heart of the hill. They found that this large cave was entirely man-made, the walls still bearing the marks of the ancient tools that had hacked at them centuries before. Thus on their first day they found what has since proved to be the biggest of all the caves made at Rio Tinto by the silver miners of the pre-Roman period in those places where the seams of ore had been particularly wide and rich.

The two investigators found that there had been considerable rock falls and decided that it was impossible to make a detailed examination of the cave without expending more labour than they cared on clearing

*Entrance of the cueva del lago: photo circa 1880.*

away the debris. So they returned to the surface where they found a shaft which Delgado descended and from the sides of which he cut out 25 pounds of mineral for assay purposes. Later the priest hired a local labourer to do the heavy work and returned with him to the hill to dig out another load of ore. The two samples were then sent to Mendoza for testing, accompanied by Delgado's comments on the partial remains of furnaces and buildings which he had observed among the rocks and dense brushwood.

The priest was shown another of the great caves dating from ancient times, the *cueva del lago*, so called because of the lake of highly vitriolized water which had formed in it over the centuries. Here, too, was a spring which was the source of the river which the locals called the *río tinto* because of its wine-red colour due, Delgado was sure, to the spring having its origin in a body of iron and copper.

*The river Tinto.*

The priest developed a great interest in the river and spent much time talking about it to the villagers who explained to him that it was devoid of fish and undrinkable, though useful medicinally if applied sparingly to some diseases of the eyes and body such as hydatid cysts and herpes. Their claims that iron would disappear if placed in the water he tested for himself and found to be true. The water had another property which he would not entrust to paper but declared he would keep secret until he could report it personally. (One wonders what the mystery was: had he perhaps noticed that iron not only disappeared in the river's water but was also converted by the immersion into copper, and preferred not to say this in writing but to confront his superiors with the actual proof in his hands?)

He reported, too, on the thick red deposit of silt that was continuously accumulating on the river banks. This was copperas, rich in iron sulphate, which every August and September had to be collected by the people of the neighbourhood for delivery to the archbishop of Seville as a customary tribute. This commodity was used in dyeing, tanning and ink-making and presumably when it reached Seville was sold to augment the archiepiscopal revenues.*

As this inspection continued Delgado found a number of drainage adits and investigated more than fifty shafts. He must have descended them by rope for the deepest was 90 feet, too deep for a ladder to be used. The priest was clearly an observant man for he noticed which shafts had been made for ventilation purposes, which for exploration and which for mineral extraction. In one of them he found a small quantity of lead which he thought additional proof that the ancient miners had worked with the principal object of obtaining silver, and took as a corroboration of this the villagers' tales of how this hill had once yielded vast quantities of gold and silver taken as tribute by imperial Rome.[2] (Apparently their folklore did not then include stories of King Solomon; or if it did Delgado kept silent on the subject.) The priest grew more enthusiastic and more zealous in his search as his excitement grew. He had no doubt when he wrote his full report on 12 August that here was one of the greatest mines of old Iberia which could still contribute to the wealth of sixteenth-century Spain. His enthusiasm was probably not shared by Aguilar who did less and less of the prospecting work as the weeks went by, preferring to leave most of it to his more energetic companion.

No doubt both men were glad to get away from the intense August heat of Rio Tinto, though Delgado saw their departure not as an escape from an onerous task but as a necessary step in his work of re-establishing the mines. He handed his report to Francisco de Mendoza, who placed it before the king only four weeks after it was written. King

*It was still collected by the local people until the end of the nineteenth century but by then it was no longer rendered to the archbishop but sold directly by the villagers to artists' colourmen. It is now collected as ferrous sulphate from the mines' cementation tanks and sold to fertilizer companies.

Philip ordered that Delgado's ore samples should be immediately assayed and that the priest should return to Rio Tinto to collect still more. Mendoza was despatched to Flanders at this time to lead a political mission which meant that Delgado lost the ear of the influential man who, even if not his ally, was at least well-known to him. Thenceforward he had to deal directly with the more remote council of finance.

Back at Rio Tinto, Delgado collected the samples, sent them for testing and then waited. As the months went by he sent a series of letters to the council praising the potential of the mines, urging action and asking for money to enable him to start mining operations. '. . . for I am a priest and not possessed of any means.' His last letter was sent from Rio Tinto on 10 June 1557 and complained of the way in which his earlier letters had been ignored. To encourage the council to reply he enclosed three small pieces of silver which he had produced by his own efforts. But again he met with no response. He decided to make the journey to Valladolid where the king was in residence in the hope of influencing events by a personal interview. But within a few days of his arrival in August he died there with nothing accomplished.

In the margin of Delgado's last letter the king noted: 'He is dead. This to be sent to don Francisco [de Mendoza] for him to look into the matter and advise me whether the place and the mine are as stated.'[3] As a result of the royal instruction Mendoza, who had recently returned to Spain, made another journey to Rio Tinto. If he wrote a report on this visit it does not seem to have survived and the council of finance allowed the subject to lapse.

The failure of Philip the second's government to exploit the rediscovery was not necessarily so foolish as it had appeared to the dead priest. The explorations had been restricted to *cerro* Salomón and the heavily worked and largely exhausted seams of antiquity. The minerals left in such Roman workings as remained unflooded by water were probably scarcely worth the expense of excavation. True, there were other rich veins buried under the soil of *cerro* Colorado and the other hills around, but of these Delgado and his contemporaries knew nothing. All their investigations centred around *cerro* Salomón, and the possibility of finding fresh mineral veins there must have appeared fairly remote to the men who would have to authorize the expenditure. To drain the floodwaters in the deeper galleries simply to see what they contained, or in order to dig below them, would have required a heavy investment of capital which might never have been recovered. Nonetheless, Delgado's work was not completely fruitless for he had made the existence of the mines widely known and his enthusiasm posthumously inspired later adventurers to dream of re-

*King Philip the second.*

opening them; a dream that one day was to be realized.

Meanwhile the government continued with its policy of trying to encourage mining within Spain. In January 1559 all former mining concessions from the crown were revoked and new grants authorized to anyone willing to undertake mining operations. As a result there was a flurry of activity amongst would-be mine concessionaires throughout the country, though very few of the new claims which were registered were ever worked. In September 1569 a prospector called Juan de Cabrera registered with the chief office of accounts in Madrid a claim 'to an ancient cave of red mineral in the hill beneath the old castle in the district of Zalamea'. The following January he lodged additional claims with the royal officials in Guadalcanal in respect of a number of slag heaps at Rio Tinto (doubtless he was thinking of resmelting the old slag to extract any residual metal) together with three ancient shafts. A few weeks later Francisco Pérez de Canales and Diego Blanco of Valencia registered a joint interest in the cave called Salitre and all the slag heaps of Rio Tinto, and yet another prospector Juan del Valle, laid claim to an old shaft. So many registrations at the same time could not have been coincidental; there was a race of some sort. But the four competitors made no attempts to start mining seriously and their claims lapsed when they failed to apply for the appropriate warrants.[4]

All this time, despite the country's great wealth and the spectacular grandeur of Philip the second's reign, expressed in superb architecture, great literature and the glory of foreign conquest, the high cost of war gradually pauperized the Spanish people and produced no fewer than three decrees proclaiming the bankruptcy of the realm. In 1598 the death of the great king placed on the throne his son, the incompetent Philip the third (1598–1621). Despite the financial crises of his reign the dead king had struggled against inflation and economic disorder more successfully than might have been expected, but his departure from the scene was the signal for the plundering of the country's depleted resources by his son's ministers and favourites. Eventually the crown was reduced to sending officials accompanied by priests on door to door collections throughout the country in an attempt to get contributions to meet the government's day to day expenses. The deterioration continued under Philip the fourth (1621–1665) a weakling who came to the throne at the age of sixteen and whose powerful favourite the *Duque* de Olivares pillaged the treasury while the government sank deeper and deeper into debt as a result of its involvement in the thirty years' war.

Neither governments nor individuals have normally been prepared to invest large sums of money in commercial enterprises during periods of severe economic dislocation, so it is not surprising that despite the crown's attempts to encourage mining little was actually done. In 1624 Olivares established a council of mines which issued fresh regulations for registering claims in an attempt to rekindle the enthusiasm of prospectors. In the course of the next five years over 13,000 claims were registered from all over Spain, of which 8,000 related to former Roman mines, though few of the claims were ever worked. The great family of international financiers, the Fuggers, attempted to reopen the mines of Guadalcanal between 1632 and 1636 but gave up after losing vast sums of money; and where they failed few could hope to succeed. However, the new council of mines studied the reports on the Rio Tinto mines written by Delgado seventy years before and decided that serious consideration should be given to re-opening them as a crown enterprise.

Consequently Philip the fourth issued a commission to Gregorio López Madera of the council of Castile and of the council of mines requiring him to go to Rio Tinto and take with him Tomás de Cardona the master of the royal household and *fiscal** of the council of mines. There they were to inspect the hill and its shafts and caves, and also the nearby slag heaps. They were authorized to take as many labourers, assayers and smelting workers as they thought they would need to prepare reports on the quantity and quality of the mineral, the probable costs of mining it and of producing a metal suitable for minting coins, and the possibility of making from it an alloy to replace the tin used at the crown's Seville foundry for making artillery, shot and bells. López and Cardona submitted accounts showing 34,000 *maravedíses* (£119) expended on the assaying and the smelting in the course of which their men produced a white metal by treating material from the slag heaps.

The original reports seem to be no longer in existence, but the members of the council of Castile obviously discussed them for they put a series of supplementary questions in writing to Cardona about the metal made from the slag. In reply he assured them that he had discovered a completely new metal†, never previously known, quite suitable for making coinage, cannon balls and church bells. After further consideration the council decided that it could have no value and the possibility of making it or re-opening the Rio Tinto mines was dropped.[5]

During the rest of the seventeenth century a number of private individuals registered mining claims at Rio Tinto but with one exception failed to follow them up. In 1637 *Capitan* Francisco Moreno de Busto registered the cave of Murcielagos (that is, 'of the bats') and that of Salitre 'in the hill of the old castle of Solomon'.[6] Alvaro Alonso de

*The *fiscales* were men experienced in mining, whereas the majority of council members were lawyers and academics.

†This white metal is today known as speiss (although at Rio Tinto it is called *blanquillo*), an alloy of iron with a high content of arsenic or antimony. It is extremely brittle and has little, if any, use.

Garfias claimed the right in 1661 to precipitate sulphate of copper and to convert iron into copper by immersion in the river;[7] plainly he had either conducted his own experiments or received information from the local people there. In 1680 *capitan de caballería* Sebastian Vallejo was granted the right to excavate minerals and actually began the work of clearing out a Roman water adit in order to drain the mines; but a sudden rush of water drowned his workmen and he himself died shortly after.[8] In 1695 Roque de Salas y Ulloa revived the idea of using the river water for fabricating copper and was granted a licence to do so.[9] Then Juan de Guevara in 1714, followed by Nicolas Vaillant in 1719, applied for and received concessions to work the mines of Rio Tinto, Guadalcanal, Cazalla, Galaroza and Aracena,[10] though they too failed to exercise their privileges.

Thus despite all the various projects, 168 years passed from the time that Philip the second's search party found *cerro* Salomón, riddled with the shafts and galleries of the ancient miners, until the appearance in 1724 of another adventurer, a man who was prepared to take risks and back his judgements with action. But though during those 168 years all the projected enterprises came to nothing, the failure is understandable. The reason in great part lay in the remoteness and inaccessibility of Rio Tinto and the large amount of capital that would have been necessary to start production on any scale. In part, too, it lay in the nature of sixteenth and seventeenth-century Spanish entrepreneurs: aroused to enthusiasm by dreams of vast profits to be had for the taking, nearly always untrained in scientific techniques, they soon lost interest in their schemes when they encountered difficulties and abandoned them in favour of new dreams inspired by fresh mirages. Yet the projects had one effect. They served as a constant reminder of the mines' existence; and, by exciting the imagination of men with thoughts of fabulous wealth, ensured that given the right circumstances a man, or a group of men, bolder and more practical than the rest, would take up the challenge and prove that the difficulties could be surmounted.

✠

# MANIFIESTO POR D. LIEBERTO VVOLTERS VONSIOHIELM,

CAVALLERO DE NACION SUECO,

EN QUE SE HAZE PUBLICO EL Aſsiento, y contrata que ha hecho con ſu Mag. por el tiempo de treinta años, de las Minas de Oro, y Plata de Guadalcanàl, Rio Tinto, Cazalla, Aracena, y Galaroſa, en las Provincias de Andalucia, y Extremadura;

Y EL PROYECTO, Y COMPAñIA

QUE SE ESTABLECE PARA QUANTOS quiſieren entrar en ella à beneficiar ſus intereſſes, con las calidades, y condiciones que ſe expreſſan.

CON LICENCIA: En Madrid, Año de M.DCC.XXV.

¶ *Hallaràſe eſte Papel en caſa de Juan de Moya, Mercader de Libros en la Carrera de San Geronimo, junto al Marquès de Campo-Florido.*

3

# *Congenial Souls*

## The struggle to reopen the mines: 1724–58

The beginning of the eighteenth century saw Spain turned into the battlefield of Europe in a war to determine the balance of power and the succession to the Spanish throne. For eleven years it suffered the humiliation of foreign armies tramping to and fro across the country, only to find on their departure that it still had to face the horrors of civil war. With the final advent of peace in 1714 the treasury was exhausted and the organization of government in complete confusion. In the aftermath of inertia, industry and commerce were virtually at a standstill. Almost paradoxically these conditions attracted to Spain a number of foreign entrepreneurs and adventurers hoping to make their fortunes from the country's economic paralysis.

Amongst them was a Swede from Stockholm called Liebert Wolters with an interesting background. His father, Abrahm Wolters, was a German from Danzig who had entered the service of the Swedish crown as a commissary, organizing the provision of stores and food for the Swedish army and navy, and who had married Ester the daughter of Henrik Gerdtsen, formerly the commander of a Dutch fleet of twenty-four vessels, raised in 1644 to assist the Swedes to repulse Danish attacks on their shipping. In recognition of his services Gerdtsen was made a Swedish admiral and knighted, whereupon he took the name of von Sjöhjelm (meaning 'sea-helm'). Abrahm Wolters and Ester von Sjöhjelm had seven children of whom the eldest was Liebert. When his mother died on 12 October 1685 the boy was still only twenty and serving as an ensign in a Dutch regiment.

Liebert Wolters continued in the army, never marrying, and took part in the war of the Spanish succession, taking his discharge in Spain in 1713 at the age of forty-eight. At the time various attempts were being made to salvage the Spanish and French treasure fleet sunk in Vigo harbour in 1701 by a combined British and Dutch force. These salvage efforts were completely unsuccessful since the ships lay in 60 feet of water and the divers, without any diving equipment, were unable to stay long on the sea-bed. Wolters became interested in the problem and devised a leather diving-suit and helmet with a long hose attached, and reaching to the surface. This enabled him to descend to the required depth and stay there for relatively long periods.

*Title page of Wolters's company prospectus of 1725.*

He therefore asked the Spanish government for the award of a salvage contract for the sunken Vigo fleet, and in 1719 this was granted, the terms being that he paid his own expenses and received one third of the value of all goods recovered. He had some success with his operations and the French government was sufficiently interested in the suit he had invented to request a report on it. (A Spanish version of this is in the British Museum.)* Following the Spanish custom he used his mother's surname after his father's but the Spanish documents in which he occurs shows that it was quickly distorted to something which the Spaniards could write and pronounce more easily. Consequently he became known as Liebert Wolters Vonsishielm, or variants of that name.

When his Vigo contract was due for renewal in 1724 he submitted with his petition another asking for a concession to reopen and work the mines of Rio Tinto, Guadalcanal, Cazalla, Aracena and Galaroza in the regions of Andalusia and Estramadura, phrasing his request in almost exactly the same words as those used by Vaillant in his application of 1719.[1] It would seem that through the good offices of friends in the royal service he obtained access to finance council records concerning these mines. To make his request more attractive to the government, Wolters undertook to bring to Spain a number of skilled miners from Sweden together with the drainage-pumps necessary to dewater the ancient flooded galleries.

Despite the prevalent hostility to foreigners, much of which was due to the recent war, the government of Philip the fifth realized that if mining for minerals was ever to be done on any scale it would be necessary to permit non-Spaniards, able to recruit skilled foreign labour, to undertake the work. The administration was certainly anxious to encourage the industry and had removed a ban on the export of metals in the hope of promoting the reopening of some of the country's long-abandoned mines. Moreover foreign miners were not an unheard of phenomenon in Spain. Numbers of Germans had been employed there in the past: ninety-three had been recruited to work at Guadalcanal when an attempt had been made to restart those mines in 1556; others had been taken on at Almaden in the early seventeenth century† and also at Linares.[2] Consequently the economic climate was right for Wolters's application and precedent ensured that his foreign nationality was no bar. Thus in 1724 in response to his petition he was granted a contract for a term of thirty years.

The next year he published a lengthy prospectus containing his proposals for establishing a company and for raising the working capital he needed.[3] He declared that his intention was to produce precious metals and described his mines as being as rich in gold and silver as any in Europe. The prospectus was optimistic to the point of

*John Lethbridge is normally credited with the invention of the diving-suit in 1715, but plainly Wolters also has a good claim to consideration.

†The Almaden mercury mine is still operating and is the largest in the world.

being dishonest: it stated that the Guadalcanal mines had provided the Fuggers, the famous German banking family, with fortunes; and that the Rio Tinto mines had formerly yielded 2,700 ducats of gold (worth almost £1,500) a day. In fact when the Fuggers held the lease of Guadalcanal from 1632–6 they wasted large sums of money in unsuccessfully trying to drain those mines, and Rio Tinto had certainly provided no gold since the days of the Roman occupation.

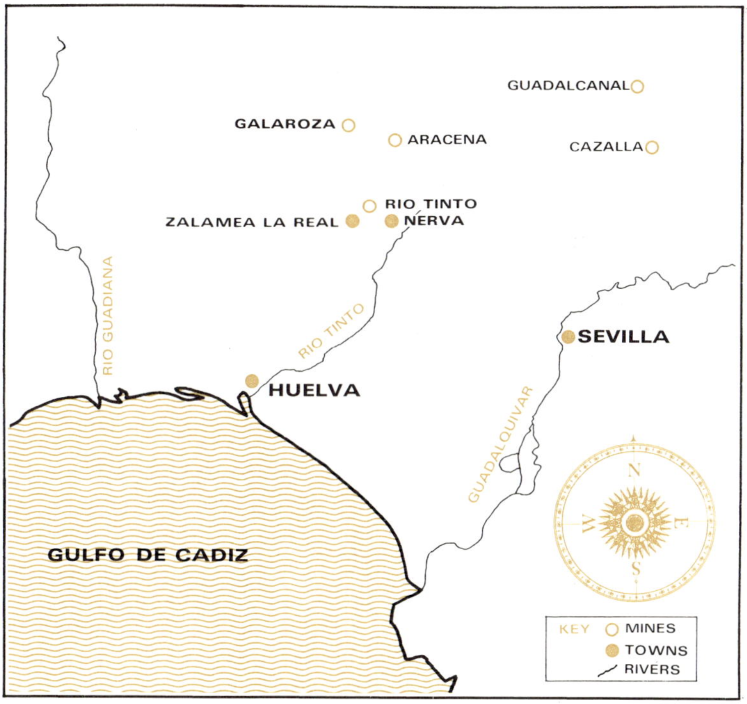


Map of the Rio Tinto area.

The articles of the company, la Compañía y Asiento de las Minas de Guadalcanal, Cazalla, Aracena, Galaroza y Río Tinto, which was one of the first joint-stock companies in Spain, provided for the issue of 2,000 shares of fifty doubloons (approximately £31) each, of which Wolters received 700 without payment in return for assigning his contract to the company. Shareholders were required to pay five doubloons (£3 2s 6d) for each share on application, ten doubloons (£6 5s od) on 31 May 1726, a further ten on 31 July of the same year and the remaining twenty-five when called on. Twenty shares carried the right to one vote though Wolters was restricted to a maximum of six votes for his 700 shares. By royal approval the articles designated him president of the company with the right to nominate his successor, and allowed him to appoint six of the ten board members. They also provided that he alone had power to appoint, to decide the wages of and to dismiss all employees. The terms of the grant also stipulated that one-fifth of all profits had to be paid to the king with the remainder distributed between the shareholders on a pro-rata basis.

The company was formally established in Madrid on 4 September 1725, and immediately became the target of a number of derisive pamphlets written by Spanish authors who poured ridicule on Wolters's promises of huge profits and described him as a liar, a swindler and a heretic (the latter on account of his Lutheran religion). These attacks resulted in a slower sale of the shares than Wolters had anticipated. When the first general meeting of shareholders was held on 5 November less than half the available shares had been taken up and those who had bought them had been induced by the pamphleteers to doubt the wisdom of their purchase.[4]

Wolters failed to get the confidence of his shareholders and at this first meeting they voted to remove him as president and replaced him with the *Marqués* de la Paz.[5] Another meeting was called for 26 November when de la Paz was voted out of office and succeeded by the *Conde* de Cogorani, and eight shareholders were elected to the board instead of the four for whom there was provision. Cogorani warned the meeting that it must be careful not to infringe too much on Wolters's privileges as this would transgress the royal prerogative under which the company's articles had been issued. The shareholders, however, were convinced that they had been defrauded: they refused a request from Wolters for money to enable him to start work at the mines and resolved to commission a mining expert called Robert Shee, who appears to have been British, to investigate the properties and to forecast their prospects.[6]

By December precisely half the shares had been allocated and Wolters felt sufficiently confident to write to friends in Upsala to recruit fourteen Swedish miners and to purchase two pumps. Early in January he started his circuitous journey to Rio Tinto by way of Aracena, Galaroza, and Guadalcanal. His first report to the shareholders was written from Rio Tinto on 18 January 1726 stating that he had taken formal possession at each of the mines on behalf of the company but that it was at Rio Tinto that he intended to concentrate his efforts. On his orders labourers had already started to cut wood and to make bricks and tiles for the houses to accommodate the Swedish workmen. Wolters had discovered a number of windmills within the mines' boundaries which were paying an annual tribute of 15,000 *reales* (£156) to the king, and he believed that under the terms of the contract he was entitled to take possession of these for the company but thought it wiser not to do so as he had no wish to prejudice the king's rights. Wolters reported, too, that he had with him two *arrobas* of ancient slag from Guadalcanal and some ore excavated at Rio Tinto which he was about to assay. His expectations were such that he asked the board to appoint an inspector and also an accountant and to send them to him by March at the latest since this was when he expected to start smelting operations.[7]

Robert Shee arrived at Rio Tinto in March to make his inspection. His report to the shareholders, dated 23 March 1726, was lyrical. The Roman ruins he described as those of the long lost (in fact entirely mythical) celtic city of Gran Britania.

Rio Tinto village (the modern Nerva) where Wolters had set up his headquarters, Shee wrote, had some eighty inhabitants and was part of the administrative district of Zalamea la Real. The town of Zalamea la Real itself, some six miles from Rio Tinto *pueblo*, had over a thousand people, an abundance of good water, fertile soil and an excellent situation and climate. Shee declared its inhabitants to be industrious and prosperous, producing considerable quantities of hides, honey and wax, and rearing pigs in large numbers.

Plainly impressed by the extent of the ancient mining operations at Rio Tinto and the great slag heaps, Shee reported on a number of ancient shafts and galleries and the condition of some of the water

*Interior of the cueva del lago: photo circa 1880.*

adits which he found. Along one of these he was able to walk upright 264 yards into the heart of *cerro* Salomón where he discovered a large lake of highly vitriolized water some 78 feet deep and so clean and translucent that he could distinguish the bottom; obviously the *cueva del lago*. From his visual inspection of the walls of some of the galleries he decided that the principal contents of the ores were gold, silver and iron, and the latter, he thought, would prove as good as any he had seen in England. He had been told a local story that there was a great hoard of treasure hidden in the mines by the Saracens and awaiting discovery, and he sent back with his report some gold coins he had picked up as a possible corroboration of this folktale. Shee's conclusion was that the shareholders could expect good dividends for themselves and great benefits for Spain from their enterprise.[8]

When the report reached Madrid it produced a rush for the remaining shares and a general scramble amongst the shareholders for positions as officers of the company.* In the five months from April to August there were no fewer than twenty-eight shareholders' meetings all of which were extremely acrimonious. But whereas before the members of the company had been indignant over what they thought was a fraudulent prospectus, now they were excited by the prospect of easy wealth and moved by jealousy and greed. Hostility was no longer directed at Wolters as a lying heretic but as a foreign adventurer who held in his own hands over a third of the shares and was about to enrich himself as director of a mine which looked like Spain's greatest single asset. At these meetings shareholders vied with each other for election as officers and for the award of fees from the company's funds. Almost every gathering was devoted to trying to reverse the decisions taken at a previous meeting and to secure the election of new officers. Streams of petitions were sent by the discontented to the king seeking redress of their grievances.

Wolters returned to Madrid to await the arrival of the overdue machinery and miners from Sweden and to try and make his position more secure. On 3 September he obtained a royal decree excluding from all part in the company's management the eight elected board members and two of the most vociferous and quarrelsome shareholders, the *Marqués* de Villadarias and Vicente Gutierrez Coronel. All the resolutions moved by these two men and adopted at general meetings were annulled by the decree, which in addition reappointed Wolters as president and summoned another general meeting to elect four board members in conformity with the articles. It also referred to the fact that the miners and the drainage-pumps had just reached Madrid from Sweden and ordered that Wolters was not to

*The Rio Tinto tradition that many ladies of the court, enthusiastic over Wolters's good looks, bought the shares is disproved by the list of original shareholders which shows that all the participants were men.

be hindered any more by shareholders in his attempts to start the drainage operations and to commence work.[9]

However, on 8 October Wolters wrote to his accountant at the mines complaining of the inability of the shareholders to reach decisions and of their refusal to vote him any money. Until they did so he could not afford to transport the pumps to Rio Tinto to begin operations. Moreover the Swedes who were also waiting to make the journey were entitled to wages from the time of their arrival and he had to pawn his own personal jewellery to pay them.[10] In desperation he petitioned the finance council to intervene in order to resolve the continual quarrelling. On 17 November another royal decree was published which assigned rules for the conduct of meetings, defined the spheres of activity of the company's officers and detailed the rights of Wolters as contractor.

In a fury Cogorani organized a hastily summoned meeting of shareholders for the next day which authorized a protest to the government and a demand for the removal of Wolters from the company and the award of his contract to the other participants. Their letter was delivered by Cogorani and some of his associates who were delighted when told that the finance minister was sympathetic to their demands and would make a formal reply in due course.[11] Wolters had fixed the date for the start of the drainage operations as 18 December but on his return to Rio Tinto decided that there was little point in installing the machinery and starting work while the future of the company and its operations was in doubt and he resolved to await the government's reply.[12]

The latter came in the form of a royal decree on 4 July 1727 in which Joaquin de Barrenechea, the finance minister, proposed an ingenious solution to the difficulties. The company was dissolved and two new companies created. One consisted of all the shareholders save Wolters and was given the mines of Guadalcanal, Cazalla and Galaroza; and the other had only Wolters as a shareholder and was awarded the mines of Rio Tinto and Aracena. The contract in effect was divided into two parts, though its various conditions were still to apply and it was to be allowed to continue in force, apart from the new arrangements, for the full course of years.[13] Thus the shares of Wolters's new company, which called itself la Compañía de las Minas de Río Tinto y Aracena, consisted solely of the 700 originally allotted to him free of charge while the 1300 other shares were transferred to the second of the new organizations which took the name of the Compañía Española. Wolters had managed by this time to recover his expenses from the capital fund but since he had received no other payment he was awarded the smaller of the two machines imported from Sweden and

500 doubloons (£313) to be paid by the other shareholders in return for his relinquishing all claims against them; while they were ordered to pay all salaries owing, awarded the larger of the two draining machines and allowed to take their pick of the foreign workmen brought into the country by Wolters.[14]

The new arrangement had obviously received careful thought, for it was not only the best way of ending the incessant quarrels but also the most effective way of ensuring that the mines named in the original contract were worked, for Wolters plainly intended to concentrate on the Rio Tinto mines and ignore the others. A copy of the decree reached Wolters at Rio Tinto shortly before his death on 26 July and knowing its provisions he nominated his nephew Samuel Tiquet, who had been one of the band of miners who had joined him from Sweden, as his heir and successor to the contract.[15]

It did not take long for the news of Wolters's death to reach Zalamea la Real and immediately the town's *alcalde* (the mayor and magistrate) went to the mines and insisted on making an inventory which included not only the dead man's property but also all the mines' records, machinery and money, placing a prohibition on their use until such time as the will had been proved. This turned out to be the first round in a dispute that was to be conducted with great bitterness during the next sixty years between the authorities of Zalamea la Real and the management at the mines, in which the civic officials were to maintain that their rights included jurisdiction within the mines' area while the mining men claimed that, under Spanish law, crown property even if leased out to private individuals was exempt from the control of local civic authorities.

This demonstration of municipal power brought all work to a standstill. Hastily Tiquet applied for a royal order forbidding the *alcalde* to exercise his judicial powers at the mines and asking for confirmation of the privileges granted to his uncle under the recent decree. With his letter he enclosed copies of his uncle's will and original contract, and a letter signed by sixteen Swedes employed at the mine saying that Tiquet had recently arrived with a party of six of them from the mines of Sahla in their own country having been sent for by Wolters because of his metallurgical skills.[16] As proof of his experience in this field Tiquet also enclosed two certificates from the manager and the chief assayer of the Sahla mines stating that he was an expert in mining and assaying, plus a testimonial from the Rio Tinto workmen certifying that he was experienced in smelting and refining gold, silver and other metals, and a letter from the five superintendents of workmen at the mines which also testified to his abilities.[17]

For nearly four months Tiquet and his men could do nothing save kick their heels until the day when a messenger arrived from the Escorial to hand to the Swede a copy of a royal decree dated 17 November, granting Tiquet all the rights and privileges of his uncle, and another of a warrant which had been despatched to Zalamea la Real raising the *alcalde's* embargo and completely forbidding any town official to interfere at the mines in the future.[18]

At last the way was clear for work to begin in earnest. Tiquet and his small band of workmen settled down to apply themselves to fourteen years of uninterrupted and laborious toil. At first, draining the mines was given top priority but the going was slow. Eventually the idea of reopening the long-abandoned shafts of antiquity was dropped, as one pump, few men and little money proved insufficient to enable him to tackle the immense task successfully. But the pump was kept going and somehow Tiquet discovered the possibilities of using the highly vitriolized floodwaters of the mine to produce copper by the immersion of scrap iron.[19] The water raised by his pump was channelled into tanks, and such small quantities of scrap iron as Tiquet could afford to buy were brought to Rio Tinto on the backs of mules, often over great distances: on one occasion a ton and a half was transported from Biscay, a journey of some 400 miles. The iron was placed in the tanks and the cupriferous waters poured onto it, blisters of crude copper forming on the iron as it was consumed. These were broken off and taken to a furnace for smelting, the copper so produced then being carried to Seville, again on mule-back, for sale to the merchants there.

As the small band of Swedes was reduced by death or departure for their homeland, they were replaced by a few Spaniards and a little foreign labour, but profits were apparently never sufficient to increase their number above the sixteen men employed at the mines when Tiquet succeeded his uncle. In fact the operations were losing money and the contractor was gradually selling his shares to raise working capital. As far as he could he functioned on credit, but it was hard to come by and he had little security to offer.

During these frustrating years trouble was building up for Tiquet as a result of the activities of the shareholders of his uncle's former company; trouble of which he could not have been aware until the blow fell. The *consejo* of the new Compañía Española made a contract on 12 December 1727 with a '*Dona* María Herbert de Powis' whereby she undertook to drain their mines of Guadalcanal in return for her expenses and a payment of two-fifths of the net produce of the mines during the balance of the period of their company's contract.

Lady Maria was in every way a remarkable woman, possessing

boundless ambition and energy. Her grandfather, William the Marquess of Powis, had taken his wife, son and five daughters to the melancholy French court of the exiled British king James the second following the 'glorious revolution' of 1688, where his fidelity to the house of Stuart and Roman catholicism, and his services as lord steward and chamberlain, were rewarded with the additional Jacobite titles of Marquess of Montgomery and Duke of Powis. He died and

*Lady Maria Herbert: painting on the ceiling of a room in Powis castle.*

was buried at Saint Germains in 1696, an attainted outlaw. William his son, however, using the title of Lord Montgomery, had by then returned to England where he spent a period in the Tower of London and then married a Mary Preston who bore him two sons and four daughters, one of whom was Lady Maria Herbert. Because of the old marquess's attainder the family was no longer as wealthy as it had been and, because of its adherence to the Roman catholic religion and the exiled king, it had little hope of participating in those offices of profit which William the third awarded to his British and Dutch supporters. Since Montgomery and his family found London society uncongenial in their straitened circumstances they often went abroad especially to France.

These were the circumstances that must have led the young Maria Herbert to decide that England was no place for her. The obvious alternative, a life of poverty and the charade of monarchy in the palace of Saint Germains, also had no appeal. She looked around for the first and most basic requirement of escape: a wealthy and middle-aged husband who was likely to provide her with a reasonably comfortable and early widowhood.

She found him in Count Gage. His counthood was a continental title of some sort for Joseph Gage was originally a Sussex gentleman but he had the advantage of an immense fortune derived from the American slave trade and cotton plantations in Mississippi. His schemes matched Maria's, and their contemporary Alexander Pope the English poet called them mockingly 'congenial souls'. Certainly they were a good match for each other in their avarice and ambitions though Lady Maria (who, following the continental custom of the time, still used her surname of Herbert after marriage) provided much of the driving force. They began by offering the Poles £3 million to be elected king and queen of Poland, the throne of which was vacant and where there was a search in progress for a new monarch. When their offer was rejected they made another bid, this time for the Italian kingdom of Sardinia. Again rejected, the couple decided to go to Spain, buy a number of disused gold and silver mines, reactivate them and increase their fortune even more. They began in Asturias in the north of Spain and appear to have had excellent success, Lady Maria acquiring considerable mining expertise in the process.

Getting the contract with the Compañía Española and acquiring an interest in her own right in the potentially rich mines of Guadalcanal provided a supreme test for Maria, and she rose to it splendidly. Leaving Gage to look after their northern interests she went to the mines and supervised the construction of the drainage machinery, imported French experts and recruited great numbers of workmen;

and despite all the difficulties experienced in the past with attempts to drain these workings she was at last triumphant. In June 1732, after twelve years of work and the expenditure of much of her husband's money, the enterprise was completed and the necessary technical certificates issued affirming the satisfactory state of the mines. The members of the *consejo* of the company made a call on the shareholders for funds to reimburse her expenses but the money failed to come in.

Immediately she initiated a legal action for the dissolution of the Compañía Española and a grant of the mines for herself. It took eight years to get what she wanted but in March 1740 a royal edict conveyed the ownership of the Guadalcanal mines and all its associated properties to her. In the next month she asked the council of finance for a royal charter similar to that under which the dissolved company had operated so that she could work the mines. This was granted and conveyed to her all the privileges granted to Wolters in 1725 (for this was the contract under which the Compañía Española had authority to mine and manufacture metals) together with an extension of thirty years from 1740. Her subtle mind immediately saw an unexpected opportunity; for her charter repeated the wording of that of 1725 and, since it referred to all the mines granted to Wolters, she set out to persuade the finance council that she should be allowed to take possession of these. As a result she was conceded a royal licence authorizing her to take over not only the mines of Guadalcanal but also those of Cazalla, Galaroza, Aracena and Rio Tinto.[20]

It is not evident what arguments or persuasions she used. She may perhaps have been helped by influential allies though it seems more probable that her own undoubted abilities and her possession of great wealth, which would provide her with all the working capital that would be required, tipped the scales in her favour. The government of Philip the fifth was still anxious that Spain's mineral resources should be exploited to the utmost and since in this case the choice was simply which of two foreigners should be supported it would be natural for the council of finance to prefer the one who was so demonstrably capable.

Samuel Tiquet of course protested, quoting his own grants and confirmations but in the following October another royal decree stated that the long-dead Wolters had forfeited all rights by failing in his lifetime to carry out the terms of his 1725 contract and that consequently his heir was incapable of inheriting his interests. Tiquet appealed and demanded clarification of the reasons why Wolters was held to have voided the contract by non-performance; and although there was a long tradition in Spain, reinforced by decisions in the courts, that mines were royal property and disposable more or less at the whim of the crown, clearly it was felt necessary to justify the government's arbitrary and apparently high-handed decision. But when the explanation came in March 1743 it was palpably absurd.

The dismissal of Tiquet's appeal was accompanied by the statement that Wolters had failed to meet the stipulations of the contract granted to him in June 1725 because he had never drained and developed the mines of Guadalcanal, Rio Tinto, Cazalla, Aracena and Galaroza as envisaged in the grant. Therefore the grant was void. For good measure the decision stated that there could be no further appeal against it.[21] This did not prevent the angry Tiquet from pointing out that Wolters could never have drained all five mines since three of them had been taken away from him by a royal decree which had specifically safeguarded his rights and those of his heirs at Rio Tinto.

Reluctantly the Swede packed his bags and took himself to Madrid while Maria Herbert made the journey to Rio Tinto and took possession. She found that Tiquet had recently begun the manufacture of copperas and vitriol which were used in making such things as medicines and dyes. Spanish production of these commodities was insufficient to meet even the demands of the growing hatmaking industry there and much of what was used had to be imported, so she decided to expand output of these commodities at Rio Tinto at the same time as operations were put in hand to drain the underground workings.

Presumably she continued to employ Tiquet's foreign employees and added to them some of the men she had employed at Guadalcanal. The town officials of Zalamea la Real continued to pester the foreigners at the mines but she left her supervisors to deal with them, for her other interests at Guadalcanal and in the north occupied much of her time. Shortly after she took control of Rio Tinto, Cristóbal Ignacio de Montilla y Quiñones, the deputy lieutenant of Seville, was ordered to go to the mines to collect samples of the copperas and vitriol being made there on behalf of the board of commerce and mines. When he arrived at the workings at the head of a column of troops he found the mines and their buildings completely deserted. Mystified he helped himself to the samples he required and then departed. Later he discovered that when the workmen had seen the soldiers coming they had run away to hide in the hills, believing that they were about to be arrested for having imprisoned the *alcalde* of Zalamea la Real in a shed at the mines to encourage him either to stay away or show them greater respect![22]

In Madrid, Tiquet was busy lobbying royal officials in an attempt to arrange another hearing. In a written memorial he pointed out that the vitriol and copperas being produced by Lady Maria's establishment were far inferior in quality to what he had produced and that in fact no one in Spain had been able to improve on his achievements in this field. If the Rio Tinto mines were returned to him he promised to boost production to 12,000 *arrobas* (approximately 150 tons) of copper extract a year by 1762 by employing thirty extra workmen and erecting new plant.

He obviously made a good impression and the accession of a new king, in the person of Ferdinand the sixth, in July 1746 was the stroke of luck he needed. Ferdinand came to the throne intent on pursuing peace and promoting manufacturers. Within a month another royal decree was issued depriving Lady Maria Herbert of the mines of Rio Tinto and Aracena, and bestowing them on Samuel Tiquet for a further period of thirty years. When the Swede returned to take possession of the mines he found Lady Maria had been there before him and had discharged the employees, sold all the stocks and stores, disposed of the mules and horses and in an outburst of rage had destroyed the drainage-pipes and tanks and demolished all the buildings. Nothing remained of all the work undertaken since 1725 and Tiquet was faced with the task of beginning over again.[23]

From the point of view of the future development of the Rio Tinto mines, Maria Herbert's loss of her contract was perhaps unfortunate. As a mining administrator she was without equal in Spain. She had capital resources that enabled her to get results and possessed abilities far superior to most of the men with whom mining brought her into contact. A woman of action with a reputation for greed and a need for power, the thwarting of her plans to acquire a throne had directed her energies into other channels. If her luck had been better she might have left a reputation as one of Europe's more interesting monarchs. As it is she is remembered, if she is called to mind at all, as a notorious eighteenth-century adventuress who married and domineered a man of little merit, and together with him was ridiculed in verse by the smart-society poet Alexander Pope:

> The Crown of Poland, venal twice an age,
> To just three millions stinted modest Gage.
> But nobler scenes Maria's dreams unfold,
> Hereditary realms and worlds of gold.
> Congenial souls! Whose life one Av'rice joins,
> And one Fate buries in the Asturian mines.*

They were indeed 'congenial souls' in their joint ambitions and in their shared lust for gold but the same could be said of all the adventurers, Spanish and foreign, who were driven by covetousness and the prospects of making fortunes from the mineral wealth of Spain. The eighteenth and nineteenth-century history of Rio Tinto

**Moral Essays*, Epistle 3

was dominated by such people all of whom would have recognized these same characteristics in each other. It was only polite society that differentiated, accepting such behaviour as natural in the *bourgeoisie* but mocking it (perhaps from jealousy of their success) in the case of Maria and her Sussex count.*

Samuel Tiquet, once again in charge at Rio Tinto, needed more money to make good the damage done by Maria Herbert on her departure, and was given permission to start a new company which would embody his former company with its 700 shares, many of which he had sold or given away, and which would have the right to issue a further 700 shares to raise the necessary capital for development at the mines. The new stock was to be offered in the first place to current holders of the 700 contractor's shares but any not taken up by them could be issued to new subscribers.[24]

A general meeting was held in Seville in November 1747. Tiquet told those present that he estimated the cost of manufacturing vitriol and copperas in the coming year as 154,000 *reales* (£1,604) which should yield a profit of 130,000 (£1,355) rising by 1762 to 700,000 *reales* (£7,292). Tiquet now held only 99 shares, the other 601 having been distributed for cash or as gifts, although the owners of only 398 shares could be located at that time. Those which had been sold, and were styled *de justicia* shares, were in a majority compared with the *de gracia* shares held by Tiquet and those who had received them free of payment, and the owners of the former made plain that they were disappointed at so far not having had any return. They were suspicious of Tiquet and anxious not to have any future profits whittled away by a greater number of participants, and carried a resolution that none of the newly authorized shares should be issued and that only holders of the old shares should be allowed to participate in the new company, with a liability for a payment of 100 dollars (£21) for each share held whether this was *de gracia* or *de justicia*.[25]

At another Seville meeting the following March Tiquet produced accounts showing that in the three months between the two meetings he had paid wages of almost 21,000 *reales* (£219). There were also debts of over 911,000 *reales* (£9,490) most of them incurred in fighting Lady Maria Herbert and obtained against his personal guarantee. He was now penniless and anxious for some remuneration and the shareholders decided that, although they could not provide him with a salary, they would agree to his keeping all profits made during the first four months' operations of the copperas and vitriol manufactory when it started and 50 per cent of all profits during the term of the contract.[26]

When Tiquet made his first call for the money due from the

*Joseph Gage died in Spain in 1750, and Lady Maria in Paris in 1774.

members of the company he asked for 256 *reales* (£2 13s 3d) for each share which should have provided 153,856 *reales* (£1,540). Less than 55,000 (£573) were paid in. At a subsequent meeting a group of shareholders proposed splitting the contract in two with Tiquet taking the Aracena mines and themselves the mines of Rio Tinto, a suggestion which the Swede promptly rejected. In despair he told them that if only he could get half a million *reales* (£5,209) there would be no problem.[27]

When recruiting staff following the ousting of Lady Maria, Tiquet took on some of his former employees together with a number of new men. One of these was a former tailor from Valencia, a young man called Francisco Tomás Sanz. Why he should employ Sanz is not evident. Neither is it plain how they met; but the tailor quickly became his right-hand man and proved an able administrator. The middle-aged Swede and the young Spaniard became close friends and Sanz served as godfather when Tiquet gave up his Lutheran religion and was baptised into the Roman catholic church. Tiquet did not change his faith simply to acquire Spanish nationality for he remained a Swede until his death. He may have done it to comply with the laws, which were occasionally enforced, requiring all foreign residents engaged in trade and commerce to be Roman catholics; or perhaps simply because he felt the need of some religious consolations after so many years away from his own country and church.

Practising the utmost stringency, Tiquet and Sanz gradually started work draining the floodwaters, making copperas and vitriol, and exploring the ancient shafts and galleries as the water subsided. Eventually, in 1750, they began to excavate small quantities of fairly rich ores for making copper by smelting. They continued with the precipitation process whenever they could afford to buy iron and, by dint of reinvesting the modest profits being made, once more began to get the business back onto its feet. The shareholders, however, were unhappy with assurances of profitable tomorrows and demanded that Tiquet start distributing profits instead of ploughing them back into the company. In 1753 they commenced a law suit on these grounds, and in the next year appointed a delegation of five to visit the mines and report on the work being done.

Tiquet was absent when the party arrived and Sanz received them. He felt obliged to show them the works but was angry at their insulting behaviour to him and the workmen. Five of the fourteen employees signed an account of the visit in which it was recorded that the delegation described Tiquet variously as a rascal, a liar and a swindler and threatened to give him a rough handling if they could catch him. The men took a particular dislike to the lawyer in the group whom they nicknamed 'the curate', both because he had been admitted to minor orders and also on account of his extravagant and simulated piety. On an underground inspection the delegation discovered what they thought were rich veins of silver whose presence Tiquet had never declared. In derision, and using some forceful language, the five workmen pointed out that they were veins of stone, sulphur and white slate. The 'curate' in anger ordered the delegation's servants to arrest the men and despite Sanz's protests they were forced to march to Seville where they were held prisoner for three days. The report later issued by the delegation alleged that the mines contained silver which Tiquet was working and doubtless smuggling out of the country to Sweden.

Sanz was asked by Tiquet to act as his representative to the shareholders to settle the dispute and the former tailor decided to carry the fight into the enemy's camp. When the delegation's bill for 6,000 *reales'* (£63) expenses was given to him he refused to pay it from the mines' funds and wrote a contemptuous report for circulation to shareholders describing the delegation's visit and the 'curate's' behaviour. This he followed up with vigorous threats of prosecution against the delegation and of possible counter-litigation against the malcontented shareholders, coupled with assurances of good dividends in the not too distant future provided the contractor was allowed to get on with the job without the distraction and expense of legal actions. Faced with these clear-cut alternatives a general meeting of the shareholders agreed to withdraw both the law suit and the accusations made against Tiquet in the delegation's report.

There followed a period of tranquility during which production rose although the shareholders still failed to receive any profits. During the period from 1747, when Tiquet had just restarted production, until 1758, the year of his death, the amount of copper manufactured rose from virtually nil to over 2,000 *arrobas* (25 tons) a year. The total produced in those twelve years was about 162 tons costing over 10,000 *reales* (£104) a ton to make.[29]

The sudden leap in production occurred in 1750 with the solving of some of the problems which had been hindering the smelting of ores. Despite the fact that Tiquet was supposed to be an expert in metallurgy there had been little success with the smelting process and most of the copper had been obtained by the immersion of iron in the vitreous-water tanks. However, in that year a new furnace called el Chorrito* was built and successfully commenced operations on the site known as los Planes, being used for smelting the excavated ores and also for refining the crude precipitated copper which had received its first smelting in an old and somewhat inefficient furnace which Tiquet

*According to the dictionary the word means either 'small water spout' or (colloquially) 'a vulgar noise made by blowing air through the compressed lips'.

called *horno romano* (in the belief that it was a survival from Roman days). The produce of el Chorrito was still a rather crude article known as *cobre negro* (black copper) which was sold to the government under contract and sent either to the Seville artillery foundry or the Seville mint for coin making, though because of its impurities it needed refining yet again on arrival.[30]

By 1753 Spaniards and foreigners were working side by side at Rio Tinto. The administration was run by Francisco Sanz with the aid of his younger brother Gabriel who had come from Valencia to serve as paymaster. The precipitation process was run by Sacre Motard, a Spaniard, assisted by Juan Wert a miner from Flanders.* In charge of the furnaces was Jorge Estér, a Spaniard from Galicia, who was assisted by two young German brothers, Miguel and Simon Forster, while the superintendent of the underground miners was Francisco Guerra, another Spaniard.[31] It was predominantly a young community. As late as 1758 the only married men were Wert and Guerra. The isolation of the mines, the hardness of the living and the fact that pay was constantly in arrears probably made most of the men disinclined to marry and raise families until there was a prospect of greater security.

Work was hard and all the men were expected to be able to turn their hands to virtually anything. Gabriel Sanz recorded late in life how at this period his duties as paymaster included being lowered down ancient shafts at the end of a rope, clutching a lanthorn and scanning the walls for signs of ores, while his colleagues on the surface hung on to the rope's other end and pulled him back up when he shouted that he was 'in danger of succumbing to the fumes of sulphur and antimony'.[32]

At the time of Tiquet's death in 1758 there were only fourteen mines' employees and they all lived close to their work in primitive buildings on the lower slopes of *cerro* Salomón. Tiquet and Sanz, together with their servant Miguel Millan, occupied a separate house which also doubled as a store and office. It was poorly furnished and devoid of all luxury save a few items of household silver. The two married men and their wives and children shared another house while the other eleven unmarried males occupied another two. The only other buildings were a covered bakehouse and a blacksmith's shop where essential repairs to equipment could be carried out.[33]

Samuel Tiquet died on 11 September 1758 when the mine was on its way to becoming an economically productive operation and the technical problems of smelting had at last been overcome. Production costs were still too high and must have been almost as much as was brought in by sales, but at least the future was beginning to look

*Wert's numerous descendants have continued at the Rio Tinto mines until the present day.

reasonably bright. In his will Tiquet, described as 'a bachelor of Stockholm of the kingdom of Sweden', left his friend Francisco Sanz all his personal effects in the house they shared: the household furniture, the little domestic silver they used and his clothing, all valued in the inventory at no more than 500 *reales* (£5 4s 2d). Millan the servant was left 300 *reales* (£3 2s 6d) and Pedro Castilla, the surgeon from Zalamea la Real who attended the community, another 300 *reales*. Apart from 'one old mule, a Galician pony and a donkey' Tiquet had no other property save for his ninety-nine shares in the company. Fifty of these he left to Antonia Vigne the wife of Jacob Keers, a Seville merchant and the company's agent in the city, in recognition of her husband's 'great care and interest . . . in my work in these mines'. The remainder he bequeathed to his mother, Esther Wolters Tiquet, living in Sweden or, if she were already dead, to his two sisters Juana and Carlota of Stockholm. He nominated his friend Sanz to take charge of the mine with full rights of administration and an allowance of 20 *reales* (4s 2d) a day for his maintenance until the expiration of the contract, together with the 50 per cent of all net profits that would have accrued to Tiquet had he lived and Tiquet's share of the stores then remaining.[34]

The accounts for 1755–8 show that the income from sales was used entirely to pay wages, to meet expenses and pay off some of the debts incurred by the company; and that Tiquet took absolutely nothing for himself by way of salary, which explains his poverty at death.[35] One cannot help but reflect that his integrity appears remarkable when compared with that of his successor, Francisco Tomás Sanz.

*King Charles the third.*

4

# *The tailor from Valencia*

## The rise and fall of Francisco Sanz: 1758–83

Two weeks after Tiquet died Sanz wrote to the finance council informing its members of the Swede's death and requesting the continuation of Tiquet's contract in his own name, those of the other shareholders and of Tiquet's relatives in Sweden. A royal order was immediately issued to this effect, which also confirmed the clause in Tiquet's will for Sanz to stay at the mines as manager, though it was provided in addition that a lawyer called Martín Alonso de Bolaños, who had been appointed in Tiquet's time as *juez conservador* to safeguard the rights of the crown and the company's shareholders, was to continue to serve in this capacity.[1]

Sanz's long rule over the mines from 1758 to 1783 coincided with a period of almost unwonted prosperity for Spain. The death of Ferdinand the sixth brought to the throne his half-brother Charles the third (1759–88), a vigorous and successful monarch with an active concept of the part he had to play. Apart from two wars in alliance with France against Portugal and Britain in 1761–3 and 1779–83, the reign of Charles was largely one of peace and of progress in the intellectual and economic life of Spain. The powers of the holy inquisition were severely curtailed; the Jesuits, who had incurred the hostility of all elements of Spanish society by their striving after power, were expelled from the country; and Spain's twenty-four notoriously inadequate universities were reformed, with the experimental method of science and medicine introduced and the writings of Bacon, Descartes, Locke and Kant appointed as text-books. Taxes on sales, including food, were cut by at least half, and in some cases completely abolished, and a state lottery introduced to make up for the lost revenue. New roads and canals were built and government stage-coach services introduced to improve transport and communications. Rural police forces were created in a largely successful attempt to overcome brigandage and irrigation schemes were promoted to stimulate agriculture.

Behind all this activity lay a determination on the part of the king and his ministers to restore to Spanish society a belief in the worthiness of honest labour that the riches of the new world and the corruption of former administrations had done much to kill. Charles even issued a decree declaring that there was nothing dishonourable in members of the nobility taking up work. Plainly a new wind was stirring throughout the land.

This did not prevent Charles the third's reign from having its share of riots and revolts some of which were stirred up by reactionary forces such as the Jesuits and some by the king's own haste for change. But despite these minor problems the energy of the king and his ministers was successfully directed to producing an ever increasing volume of trade, commerce and industry. This in turn stimulated and increased demand for copper for fabrication and coinage which worked to the advantage of the Rio Tinto mines and gave them an important role in the country's economy at the very start of what looked like being Spain's own industrial revolution.

The credit for boosting production at Rio Tinto, however, and enlarging the scale of operations there, belongs almost entirely to Francisco Sanz. His opponents when they came to belittle his achievements or to accuse him of dishonesty referred to him disparagingly as 'the former tailor from Valencia'. It is true that he was not a skilled miner. Under his direction his men worked with little system or order and created a fresh labyrinth of galleries whose walls crumbled and whose roofs collapsed in later years creating hazards additional to those left by the ancient miners of prehistoric and Roman times. But within a short time of his assuming command at Rio Tinto he had expanded the work force to exploit the mass of ore on the south side of the *cerro* Salomón on a much larger scale than his predecessors had ever envisaged, completely rebuilt the furnaces and works, and turned the four houses on the south slope of the hill into a substantial village, the *pueblo* of la Mina, which he planned as a settled community and not a mining-camp, a place where men could take their wives and raise families, and where they could put down roots and develop a sense of belonging.

From the moment of taking charge at the mines Francisco Sanz applied himself energetically to solving the problems he had inherited from Tiquet. When in January 1759 Bolaños the *juez conservador* submitted his first report to the board of commerce and mines he loudly sang Sanz's praises: production was rising and the income from the increased sales was being devoted to paying off the heavy debts with which the company had been burdened at the time of Tiquet's death. By restoring confidence among creditors Sanz was laying the basis for future expansion by demonstrating the financial soundness of the undertaking. But a year later Sanz and Bolaños were at loggerheads. The judge now found that Sanz was ignoring his authority and spending money without reference to him. Moreover the manager was using the new title of *administrador asentista*, or administrator-contractor, in place of the official designation of *administrador*: plain evidence to Bolaños that Sanz was trying to usurp the rights of Tiquet's family and the other shareholders and aiming at absolute rule within the mines.[2]

The judge's protests to Madrid fell on deaf ears. In its anxiety to promote mineral mining, and because Sanz undoubtedly possessed considerable talents, the government was prepared to let him cut a few corners and fell in with his suggestion to remove the pettifogging Bolaños. The post went to a Seville lawyer, Bernardo Luque. He was to enjoy a happy relationship with Francisco Sanz through the simple expedient of giving him a free hand. With the new and compliant *juez conservador* installed Sanz was able to send to Valencia for a number of his relatives to join his younger brother Gabriel and himself at the mines. Luque also agreed with Sanz that there was no need to call company meetings until such time as either the government or a considerable body of shareholders became insistent. In consequence no meetings at all were held between 1758 and 1770. Any discontent with lack of profits or any concern over the administration of the mines was therefore not voiced publicly during some twelve years and Sanz, buried in the remoteness of the *sierra* was able to become a complete autocrat. But though unworried by shareholders he was not without problems. The sheer labour of supervising the work in the office, of creating the mining village and of directing the miners at the work face must have been enormous.

It was on one of his underground visits in July 1762, to inspect a newly discovered Roman gallery in the south-east corner of the *cerro* Salomón, that he discovered a metal plate erected in honour of the Emperor Nerva. He decided to investigate the gallery but a sudden fall of rock killed two of his companions and the remainder of the men with him refused to go forward. Reluctantly he gave the order to leave, but in a twenty-two page pamphlet which he wrote about the discovery (*Memoria antigua de Romanos en las minas de Río Tinto*, published at Seville on 4 September in the same year) he reveals a little of his complex and energetic personality.

Above all else his writing shows a strong sense of nationality: the belief that at Rio Tinto he was uncovering one of the greatest mines of the ancient world for the glory of Spain. Allied with this was a self-confidence in himself as an efficient businessman, eager and ready to take advantage of the economic benefits made available to his country by 'the most mighty, catholic, pious and blessed king . . . that best of princes' who ruled 'happy Spain'. That Sanz did not mention what was undoubtedly his principal motivation, the desire to accumulate wealth, is understandable enough. But his pamphlet shows how the

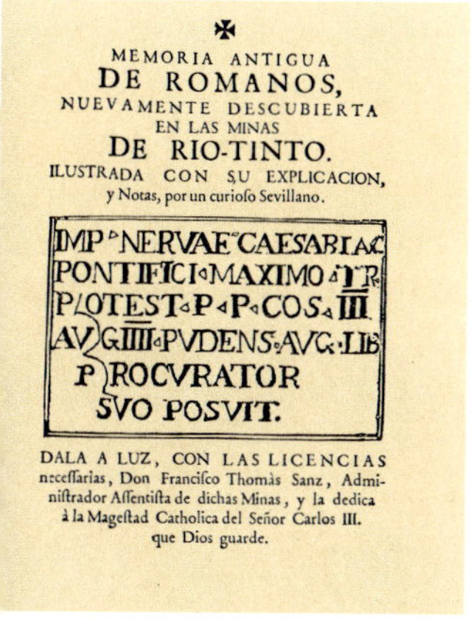
MEMORIA ANTIGUA
DE ROMANOS,
NUEVAMENTE DESCUBIERTA
EN LAS MINAS
DE RIO-TINTO.
ILUSTRADA CON SU EXPLICACION,
y Notas, por un curioſo Sevillano.

IMP NERVAE CAESARI AC
PONTIFICI MAXIMO TR
POTEST P P COS III
AVG IIII PVDENS AVG LIB
PROCVRATOR
SVO POSVIT.

DALA A LUZ, CON LAS LICENCIAS
neceſſarias, Don Franciſco Thomàs Sanz, Admi-
niſtrador Aſſentiſta de dichas Minas, y la dedica
à la Mageſtad Catholica del Señor Carlos III.
que Dios guarde.

*Title page of Sanz's 'Memoria antigua de Romanos en las minas de Río-Tinto'.*

head of one business at least felt about the policies of the new government and the effect they were already beginning to have on the country's economy. It would, of course, have been unnatural for Francisco Sanz not to have been enthusiastic about Charles since one of the king's first tax reforms cut the royal share of Rio Tinto produce from one-fifth to one-thirtieth in order to encourage greater capital investment from profits.

The difficulties with the civic authorities of Zalamea la Real were always in the background and in 1763 they came again to the fore. The town officials complained to the board of commerce and mines about the extent of wood-felling on the common waste land around the mines and its over-population with mines' cattle. The view of the townsmen was reasonable enough. Much of the prosperity of the town, and the livelihood of many of its poor, depended on the waste land within the boundaries of the *ayuntamiento* being freely available. The destruction of trees by Sanz for housing and mining purposes reduced the town's fuel supplies while the ever-growing size of the mines' herd of cattle meant that there was less rough-pasture available for the townsmen's animals. The more the mines were worked the larger grew the work-force, and the more would the citizens of Zalamea la Real suffer.

The conflict of interest between these local inhabitants who were primarily dependent on agriculture and its allied trades, and those principally dependent on industry, was inevitable. Sanz defended his use of the waste land in a letter to the board by appealing to the customary rights attached to mining enterprises, and maintained that when he cut trees he paid a fair price and that as yet his cattle had not deprived others of their grazing. If he had wished he could have pointed out that many of his workmen were from Zalamea la Real and that their wages and the mines' own purchases were contributing to the town's economy. But *Don* Francisco Sanz was not the man to truckle to those whom he regarded as his inferiors and he was not prepared to negotiate a mutual arrangement with the town over the use of the common land.

Consequently the complaints from Zalamea la Real to Madrid continued. It suited the government to let Sanz have his own way in the dispute with the town for he was apparently scrupulously paying his taxes and establishing an important mining concern that could be of great benefit to the country. But of necessity Sanz had to be asked about the complaints. In August 1768 he sent a lengthy defence to the board denying that he had abused the rights over the waste land to which he was entitled, and expressing indignation at the possibility that the value of his work was not appreciated in Madrid. He tetchily concluded his letter by saying that since he had become the contractor ten years before (he now described himself simply as *asentista* without the prefix of *administrador*) he had spent an immense fortune, which had come to him as his wife's dowry, on improving the mines and was now close to poverty. He felt that in fairness the government should give him the chance to recover this money by giving him an absolute grant of the mines for his own life and those of his sons, or at the very least a grant for thirty years after the expiry of the current contract which was due to run out in 1776. Accompanying this letter was an enthusiastic recommendation from Luque describing Sanz's achievements at Rio Tinto and urging the government's agreement to the proposals, though he noticeably omitted all mention of the rights of the company's shareholders.[3]

Faced with this request, the board of commerce and mines replied to Sanz asking for fully comprehensive accounts from 1758 onwards to help them reach a decision. The following July he sent a set of accounts covering the period from September 1758 to June 1769 which showed the manufacture of 55,816 *arrobas* (698 tons) of copper during those years, the payment in full of the royal thirtieth, and profits from sales of almost 417,000 *reales* (£4,344) which had been in part reinvested in the mines and in part used to buy back 414 shares from various shareholders. (This was allegedly done so that the eventual distribution of profits to individual shareholders could be much larger than it would otherwise have been.) Debts then stood at 50,000 *reales* (£521).[4]

The members of the board viewed the figures suspiciously: in their reply they demanded more detailed figures, together with reports on the actual works Sanz claimed to have carried out, with a view to printing the accounts for distribution to shareholders. Meanwhile Tiquet's Swedish relatives, who were annoyed that they had so far not received any income from Rio Tinto, arranged for their country's minister in Spain, Baron von Friesendorf, to call a special general meeting of shareholders in Madrid. When it took place in April 1770 Daniel Sheindemburg, the almoner of the Swedish mission, proposed on the family's behalf that a petition be sent to the king requesting an extension of the contract in their names for a further thirty years so that they could reimburse themselves for the outlay and losses of Wolters and Tiquet. The other shareholders rejected the proposal. Though everyone wished to see some return for their money in the near future they were unable to agree on any unified course of action to protect their interests.[5] With the meeting safely behind him Sanz returned to run the mines in his own way, safe for a few more years from outside criticism and interference, and doubtless confident that in view of his success in setting up one of Spain's major enterprises he could rely on the government to let things go on much as before.

Certainly in the ten years between Tiquet's death in 1758 and the making of inventories in 1768 there had been great strides forward at Rio Tinto. In the former year there had been only six buildings on the lower slopes of *cerro* Salomón. By 1768 Sanz had converted this small hamlet into the substantial *pueblo* of la Mina (which later became known informally as Rio Tinto when the larger *pueblo* of that name became Nerva). Though still in the process of being built it contained forty-seven houses for senior workmen and their families (the labourers still lived in the *pueblos* round about); and to serve the new community there was a slaughter house, communal ovens for baking bread and a shop for the sale of wine, vinegar, oil and all the other necessities which had to be brought into the village from outside. Sanz had also built a small barracks for ten armed watchmen, necessary not only to deal with any trouble caused by irate townsmen from Zalamea la Real but also to protect the mines and their supplies from the brigands who still roamed the *sierra*. There were stables, too, for the mines' fourteen horses and three mules. (More were unnecessary: at this time the Seville copper merchants took their own mule teams over the hills to Rio Tinto in order to carry their purchases back with them.)

Although the mines' village had at first no church of its own, there existed during the early years of Sanz's administration a small chapel, the hermitage of San Roque. Whether this building was ever used by a hermit is doubtful. It was probably erected by Sanz simply as a private chapel to house a statue of San Roque whom Sanz had decided would be the patron saint of his village. Although private devotions would have been made there, the community still depended on the parish church at Nerva. (This chapel still exists and services are occasionally held there, although the effigy of the saint is now in the modern church of the village of el Valle.)

*The statue of San Roque from Sanz's hermitage. (Now in the church of Santa Barbara, Rio Tinto.)*

But before an inventory of the mines' possessions were made in 1783, Sanz had erected on the edge of the central square of the new town he was building, a full-sized church, complete with a belfry holding three bells and a clock, and capable of holding 600 people. Such a large building must have been intended as an open display of his confidence in the future prosperity and growth of the mines. In 1769 he added to the *pueblo* an additional five workmen's houses and erected a mansion for himself. Also on the slopes of Salomón, not far from the houses but on higher ground, he built in those ten years from 1758–68 four new smelting furnaces which he named San Francisco de Paula, San José, San Francisco de Borja and Nuestra Señora de los desamparados (Our Lady of the helpless), and two new refining furnaces to produce purified copper which he called San Gabriel and

Nuestra Señora del rosario (Our Lady of the rosary). These, together with a number of stone (and hopefully thief-proof) stores for tools, gunpowder, grain and other supplies, made up the surface works of the Rio Tinto mines.

Three shafts with their respective galleries leading off them had been brought into active operation during the same ten years: San Christof, otherwise San Cristóbal, Santa Barbara and Santa María. (The name Christof, used as frequently in those years as the Spanish form Cristóbal, suggests that this shaft may have been constructed by some of the Germans or Swedes that Sanz employed.) They were on the south side of Salomón near to the works and the *pueblo*. At the head of each was a windlass and ropes for raising the broken ores from the bottom of the shaft to the surface, and a dumping ground consisting of a square clearing made by cutting down the vegetation on the hillside. From the dumping grounds to the smelting furnaces were constructed iron tramways over which the ores were conveyed in hand-pushed carts.

Within the mines' area, two cart roads had been made and twelve bridges constructed. Two large gardens employing four full-time gardeners had been laid out and two orchards planted to supply the new *pueblo* with vegetables and fruit. A plantation of oak trees had been started to provide timber for the community and the mines. When Sanz had taken charge the number of workmen had stood at fourteen. Ten years later it had risen to 400, while another nine years on, in 1777, the number of workmen had increased to 780.

In subsequent years Sanz pulled down the smelting and refining buildings and rebuilt them with their foundations six feet lower to give them greater strength; four ancient water conduits were cleaned out and repaired, and two new conduits constructed, to drain off the floodwaters from the lower levels of Salomón and thus provide access to more Roman galleries. As production increased Sanz bought 120 pack-mules to carry the extra copper he was making to Seville where he was hoping to create new markets. As the population increased so it became necessary to construct larger ovens for making the bricks and tiles with which to build new houses. When six small quarries of jasper were uncovered Sanz engaged a lapidary to cut the stones, taking yet another opportunity to increase his profits. The policy of tree planting was assiduously promoted: by 1783, the time of Francisco Sanz's final departure from Rio Tinto, the plantations consisted of 1,500 fruit trees, 5,000 oaks and 13,500 black poplars.[6]

Sanz's achievements were not unnoticed in Madrid and they were acknowledged even by the officials of Zalamea la Real when they submitted their complaints about him to the government. In a letter sent to the king in 1777, complaining that Sanz had illegally extended the mines' boundaries by one and a half miles, they estimated that since 1758 he had paid some 12 million *reales* (£125,000) to men from Zalamea la Real as wages and they assured the king that the mines were 'the happiness of our town'. In a way which the signatories had not expected they caused Sanz some difficulty; for an officer of the finance council, on seeing their letter, wondered whether perhaps the time had arrived to re-examine Sanz's earnings with a view to establishing whether enough was being paid in taxes. A memorandum was submitted to the council members pointing out that if Sanz was treating the men of Zalamea la Real so generously he must certainly be taking for himself more than the 20 *reales* (4s 2d) a day which showed in the accounts as his wages for the past nineteen years. In consequence it was decided to look more closely at the mines' finances and this contributed to the difficulties in which Sanz eventually found himself.

Nonetheless he still stood in good stead in certain government quarters, the Zalamea la Real officials in 1777 commenting that it was to his credit that all the senior posts at the mines were now held by Spaniards as he had dismissed the foreigners in his employ; and a finance council memorandum of the next year echoed the sentiments of the townsmen, though for more practical reasons. 'He is a Spaniard and as a result the labourers and masters of the works are Spaniards; and through him has been obtained the secret of refining which was a private process of the brothers Peringer, German master-workmen, whom he was obliged to dismiss.' Thus by using skilled foreigners to teach his Spanish workmen he was providing Spain with men of its own trained in important industrial techniques. The memorandum anticipated that the day would come when, as a result of Sanz's work at Rio Tinto, the mines there would serve as a national school of mining and metallurgy in which Spaniards could be trained to exploit their country's own mineral resources without the need to depend on outsiders.[7] That this did not happen, largely because of the incompetence of later governments, was one of Spain's misfortunes.

While so dramatically altering the state of affairs at Rio Tinto, Francisco Sanz took care to look after not only his own interests but also those of his family, giving lucrative posts to his brother, son, son-in-law, and various nephews; he would have been atypical of his time if he had not. His brother, Gabriel, for example, who had been employed by Tiquet in 1750 at Francisco's request to take charge of the books, had in 1756 been given the title of accountant and paymaster at a salary of 250 ducats (£138) a year. Both title and salary proved empty for the number of employees never rose above sixteen and there was never sufficient money to pay him his full wages. However, when

Francisco Sanz succeeded Tiquet, Gabriel was given the additional posts of notary, postmaster, chief correspondent and director with a salary of 500 ducats a year which henceforward was paid in full.[8] (Gabriel, though doubtless a rogue as the rest of the Sanz family proved to be, was an interesting character. His spare time he devoted to scientific experiments and to writing and publishing papers on such subjects as 'how to discover marl', 'the use of flax' and 'the management of bees'. His paper on 'a new way of producing copper by a chemical process' does not seem to have survived, though the comment of Pedro Henry, the professor of mathematics at Seville who read it at the government's request, is still preserved. 'The man', Henry wrote, 'is a bare-faced charlatan' [*un charlatán presuntuoso*].)[9]

Once again the Swedish minister in Madrid, now Baron von Sparre, started objecting to Sanz's behaviour on behalf of Tiquet's family. He lodged a formal complaint that Sanz had usurped their rights and exceeded his powers, alleging that he was behaving as if he were the contractor whereas he was legally only a managing clerk appointed by Tiquet's will to administer the mines on behalf of the dead man's family and other shareholders at a salary of 20 *reales* a day. He had never submitted any detailed accounts and the only information they had ever received had been the highly dubious accounts printed in 1769. The Swedish minister probably expected little sympathy from the Spanish government and even less action, and he decided to send the mission's almoner, Daniel Scheindemburg, to Rio Tinto with an engineer called Storn who worked at the Almaden mines. Their instructions were to prepare a report about the work in progress. After their arrival the two men took lodgings in an inn and then made a number of expeditions onto mines property and tried to see what was happening underground. On Sanz's orders they were ejected whenever they were spotted and Luque sent an indignant protest to the finance council accusing Scheindemburg of causing trouble and questioning the workmen. Defeated in his main object, the Swedish diplomat sent for one of the Peringer brothers, who were nursing a grievance against Sanz for their dismissal, and in the comfort of his inn made copious notes of the stories the German told him as to how great profits were being made from unrecorded sales.[10]

On his return Scheindemburg gave a copy of his notes to the board of commerce and mines and demanded that they consider the question of the ownership of the mines of Rio Tinto. (The Aracena mine which had formed part of the grant was no longer of interest to anyone and could have been considered to have technically reverted to the crown because of failure to work it.) In the summer of 1777 the board got down to examining the original contracts of 1725 and 1746, and its members realized with a jolt that the 1746 grant had officially expired the previous year so that legally the mines were once again the property of the crown. Moreover, it seemed to them that a good case existed for interpreting the clauses of the contract as having been intended for thirty years or the life of the grantee, whichever was the shorter period, and that Tiquet's ownership had died with him.

The officers of the board had doubts whether the mines could be worked directly by the crown in the future and suggested that Sanz should be given a thirty-year contract. The finance council disagreed strongly. In its recommendation to the king it expressed strong doubts about Sanz's honesty, disbelieved his claim that debts dating back to the time of Tiquet were still outstanding and pointed out that this was incompatible with his other allegation that he had used all the profits to expand the mines and pay off creditors. Clearly, whatever he had done with the money he had made he had not looked after the interests of the shareholders. Consequently the council's officers suggested that he be compelled to produce verifiable accounts for the years 1758–76, that an inventory be made of the mines' properties by persons acceptable to the Swedish minister, the shareholders and the creditors, and that an administrator other than Sanz should be appointed to run the mines for the crown.[11]

While the finance minister considered the problem through the winter and the spring, an attempt was made to murder Sanz by two townsmen of Zalamea la Real for reasons that are not apparent. While the men were awaiting execution in Seville they received the sacrament of penance from Andrés de Santa, a bare-footed carmelite. Sanz was moved to fury and was accused by the townsmen of ordering his men to find the friar and to beat him up. The *juez conservador* protested to the finance council at the same time that a number of townspeople were trying to lay hands on Sanz to do him harm.[12] Nothing further seems to have come of the episode and soon there were other things to think about, for on 3 August the long-awaited royal decree was issued.

Faced with the two proposed alternatives of giving Sanz a contract for thirty years or removing him and taking the mines under the crown's control, the king and his minister had liked neither. As they saw it, the mines had become so valuable that it was essential that never again should the crown lose control, and yet there was no obvious capable successor to Sanz who could be appointed to run them. In the circumstances it was decided that, though the government would work the mines in future, Sanz would be retained as the royal manager while his suspected propensity to dishonesty would be controlled by removing his closest associates. Consequently Bernardo

Luque was dismissed as *juez conservador* after twenty-two years loyal, and presumably well-rewarded, service to Sanz and replaced by a Seville lawyer called Martín de Ulloa who was acceptable to the Swedish minister and who was given wide powers of reorganization at the mines. After twenty-seven years at the mines Gabriel Sanz, too, was dismissed as was José, Francisco Sanz's son, who held the office of comptroller. Francisco's salary as administrator for the crown was fixed by the decree at the respectable salary of 900 ducats (£495) a year, and shortly after he applied for, and was granted, a royal order that it would be paid monthly from the mines' revenues with preference over all other debts.[13] Apart from his younger brother and his son, his other relatives were left undisturbed in their positions, a mistake which was to cause considerable trouble for the crown at a later date. And though Luque quit Rio Tinto, Sanz's son and brother stayed on and received unofficial pensions from the mines' funds until his final fall from power.

(Gabriel Sanz, despite his unemployment, was quite happy for a while. His income from the mines' petty cash was sufficient to support his large family while devoting himself to his scientific experiments. It was not until his elder brother was dismissed as administrator in 1783 that he really felt the pinch and even then other members of the family at Rio Tinto helped him with money. Fortunately the younger of the Sanz brothers was a born optimist. For at least eighteen years after he had been sacked, Gabriel wrote regularly to the council of finance asking if there was any hope of alternative employment in the king's service, accepting with apparent equanimity a string of replies rejecting his requests, written in such forceful language that only the most thick-skinned of men could have persevered. Not until 1797 did he give up hope and even then he asked the finance minister, Pedro de Lerena for one more favour: the right, in view of his past services, to be allowed to wear an official uniform. One similar to that worn by the French engineers then administering the mines of Guadalcanal was what he would have preferred.[14] No reply seems to have been sent to him.)

The 'verifiable accounts' for which the finance council had called were produced by Sanz but how much reliance could be placed on them was questionable. From his later conduct it is probably not incorrect to assume that they under-recorded both copper production and sales. However, the government officers, *faute de mieux*, had to accept them as a basis for their calculations, and from them they estimated that debts at Tiquet's death in 1758 stood at 601,000 *reales* (£6,260) of which 542,000 (£5,646) had been paid off by 1778. Total copper production in the twenty years they estimated as 125,295 *arrobas* (1,566 tons). This gave a yearly average of 6,284 (78 tons) compared with the 2,051 *arrobas* (26 tons) manufactured in Tiquet's last complete year of operations. Between 1758 and 1778 the crown had received as its royal thirtieth 3,861 *arrobas* (48 tons).[15]

The new inventory that was made at the time of the reversion of the mines to the government was the cause of considerable controversy. The new comptroller who made the inventory was probably over anxious to favour the crown, and though he and Sanz agreed on the valuation of stocks and stores they disagreed over most other items. When it came to valuing things such as waterpipes, pit-props and similar articles the comptroller was prepared to include only those items actually being used in draining and working the mines, and refused to consider those used in searching for ores and which represented a considerable outlay, though probably not as much as Sanz claimed. On the basis of the inventory and accounts the finance council concluded that an operating profit had been made over the twenty years of 1,030,000 *reales* (£10,729) and, setting against this the 727,000 *reales* (£7,573) claimed to have been paid in reducing Tiquet's debts and buying-in shares, they came to the conclusion that Sanz had made a minimum profit of 303,000 *reales* (£3,156) which had gone into his own pocket.

On the 14 January 1779 a royal decree declared that the company's books of account would be closed on the last day of February and the calculated profit allocated amongst the shareholders. Since some of these had not been heard of since 1748 it was ruled that in the event of their not coming forward by 28 February then their shares would be ignored in the distribution of profits.[16] But though the government may be given credit for its good intentions, its knowledge of Sanz was sadly inadequate: he managed to produce such confusion by his manipulation of the accounts and by arguing legal points that the books were never officially closed and the shareholders never received a distribution of dividends.

But at the time of the publication of the decree, respect for his mining expertise and belief in his reasonableness was still such that the document included a clause stating that, in addition to his salary as the royal administrator, he would also receive 5 per cent of the net profits. Possibly it was thought this would make him more diligent in caring for the king's interests. However, it did not take long for the government to discover their mistake.

Sanz, in response to requests for more information, produced so many and such varied explanations relating to his management of the mines during the previous twenty years that the council of commerce and mines could not cope. It set up two committees to examine the affairs of the Rio Tinto mines, one composed of civil servants and the other of judges. Since the committees could not agree on their respective spheres of interest and were constantly complaining of each other's activities (one of the judges resigning in protest) the minister of finance ordered them to hold joint meetings. The arrangement did not lessen the antipathy or make for much greater efficiency and the enquiries dragged on for eight years until 1787,[17] by which time other events and decisions made the committees' deliberations of little consequence.

At first, part of the committees' difficulties lay in the fact that Sanz was still at Rio Tinto administering the mines and, by taking advantage of the problems of communications between there and Madrid and by pleading overwork in looking after the king's interests he was able to delay his evasive replies to the committees' demands for information. The new *juez conservador*, Martín de Ulloa, was no help to the Madrid officials. Within a short time of his appointment Sanz and he had become bosom friends; their friendship (it was alleged by one mines' official) being cemented by the gift of a share of the illicit profits being made at the crown's expense.[18]

In the January of 1780 Sanz reported to the board of commerce and mines that there had been a large-scale inundation of floodwaters. There was no loss of life but he estimated the damage at 148,000 *reales* (£1,542) and claimed that 504 *arrobas* (6 tons) of unrefined, and 320 *arrobas* (4 tons) of fine, copper stored in underground galleries had been lost. Questions put to mines' employees by the Madrid civil servants showed that flooding had occurred but that the estimate of the damage was grossly exaggerated. At least one civil servant thought that such a large scale inundation proved that the accounts, which showed a high level of expenditure on drainage works, were false. It also seemed to him to be strange that copper should be taken underground for storage instead of into the warehouses. Doubtless he was right to suspect trickery but the council found the matter so complicated that it was decided it would be easier to accept Sanz's report and let the matter rest.[19]

Then in 1782 Joaquin Buiza, a government official who had been sent to the mines' counting-house four years earlier as part of the general reorganization, reported to Madrid a deficiency of 8 tons of copper from a consignment of 11½ tons stored in a warehouse and marked for the king's use. He also accused Sanz of sending 5 tons of the missing metal by mule-back to Almonaster where it was sold to José Jiménez, the parish priest of Santa Cruz in the city of Ecija, who had resold it to some bell founders. They had used it to make nine bells for three monasteries. Buiza had constructed what must have looked like a watertight case against the administrator and a committee of enquiry was appointed. But Sanz moved quickly to turn the scales. He ordered Buiza's arrest and incarceration in the mines' lock-up on a counter-accusation of the theft.

The committees found it impossible, because of the confused state of the mines' accounts, to discover any greater deficiency than 1¾ tons of copper, though Sanz as administrator was held responsible for its loss and ordered to make up the missing amount. Instructions were given that copper for the king's use should henceforward be kept in a separate warehouse with a lock that required three keys to open it, one of which was to be held by the administrator, one by the accountant and another by the *juez sub-delegado* (as the representative for the absent *juez conservador*). The individuals named in Buiza's account of the theft all pleaded ignorance and the enquiry made little progress.

Over a year after his arrest Buiza was still imprisoned and protesting against the lack of a trial. In two letters to Madrid written in August 1783 he demanded an immediate hearing. He now admitted that he had a prior knowledge of the theft but claimed the only beneficiaries of it had been Sanz and the accountant. To try and convince the committees that he was telling the truth he volunteered information about some of Sanz's other activities. He described how the administrator bought, in Rio Tinto, smuggled gunpowder for blasting for 4 *reales* a pound while charging the finance council the Seville price of 4½ *reales* plus fictitious transportation costs. On another occasion he alleged that thirty pounds of mines' gunpowder had been sold to a priest by Sanz who had pocketed the proceeds. Buiza also claimed that two sets of accounts were maintained: a false set being sent to Madrid while, for Sanz's use, an accurate set was prepared which was periodically destroyed, in proof of which he said he had once seen the latter being torn into little pieces by Sanz when he no longer needed it.

Buiza claimed to have witnessed on two occasions sales of copper which did not get recorded in the official accounts and for which Sanz received 80,000 (£833) and 176,000 *reales* (£1,833) respectively. He also accused his chief of obtaining wages from the finance council for non-existent workmen and of inserting his own household servants on the miners' wage lists. According to Buiza, when Sanz's son José was married he was given 1,200 dollars (some £480) from the mines' money, while Sanz's brother Gabriel was receiving 20 *reales* (approximately 4s 2d) a day from the same source.

The committees of enquiry decided on further investigations and Sanz wrote to Martín de Ulloa the absent *juez conservador*, at Gibraltar telling him to return immediately to Rio Tinto. On his arrival the judge held a conference with the administrator as a result of which he decided to search Buiza's house. In a report to the committee Ulloa claimed to have found there a set of pick-locks and copies of keys to the warehouses. The investigators were unable to reach a decision on the matter. The lawyers amongst them suspected Sanz and the civil servants Buiza. Once again it was decided that the affair was too complicated for a decision to be reached.[20]

Presumably Buiza was eventually set free without trial. Whatever the prisoner's participation in the theft may have been, it would seem certain that Sanz was engaged in at least some of the dishonest activities of the sort alleged; for in 1852, when the gallery known as San Luis was being re-opened under the impression that it had not been worked since Roman times, a hidden store of over 2,600 *arrobas* (32½ tons) of copper was found together with a small quantity of gold and silver bullion, and also coins and implements indicating that this gallery had been in use in Sanz's days but had been abandoned as the result of a serious collapse of rock. The fact that this place was being used for storage was never mentioned by Sanz. Nor did he report the loss of this particular quantity of copper at any time or reveal that he was mining for gold and silver.[21]

From the time of his appointment as royal manager Sanz was continuously involved in disputes with the council of finance officials known as the directors of revenue in connexion with his claims for expenditure on mines' maintenance and his sales' figures. At the end of 1780 he claimed his 5 per cent of the profit on sales which he had estimated as 85,645 *reales* (£892) for the two years 1779–80. The directors refused to pay him: they had not received his accounts for 1780 and for the twelve months previous to that they calculated that there had been no profit at all but an overall loss of 6,200 *reales* (£64). By 1783 the accountant's books were in such confusion that the directors insisted on a new accounting system and the making of a new inventory. When completed, the latter gave a total valuation for the mines of nearly four million *reales* (£41,667). At Sanz's request the directors agreed to add to this another million (£10,417) for the value of copper contained in unsmelted ores, but rejected the idea of including an additional ten million *reales* (£104,167) which Sanz claimed represented the cost of underground work on three different levels during the thirty-six years the mines had been in his care.[22]

The broken ores awaiting smelting were estimated to contain 24,000 *arrobas* (300 tons) of crude copper: the equivalent of four years' production which, when compared with Sanz's statistics, tended to confirm the suspicion that he had been sending to Madrid gross understatements of his annual production. Yet another team of accountants was asked to examine his accounts in the light of the new inventory. They decided that between 1758–82 Sanz had probably helped himself to profits which he had never disclosed of the order of eight million *reales* (£83,334).

In February 1783 there were a series of disputes at the mines between the Sanz faction and the civil servants sent in by the finance council. Both sides were accusing each other of obstruction and non-cooperation. Sanz declared that the mines had run out of cash and refused to pay any wages though he continued to incur debts. After fourteen weeks the directors of revenue reluctantly agreed to send him half a million *reales* (£5,208) in order to allow the resumption of work but at the same time they recommended to their minister that he should be replaced.[23] In August only three weeks later the king signed a royal decree: almost unbelievably the former tailor from Valencia was dismissed.

The years of Sanz's rule at Rio Tinto were important for showing that mining there was a practical proposition if carried out with energy and intelligence despite the twin problems of isolation and high transportation costs. The final figures prepared for the finance council showing the alleged production of all copper during his administration, understated though it must be because of his fraudulent activities, still indicate the nature of his success. In Tiquet's last year 2,501 *arrobas* (31½ tons) of crude copper had been produced. In the first six years of Sanz's management the average output recorded was 5,300 *arrobas* (66½ tons) which rose consistently during the following period to reach an average of 8,000 *arrobas* (100 tons) in the last six years. However, these figures are certainly underestimates. In the six years after Sanz left Rio Tinto (1783–88) the average quantity of crude copper produced each year was approximately 12,800 *arrobas* (160 tons).[25] Since production was not likely to have increased in those years, only to have been more honestly recorded, it looks as if Sanz may be credited with the misappropriation of about one-third of the copper produced in his time as royal administrator.

Sanz's dismissal was not the end of Sanz, however. He continued to be a nuisance to the government for another seventeen years. He began by going to Seville and somewhat audaciously called a meeting of the shareholders of the former company at which he proposed a legal action to try to recover from the crown the ten million *reales* which he still claimed he had spent on mining and exploration, and which he now declared ought to be divided among the shareholders. But they did not have his enthusiasm and most of them agreed to sell him any rights they might possess. Simultaneously Tiquet's family began two legal actions. The first (which was quickly dropped) was for a declaration that they were the legal owners of the mines and that the reversion to the crown was illegal. The second was against Sanz and the crown together, and sought judgement for a share of all profits made from the time of Tiquet's death.

In 1785 the parties reached a compromise settlement. This was, that while Sanz agreed to pay the Swedes 45,000 *reales* (£469) for any rights they possessed, the council of finance was to supply them 375,000 *reales* (£3,906) worth of salt as compensation for the loss of any profits to which they might have been entitled. They were luckier than they deserved, for their shares had been practically valueless when they inherited them from Tiquet and if Sanz had not been on the spot it is doubtful if mining could have continued successfully. Sanz and the crown agreed that he should receive 700,000 *reales* (£7,292) for the value of all the tools and goods at the mine at the time of the reversion. There was an attempt by some royal officials to have this payment cancelled on the grounds that the inventory was inaccurate, but Sanz obtained a royal decree confirming it and declaring that the inventory and valuation could not be faulted.[27]

A year later Sanz reached another settlement with the crown in which he gave up his claim for 5 per cent commission on profits during his period as royal administrator in return for a payment of 36,000 *reales* (£375) a year for three and a half years. Encouraged by this he petitioned for a pension, scornfully rejected an offer of 12,000 *reales* (£125) a year and finally accepted one of 24,000 (£250).[28]

Success went to his head: he commenced another action for the ten million *reales* alleged to have been spent on the mines, plus the delivery to him of some lead cauldrons used to make vitriol which had been inadvertently omitted from the inventory. Thereupon, the few remaining shareholders also decided to bring a claim for the joint ownership of everything listed in the two inventories of 1779 and 1783.[29]

It was ruled in February 1787 that everything at the mines at the time of the reversion had belonged to the crown save for the metal in the warehouses, the value of which would be divided proportionately among the owners of the company's shares. But it was decided it would be necessary in the interests of fairness to hear further claims from anyone owed money by the company before the payment for the metal was distributed. These claims came in such great numbers that in May 1789 they were referred to the chamber of justice for a decision. Twelve months later the judges asked the directors of revenue to help sort out the confusion but the directors declined, explaining that they were already overworked.[30]

At the suggestion of the judges a royal decree was issued in May 1791 setting up a tribunal to examine the claims and the opinions expressed about them by the judges and various government departments. The three members were named as Rosendo Saez de Parayuelo *Conde* de Montarco, Francisco Angulo the government director-general of mines and Sanz himself. However, in February of the following year Sanz wrote to the minister of finance stating that despite his endeavours the tribunal had not yet met. He added that he was living in great poverty and that though 5,000 dollars (£2,000) had recently been paid to him as an advance on his own admitted claims as a shareholder, he would be grateful for a further advance of 20,000 dollars (£8,000). Five days later the king (now Charles the fourth) ordered the immediate meeting of the tribunal.

When it met Sanz did not take his seat but was represented by a lawyer, José Moscoso, who took the opportunity to renew Sanz's claims for expenditure on mining and exploration and for compensation for the goods handed over to the crown. He suggested that the tribunal should recommend as compensation a new contract for the mines for a minimum period of twenty years to be given to Sanz and his heirs, or else an award of five million *reales* (£52,083) for the alleged expenditure on underground work plus another 120,000 (£1,250) to compensate for errors in calculating the true value of the unsmelted ores at the time of the transfer to the crown. If the latter course was preferred Sanz would be prepared to accept two million (£20,834) in cash straightway, and the post of government tax collector for the towns of Valverde del Camino, Zalamea la Real, Aracena, Castillo de las Guardas and Aznalcollar, until he had recovered the remainder of the money, together with interest at 4 per cent per annum, from the taxes of those towns.[31]

Montarco and Angulo rejected both proposals out of hand. However, so many representations were made by people claiming to be shareholders or creditors, and so much time was spent trying to get Sanz and his attorney to produce original documents, that over six years passed before the tribunal's report was ready.[32] When it was submitted to the minister in July 1798 it rejected all Sanz's claims and submissions, doubted his honesty when running the mines and commented acidly on his lack of truth before the tribunal. Amongst its findings was one declaring him to owe to the crown and the remaining shareholders 630,000 *reales* (£6,563).[33]

On 2 October 1800, while still waiting for the government's official reaction, Francisco Sanz died. It was not until November 1802 that a royal decree ordered the implementation of the report's recommendations and instructed his heirs to pay the sum owed to the crown. Immediately his family initiated new legal moves in an attempt to delay payment and claimed for themselves the freehold ownership of the Rio Tinto mines. The claim was palpably absurd but in the troubled times in which Spain found herself the case drifted on and was never settled. The family never paid the money and the pretence of ownership became part of its traditions; so that when as late as 1873 the state sold the mines to the Matheson consortium the Sanz heirs protested against the government disposing of their property.[34]

*Reconstructed view of the pueblo of la Mina (Rio Tinto) in the latter eighteenth century.*

5

# The new broom

## Director General Angulo: 1783–98

Whoever had been appointed to succeed Sanz in 1783 would have encountered difficulties, for the dismissed administrator had been popular with the workmen and had left behind him a number of his quarrelsome relatives in senior positions. It was probably for this reason that the council of finance selected for the job an autocratic and waspish-tongued official called Manuel de Aguirre, mistakenly thinking that this particular despot could impose order on the administrative confusion left by Sanz and instil a sense of discipline into the workforce. In fact few men could have proved less suitable than Aguirre who was a man of violent temper and abusive language, with little ability as an administrator.

Francisco Sanz had maintained an office in Seville where copper-traders could place orders and settle their accounts. Aguirre closed it down thinking it an expensive luxury and seeing no reason why the merchants should not travel to Rio Tinto to make their purchases and pay spot cash. But the experiences of a few of the would-be purchasers, who lost large sums of money to mountain bandits on their way to the mines, discouraged many other potential buyers from making the journey. They preferred to refuse contracts rather than to face the robbers and the other inconveniences that the trip to the mines entailed. Consequently the government sold less copper than it could have done.

Aguirre also sold the train of mules acquired by Sanz to carry the metal to Seville, taking the view that purchasers should provide their own transport. This decision, too, was a mistake. Previously a delivery charge had been made to customers to cover the cost of transporting the copper, and the mule teams and their handlers had made the return journey carrying the mines' supplies at no extra expense. But when the mule train was sold all the mines' purchases cost more as a result of the delivery charges imposed by suppliers. Aguirre also disposed of most of the other livestock, including the herd of cows kept to supply the community with milk, cheese and meat, and retained only eleven horses and three cart-oxen.,

In the three years following his appointment he cut back sharply on all expenditure, including much necessary maintenance work. Consequently a number of the mines' drainage adits became blocked, floodwaters rose and there was a considerable contraction in the size of the underground workings with a corresponding decrease in the quantity of ores excavated. For lack of repairs, the mines' buildings and houses decreased in value between 1783 and 1786 from 1,280,000 *reales* (£13,333) to 797,000 *reales* (£8,302), a drop of 40 per cent. The gardens and orchards were neglected and declined in value from 350,000 *reales* (£3,646) to 250,000 *reales* (£2,604), almost 30 per cent. During that short time the only part of the enterprise to flourish was the woodlands which, as a result of natural increase and the skill of the foresters appointed by Sanz, rose in worth from three-quarters of a million *reales* (£7,812) to one million (£10,417).

From the time of his arrival at Rio Tinto, Aguirre's behaviour involved him in a series of disputes with all grades of the mines' employees. The ill feeling mounted until in December 1786 the workmen went on strike. They appear not only to have downed tools but also to have taken a threatening attitude towards Aguirre and, alarmed at the possibility of riots, minister of finance Pedro Lerena issued a royal decree referring to the confusion at the mines, the disorder amongst the employees and the paralysis of work, and announcing the replacement of Aguirre as administrator by Melchor Jiménez.

Jiménez, however, was not to be allowed the free hand given to his predecessor. He was informed that although he would proceed immediately to the mines he would be accompanied by the crown's senior mining official, Francisco Angulo, a former professor of chemistry who had been appointed director general of mines, and would obey him in all respects. Lerena told Angulo of the arrangement and instructed him to stay at Rio Tinto as long as was necessary to re-establish order and get the mines working successfully. Profitability was essential and henceforward the Rio Tinto mines were to be the director general's principal concern. The finance council wished to be kept fully informed as to what was happening and Angulo was to prepare detailed reports for its consideration. To help in making these he was provided with Juan López de Peñalver, an engineer, and Floriano Coetanfao, a royal pensioner. Angulo was left in no doubt that he was expected to be the new broom to sweep out the mess and the muddle resulting from the years of Sanz's dishonesty and the briefer period of Aguirre's incompetence.[1]

The four men and their servants arrived at Rio Tinto on 25 January 1787. Angulo immediately handed Aguirre written instructions from the minister ordering him to go at once to the Cadiz salt works to take up the post of accountant, and then made formal delivery to Jiménez of the mines and the attached properties. That same day an audit was started into Aguirre's accounts and work began on an inventory. Angulo left Coetanfao to supervise these while he persuaded the strikers to return to work, and then spent five days on making an inspection of the mines with Peñalver, the engineer he had brought with him.[2]

On the sixth day he wrote his minister a detailed report. It was obvious to Angulo that unless the mines could be put in order quickly future profits would be swallowed up by unnecessarily high running expenses. The shafts and galleries were in a very insecure state and capital expenditure on these and on the drainage adits was the first essential. If this was done, and certain technical problems were overcome, then the richness of the copper, sulphur and iron ores would give good profits.

One of the needs that had to be made good was a lack of maps of the underground galleries. This deficiency led to haphazard mining. To achieve greater efficiency it was necessary to have plans showing both the ancient and the modern workings and consequently Angulo asked the minister to send him a skilled mining surveyor and a cartographer.

The director general thought that the men carrying out the smelting of ores and the refining of the copper were not sufficiently expert in the work, and, although he appreciated that there was a general lack of knowledge on these subjects in Spain, it was plainly something that he would have to remedy. He found that production methods were old fashioned and uneconomic. Excavated ores for example, were drawn up from the bottom of shafts to the surface with windlasses operated by hand, a slow method and costly in wages. A machine worked by a horse and only one man would be able to raise larger quantities at a time and be less expensive to operate. The mines' carpenter could make a horse-whim for between 7,000 and 9,000 *reales* (£73 and £94) and Angulo asked Lerena for permission to have at least one constructed.

He found that there was a practice whereby all energies were concentrated on excavating great quantities of ore which were stockpiled until it was decided to curtail mining and switch to calcination. This meant that the ore was often exposed for long periods to dampness in the air which made it crumble into small pieces. These in turn choked the cavities and cracks in the calcinators which prevented circulation of the air during roasting, extinguished the fires and led to the dumping of immense quantities of half-burnt ores. He proposed a system of conterminous mining and calcination with a regular flow of excavated minerals to the calcining ground.

Angulo was worried because the ores contained so much sulphur that three calcinations were required to get rid of it and these required considerable quantities of wood, a highly expensive item at Rio Tinto, for not enough could be obtained locally and what was purchased in the countryside around not only cost a great deal but also incurred heavy transportation charges. He commented that in Hanover a method was used which removed almost all the sulphur in one calcination and, instead of sending it up into the air in thick fumes, permitted its collection and thus its sale. If it was possible he would introduce the German system at Rio Tinto. One day the policy of afforestation introduced by Sanz would ensure abundant supplies of wood but that time was still some years ahead.

He saw no reason why the output of copper could not be increased as the furnaces were standing idle much of the time, but it was essential to improve the quality of the refining. The royal cannon foundry at Seville, which relied very largely on Rio Tinto copper, found the metal had to be further refined on delivery before it could be used for making armaments; an unnecessary expense. To achieve the improvement he proposed the installation of reverberatory furnaces. Since at that very time Franz Peringer a German expert (and son of Johannes Peringer, one of the two brothers Sanz had employed at Rio Tinto) was constructing such furnaces at the Seville foundry, Angulo asked Lerena to arrange for him to be sent to the mines to install some there. Angulo reckoned that this would not only cut production costs but also allow higher prices to be charged to the Seville merchants who bought such copper as was surplus to the government's own requirements. For the benefit of these traders Angulo also proposed reopening the office in Seville which would encourage increased sales.[3]

In the following six months Angulo waited impatiently for the minister's decisions. In the meantime he had grown tired of waiting for a cartographer and had got Peñalver to start preparing the necessary maps of the underground workings, though he was making slow progress because of the dangerous condition of many of the galleries and their intense heat which made it impossible for him to stay in some parts of the mines for more than a short while. Nonetheless, when the minister of finance did reply on 7 June he told the director general that he was sending to help in the task Leopold Stonz, a German engineer who had been specially recruited in France, where he had been working, to be Spain's sub-inspector of mines, and Stontz's nephew Heinrich Schnell, a mathematician and draughtsman. Otherwise the letter was disappointing. It did not even refer to his request to introduce horse-whims and reverberatory furnaces. Angulo sat down to write once more for permission to install these.

In this letter he told his minister of another problem which had manifested itself with the coming of harvest time. For to Angulo's dismay the labourers, although in his opinion hard-working as a class of men, all quit the mines at the start of the harvest to go into the countryside around to earn the high wages paid to temporary agricultural workers. The only way that this could be prevented in future, he thought, was by improving working and living conditions and by providing social benefits for the men on condition that they did not terminate their work at the mines in this way every summer. The improvements he proposed included the building of more houses to relieve overcrowding, providing the mineworkers with the privilege of exemption from military service, the setting up of a fund (financed partly from mines' revenue and partly from workmen's contributions) to give the men sick-pay in time of illness, and the introduction of a regular system of pensions for aged former workmen. Angulo also considered it necessary to make jobs more attractive by replacing as much manual work as possible with machinery.[4] Plainly, the director general was taking the minister at his word and was intent on providing him with the blueprints for a radical transformation of the Rio Tinto mines. The changes being urged by Angulo made good sense and, although the ideas behind them were advanced for the time, they can be seen as a natural result of the national revival stimulated by the policies of Charles the third's benevolent despotism. Angulo, as one of the younger generation of highly efficient civil servants coming to the fore after thirty years of enlightened royal rule, was proposing merely a logical extension of the king's own programmes.

But royal despots, however enlightened, grow old as other men; and the reign of Charles, which had done more for the advancement of the Spanish people than any other, was drawing to its close. The king's intense activity began to decline at last and, since the system of government was totally dependent on the sovereign's genius, the whole administrative machine began to slow down as the royal hand grew slack. Moreover, ministers and senior officials began to anticipate the likely requirements and attitudes of the heir to the throne, the Prince of the Asturias, and realized that different standards would prevail with his accession. In consequence Lerena's earlier decisiveness began to disappear. With a change of king just ahead he began to display a caution in his letters to Angulo that irritated the director general.

The minister wrote suggesting that the mines' carpenter's estimate of 7,000 to 9,000 *reales* for making a horse-whim was a little high and thought that Angulo should get the *asistente* of Seville to contact two Englishmen, a father and son called Binns, who were working in that city and were experienced in erecting this type of machinery, to ask what they would charge. As for the other proposals, such as the installation of reverberatory furnaces, it would be necessary to give them full consideration which would take a little time. Angulo's request for permission to build thirty-two new houses for workmen could not be considered by the finance council without a precise idea of the cost, but Lerena thought it unlikely that authority would be given to build more than twelve in the immediate future.

To this letter Angulo replied straightway. At least twelve houses were urgently needed and he proposed that of these four should be allocated to clerical staff and eight to senior workmen in the supervisory grade. The plans and designs had already been prepared by Heinrich Schnell and the clerks' houses would cost 12,000 *reales* (£125) each to build and the senior workmen's 4,000 *reales* (£42). Without explaining the reason behind his thinking he added that he had come to the decision that it was best to separate the homes of the two classes of employees and in future the houses for the clerical staff would be built about one and a half miles from the *pueblo* at the mines. When the signal came to go ahead everything was ready and the new houses were started at once under Schnell's supervision.[5]

However dismayed he may have felt at the minister's growing coolness, Angulo went ahead with his preparations for a scheme of social benefits for the workmen. It is a measure of the man's originality that not only did he devise a plan which was imaginative and far-sighted for the time but also that he worked out the details in discussion with the workmen and secured their unanimous support. He dropped his proposals for long-service pensions since the men themselves took the view, common in that period, that the care of the aged was basically the concern of the village community in which they lived. The workmen, though, were enthusiastically in favour of a sick-pay scheme and agreeable to the setting up of a fund to which they would make contributions of between 2 and 4 per cent of their earnings according to their employment categories and be entitled when sick to receive between 30 to 60 per cent of normal pay (save for the lowest paid category of boys who would receive full payment).

Angulo proposed that any additional money needed for the fund in excess of the men's contributions could come from the profits being made on the supplies sold to the men and their families in the government-owned shop at the mines. The men asked Angulo to tell the minister that when the fund had grown in size they hoped it would also be possible to use it to provide small pensions for the widows and orphaned children of mine workers.[6] But by this time the old king was dead. The dull and amiable Charles the fourth (1788–1808) was wearing the crown and the country was being ruled by his imperious wife Maria Luisa of Parma and whoever was her lover for the time being. The new reign was producing too many problems for the politicians for them to worry about the social welfare of Rio Tinto miners, and ministers found it easier to slip into inertia rather than to advocate policies which might arouse opposition or incur disapproval. Despite Angulo's periodic attempts to raise the matter, the scheme remained unimplemented although his other request, for the exemption of the Rio Tinto employees from military service, was largely conceded when all workers other than day-labourers were freed from the obligation.

Meanwhile Angulo concentrated on production difficulties and on industrial safety. The maps of the mines' interiors which were being made by Peñalver and Schnell filled him with alarm. Hitherto no thought had been given to the positioning of the walls and the pillars left behind when galleries were being excavated, and the weight of these was bearing down onto the floors causing subsidences in the passages underneath. He ordered that in future the pillars and walls were to be immediately above and below each other; and to reduce the number of workmen killed in accidents he had stone taken down from the surface to build extra pillars and to prop up ceilings, together with vast quantities of bricks to construct walls wherever there was a danger of a fall-in from above.

The casual nature of the previous eighteenth-century work, and the complications caused by the presence of ancient workings, had produced galleries which sometimes had floors less than a foot thick. Together with his supervisory staff Angulo planned future underground extensions so that they would all have uniform heights and widths, with those passages leading off from the main galleries inclining slightly upwards. This innovation was intended to make it easier to push the small, four-wheeled handcarts he was introducing which were being made to his design by the mines' carpenter. Up till then the broken ores had been shovelled into barrels at the working face and these had been rolled along the passages into the main galleries and thence to the bottom of the shafts where their contents were tipped into bags for winching to the surface.

Angulo then commenced the construction of a main shaft to cut through all the various work-levels of *cerro* Salomón to make it easier to extract and raise the mined ores. For this purpose he chose the ruinous San Gabriel shaft. He had the rubbish and the rockfalls removed, and had work started on widening it and extending it downwards through the heart of the hill. When this was done it needed only a few short communicating passages to be cut in order to link it up with all the many galleries.

Angulo was concerned, too, by the number of accidents to borers at the work face. These men used iron jumpers (chisel-ended bars) which often snapped in the hard rock when being struck on the end with a hammer. By replacing these with copper jumpers he succeeded in cutting the number of casualties and thus made a significant contribution to mines' safety.

In one of his reports Angulo described the system of piece-work used at the mines. It had been introduced by Sanz and was universal in Spanish mines at that period but, as Angulo appreciated, it had a number of drawbacks. By this system the chief supervisor granted contracts to senior underground workers for four-week periods. Each of these senior men recruited his own team of workmates from amongst the men on the approved list, and supervised them while they cut the ores in that part of the mines described in his contract. The team was paid a price, also specified in the contract, for each ton of ore excavated which was calculated on whether the section where they were to work was likely to be easy or hard going, the rate being lower in the former and higher in the latter case. The chief supervisor aimed at fixing a price per ton which would provide the workmen with an average pay of 1 to 2¼ *reales* (2½d to 5¾d) a day for boys (depending on age), 2½ to 4¼ *reales* (6¼d to 10¾d) for the miners (according to their category) and 4½ (11¼d) for the supervisor of each team.

Because the hardness or softness of the rock in any section could alter markedly in the course of a month's contract, the men often found that their daily pay suddenly increased or decreased due to the nature of the ground and through no ability or fault of their own. When earnings suddenly went down the men and their families might have to face many days of hardship until the time came when the new contracts were drawn up. On the other hand when earnings jumped the men could earn 12 *reales* (2s 6d) or more a day. When this happened, instead of saving the money against periods of reduced earnings, they often stayed away from work for a couple of days until they needed more cash. Angulo rejected the idea of paying the men fixed daily wages as in his opinion this would lead to a serious drop in output. But he proposed the introducing of weekly contracts to smooth out the anomalies and to be fairer to the men.[7] But despite his recommendations, and the petitions of the men themselves at various times, the system continued in use largely unaltered until as late as the beginning of the twentieth century.

Angulo was also experimenting with methods of calcination. He tried covering the sides of the piles of ore to be burnt with large clods of earth in order to keep the rain off them, thus stopping crumbling, and to contain the sulphur and prevent its discharge into the air. But he found that the earth made it difficult to keep the ores burning, while it still needed only a moderately strong wind to carry the sulphur off in fumes as before. The torrential and constant rains which were a feature of the wet season at Rio Tinto also made it difficult to keep the fires going and so he had kilns built with penthouses above to protect the burning piles. The first kiln made was 14 ft long, 5 ft high and 5 ft across, and consisted of a brick base and two parallel brick walls pierced with holes at two levels. Angulo found this reduced the amount of wood needed to keep the fire burning and proceeded to build some more which were designed to hold twice the quantity of ore and with the walls 8 ft apart and with only one level of holes set low down. These kilns in turn cut wood consumption still further and increased the quantity of ores that could be used in the smelting furnaces.[8]

Jiménez, the mines' administrator, and other senior members of the staff, followed by Schnell and López, after a while developed the symptoms of the disease known as tertian ague, a form of malarial fever in which the paroxysms occur at intervals of three days. The director general decided to take his team of assistants off to Cordova for a holiday in the belief that 'a change of air would be the best remedy to shake off the fear engendered by the illness'.[9] While on holiday he met a Frenchman called Gabriel de Montrosier and as a result of their conversations decided it might be practicable to establish a manufactory for vitriol and copper as at Rio Tinto. These items had been produced at the mines from the time of Tiquet until the departure of Sanz, and it would seem that this was yet another activity abandoned by Aguirre during his period as administrator.

Angulo wrote to his minister proposing that the vast quantities of small calcined ores that were being dumped as unusable could be placed in tanks of lye (which was water containing alkaline salts and was freely available at the mines as it was the ordinary drainage water drawn off through the adits) and then subjected to evaporation. The copper residue in the ores could then be extracted by the immersion of iron in the water which would also give considerable quantities of vitriol.

'In fine, I have an idea, and experience will show the truth, that the method of reducing all ore at Rio Tinto . . . to vitriol by calcination, and subsequent working to precipitate or to separate the copper by means of old iron, would be more advantageous than the method of smelting now pursued, because then there would be a small consumption of charcoal and much iron now cast away might be utilized and even converted into green vitriol which is more valuable.'[10]

He asked permission to construct tanks and to purchase iron for trial runs but his request was ignored and he was unable to implement his proposals. (Fifty years later the operation was patented by an Englishman living in Seville as his own invention and successfully used at Rio Tinto as a way of producing copper at small expense and with a high profit margin.)

Following a long correspondence with Francisco Antonio Palacios, the manager of the Linares mines, Angulo also prepared a detailed job-analysis for all the senior twenty-nine permanent employees at Rio Tinto which resulted in a royal ordinance in May 1788 entitled 'royal instructions for the good administration, government and management of the royal copper mines of Rio Tinto.'[11] This provides an unusual description of the arrangements for running a large Spanish mine and the duties of its managerial staff in the late eighteenth century.

The administrator was paid 20,000 *reales* (£208) a year and provided with a rent-free house and wood for his domestic use. Responsible for the general running of the mines he was specifically charged with periodic inspections of the works and the produce, the making of annual inventories, signing all business documents being sent from the mines and authorizing the purchase of all tools and equipment after consulting the accountant. The sending of monthly production statistics to the director general, and the arranging of copper sales in accordance with the policies laid down in Madrid, were also part of his responsibilities. Likewise the safe keeping of one of the three keys to the chest holding the mines' cash. (The other keys were held by the accountant and the treasurer.) The administrator alone had the powers to employ and dismiss workmen and had the duty to look after their general interests, being required in particular to enforce their rights to exemption both from military service and from payment of customs' duty on goods brought to the mines for their use. He was also under instructions to provide employees joining or quitting the mines with transport for their belongings.

By this ordinance the administrator was granted the official titles of *alcalde* and *juez*. Theoretically the former appointment could be made only for a local government district and there was then no *ayuntamiento* of Rio Tinto. Angulo, however, had witnessed the incessant quarrels with the town officials of Zalamea la Real and decided that the appointment of the administrator as *alcalde* of the mines' area would both enhance his status in any future disputes and make plain that the crown rejected any claims by the *alcalde* of Zalamea to jurisdiction over the Rio Tinto mines.*

*When the *ayuntamiento* of Rio Tinto was eventually created in 1841 the title of *alcalde* passed to the head of the new district council.

Under Spanish mining laws the chief administrator of a royal mine

*The Rio-Tinto Chair*

had the powers of a judge to deal with offences by workmen, since the maintenance of discipline in a mining camp posed problems of a different quality and quantity from those of a normal village or town. When the mines were leased out to private individuals the rights were exercised by a *juez subdelegado*, as at Rio Tinto during Sanz's period as private administrator. The new ordinance in making the administrator an official magistrate, specifically gave him the right to try all offences within the boundaries of the mines' area. It is probably from this time that the Rio Tinto chair dates. This is an imposing wooden throne, carved on the back with a crude though vigorous representation of the Spanish royal arms and crown, which served as the justice-seat of the head of the mines when hearing legal cases.*

The second mines' employee in order of seniority was the accountant whose department, called the counting-house, contained three clerks. Paid 12,000 *reales* (£125) a year, he was in charge of the accounts' ledgers and also of the two post-offices in the mines' area, each of which was manned by two clerks. Third in the hierarchy was the treasurer who handled the cash, making payments and receiving the money paid in settlement of bills, and receiving 6,600 *reales* (£69) as his salary.

The fourth most senior employee was the head supervisor of the mines whose duties were described at great length in the new regulations. Plainly Angulo thought of him as a key figure. With a salary of 5,500 *reales* (£57), and a rent-free house like all the senior staff, he was responsible for supervising the work of all the men working underground and at the calcination and smelting grounds. In this he was aided by two supervisors, three superintendents of calcinations and two superintendents of smelting. It was the head supervisor who decided hours of work and holidays and worked out the details of the monthly contracts awarded to the senior miners. Subject to the administrator's instructions, he settled the wages of the non-mining workmen, and had the duty of distributing and maintaining tools and equipment, and of making monthly checks to discover deficiencies. He was also required to prevent the workmen from using the mines' tools and workshop facilities for private work unless they made appropriate payment.

At sunset every working day the supervisors, superintendents and foremen met at the head supervisor's home for discussion on the day's work and to receive the next day's orders. Dating from the time of the ordinance he was to meet supervisors and superintendents every Saturday morning to examine the lists which showed the number of days worked by each workman and the pay due to him, and sign them

*Following the creation of the Rio Tinto *ayuntamiento* in 1841 the chair was given to the *alcalde* for his official seat: an act of generosity possibly prompted by its extreme lack of comfort. After the acquisition of the mines by the British it was given by the *ayuntamiento* to the Rio Tinto Company's London office for use by the chairman. After eighty years of such use (proof perhaps of the British ability to endure sustained physical discomfort) it is now preserved as a historical relic.

if correct as the authority for the counting-house and the treasurer to make payment. This provision also ended the old arrangement whereby wages were paid at the end of each day, and replaced it with a weekly system which was presumably intended both to make work easier for the counting-house clerks and to encourage the men to work a full week without taking time off.

The rest of the senior employees consisted of a charcoal storekeeper, a warehouse keeper in charge of the copper awaiting sale, a granary keeper, a foreman in charge of the combined blacksmith's and carpenter's shop, a foreman of works and a head porter, all of whom came under the chief supervisor's control. There were also other members of the staff whose activities, though unrelated to mining, were essential to the community: two chaplains to serve the church, a surgeon to tend the sick, a schoolmaster to teach the children, and a constable whose main duty was to keep order in the taverns, and these answered directly to the administrator.

Melchor Jiménez had to resign from the post of mines' administrator in August 1788 as a result of the recurrent effects of the ague, and was given the post of chief government accountant of the province of Cordova. Surprisingly, Jiménez's predecessor, Aguirre, was appointed to succeed him, and the announcement was soon followed by Aguirre's additional appointments as *alcalde*, *juez* and, at his own request, a grant of the army rank of *comisario de guerra* by the council of Castile. Once back at the mines, Aguirre made what must have been strenuous efforts to control the bad temper which had made difficulties for him during his previous administration at Rio Tinto and thereby earned Angulo's approval.

When fresh disputes arose with Zalamea la Real about the mining community's rights of pasture and the right claimed by Zalamea's citizens to collect the annual crop of acorns for their pigs from within the mines' boundary at the place known as 'el escorial' on the north side of Salomón, Aguirre proposed the appointment of an independent land surveyor to mark out the boundaries and to arbitrate where the two sides were in disagreement. Andrés de Cañete, the surveyor to the city of Seville, undertook the task in November 1790. With the assistance of representatives of the mines' administration and of the town of Zalamea la Real, he was able to mark out boundaries to the satisfaction of both parties: a notable achievement after so many years of controversy. It gave Angulo, once again in Madrid, a great deal of pleasure.[12] For him its importance lay in the fact that henceforward there need be no further disputes over grazing lands; a matter of some concern since, to meet the needs of the increasing population of the mines, Angulo had purchased new herds of cattle.

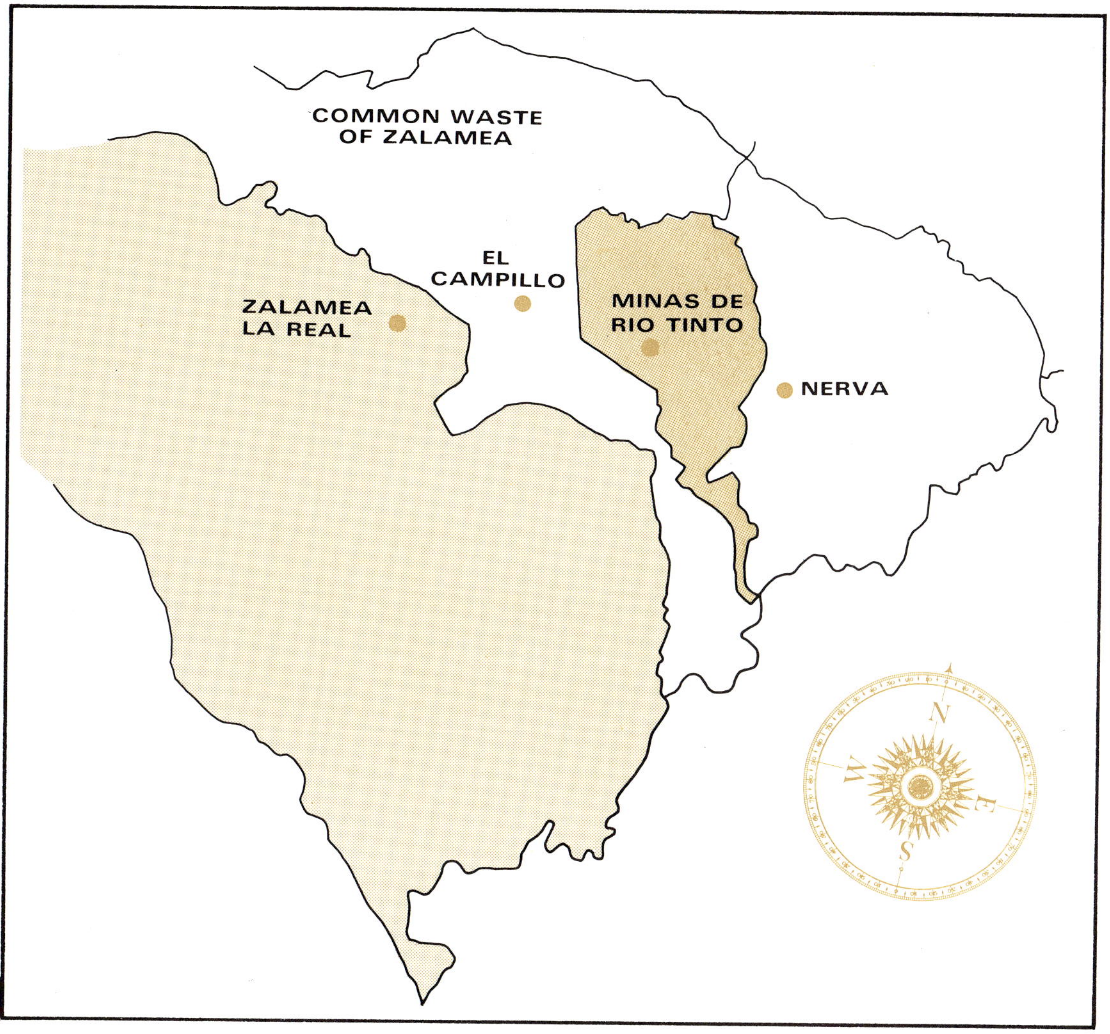


*Map showing the boundaries of the mines' area (as agreed in 1790) in relation to the district of Zalamea la Real.*

It was this growth in the number of workmen and their families that made it necessary for Aguirre to rebuild the church of Santa Barbara*, in the mines' village of la Mina on the southern slopes of Salomón, built by Sanz to hold a congregation of 600, so as to accommodate 900 worshippers.

Although by this time Angulo had returned to his office in Madrid because of pressure of work, the Rio Tinto mines still occupied much of his attention. He was keeping in close contact with Aguirre through letters and occasional visits to Rio Tinto, and urging his minister to take steps to make the mines profitable while at the same time participating in the interminable discussions and exchanges of memoranda connected with the financial claims that Francisco Sanz and his lawyer were making on the government.

*The patron saint of all users of explosives such as miners and gunners.

In 1790 Angulo received permission at last to employ Franz Peringer to install reverberatory furnaces at Rio Tinto. The German spent nine months there installing the furnaces and instructing the workmen in their use. The first results delighted Angulo. There was a great improvement in the quality of the refined copper. Not only did it cost less to produce but it became possible to charge a higher price for it to the Seville traders. Santiago Hidalgo, the director of the cannon foundry at Seville, was also enthusiastic about the improved copper being delivered there although it meant he had to dismiss a number of his own refiners as redundant. He proposed that they should be given jobs at Rio Tinto; a suggestion which Angulo rejected indignantly. It would not be sound business, he declared, and would offset the profits being made on the operation.

However, Peringer's departure to take up the post of head refiner at Seville coincided with a sharp decline in the quality of the copper being sent to the foundry from Rio Tinto. At the German's suggestion Angulo ordered the mines' refiners to report to the foundry for a further five months' training. Once there, the men showed themselves quite adequate in their work and Peringer could not understand why they had produced such inferior copper when at the mines. Mystified he sent them back, but within three weeks of their return the Seville and Barcelona foundries were protesting strongly to Angulo about the inferior nature of the copper again being delivered to them from Rio Tinto. The director general's immediate reaction was to assume that the mines' refiners were just not taking enough care. Their knowledge was plainly sufficient for they had all been trained by Peringer and their supervisor, Clemente Wert, had been awarded the title of master-workman on Peringer's nomination. From Madrid, Angulo gave instructions for one of the redundant Seville refiners to be sent to the mines with the title of chief inspector. This man's arrival at Rio Tinto solved the mystery. He found that the clerk in the warehouse was issuing the poorest quality copper for refining in the reverberatory furnaces instead of the essential red antimonial copper.[13]

For the first three years of his second term as mines' administrator Aguirre managed to remain on good terms with his staff. But in the spring of 1791 problems started once again to arise over his violent temper. Senior employees bombarded Angulo with complaints about the discontent he was causing among the workmen and clerical staff. The director general reprimanded Aguirre gently in a letter but within four weeks found it necessary to send him a strongly worded order to moderate his behaviour. For a few weeks this had the desired effect but in July 1791 a particularly violent disagreement arose between Aguirre and the mines' surgeon, Salvador Bonar, about the condition of the workmen's housing. Bonar felt it was grossly inadequate, a view which Angulo privately shared, but he knew it was difficult getting money out of the finance council for the building of new houses and that the director general was doing his best. Aguirre, however, instead of explaining matters to Bonar told him that the matter was none of his business. It was a subject which not unnaturally aroused strong feelings amongst the men with families, and the atmosphere quickly became unpleasant with staff and men showing open hostility to Aguirre and his orders.

Vicente de Letona the accountant, who had married Francisco Sanz's grand-daughter, wrote privately to Angulo suggesting that he should send a strong warning to Aguirre about his despotic behaviour and at the same time a stern reprimand to Bonar telling the surgeon to confine his activities to the practice of medicine and to looking after the pharmacy. Letona may well have been trying to curry favour with the director general but Angulo thought him a sensible young man. He despatched the two letters as suggested and also sent a third. This was to Letona thanking him for his concern and asking him in confidence for his opinion as to why there had been a sudden drop in the amount of copper being produced by the calcination method. Letona's reply was disingenuous in the extreme. He explained that Aguirre had been concentrating all his energies on trying to improve production at the furnaces, and had therefore been compelled to ignore what was happening at the calcination grounds where the men were being difficult because of the controversy over the housing. However, this fault would, in Letona's opinion, soon be corrected as the administrator had just decided to shut down four of the furnaces and reduce the amount of smelting and refining while he devoted all his time to the calcination work!

During the next nine months Letona kept Angulo posted about all the arguments in which Aguirre became involved. Then, in April 1792, Bonar the surgeon, Albores the parish priest, and Diego Dominguez the master machinist sent a formal complaint to the council of finance about Aguirre, referring to his crude and violent language and his refusal to order improvements to be carried out to the workmen's houses, many of which, they claimed, were in a bad condition and overcrowded, making it impossible for married couples 'to lead decent lives'. Angulo asked Jiménez the former administrator to visit the mines and prepare a report but on his arrival on 12 November Jiménez received another letter informing him that Aguirre was relieved of all duties and that he himself was reappointed to his duties as administrator.[14]

Angulo had been able to persuade the finance council to act quickly in the matter, but in most other respects he was unable to get the council to move as swiftly as he wanted. However, as the result of a couple of years of patient urging he was able to tell Jiménez shortly after the latter's reappointment that there was at last permission to build ninety-three new houses at the mines and Jiménez was instructed to put the work in hand.

In the first report Jiménez told Angulo that there were serious defects in all the production processes at Rio Tinto. He was already improving output at the calcination ground and thus supplying the furnaces with plenty of calcined ores, but he himself knew little of refining and smelting and his attempts 'to cultivate the little intelligence possessed by the master-workmen' met with little success. He felt obliged to ask if Franz Peringer could return to Rio Tinto with his son José to provide the men with a fresh course of instruction. The finance council agreed to this with reluctance and the comment that the chief inspector recruited from the Seville foundry was clearly not earning his pay.[15]

The officials of the directorate of revenue at the ministry of finance started to complain about the level of copper production and the mines' lack of profit. Though production was greater than it had ever been officially during Sanz's administration they calculated that the mines were losing money heavily, the loss for the year 1795 alone being almost 333,000 *reales* (£3,469). Later they calculated that during the period February 1783 to December 1796 there had been 171,800 *arrobas* (2,148 tons) of copper made which had been sold for a price of 4½ *reales* (11¼d) a pound. This brought in over 19¼ million *reales* (£200,522) but as the entire outlay on the mines, including housing and wages, amounted to well over 20 million (£208,334) there was an overall loss for those years of 850,000 *reales* (£8,854). Consequently Jiménez was forbidden to spend any money on repairs to houses and to drainage adits.

Angulo and Jiménez pointed out forcibly that the losses were directly attributable to causes outside their control. The cost of charcoal had risen four-fold between 1783 and 1796. Transportation costs had increased steeply and the finance council had not permitted these rises to be passed on to customers. Moreover production costs had risen because of lack of money for maintenance work, which had led to rock falls and inundations cutting off those parts of the mines which contained rich copper seams and could have been worked at a relatively low cost. In Angulo's view the capital outlay of those years had not been wasted. The main problem was the government's refusal to allow increased prices for Rio Tinto produce.

He made a survey of prices in Spanish copper markets in May 1796 which showed the following prices for a pound of copper: Catalonia 4 *reales* 32 *maravedises* (1s 0¾d); Corunna 10½ *reales* (2s 2¼d) for Spanish and 9½ *reales* (1s 11¾d) for imported south American copper; Saragossa 5 *reales* (1s 0½d); and Pamplona 6 *reales* (1s 3d). He accepted that the price of Rio Tinto copper had to be kept low since the government was its main purchaser and what was sold to the Seville copper merchants had to bear additional heavy transportation costs, but he thought the price of 4½ *reales* (11¼d) too low. Next month the finance council revised the prices for Rio Tinto produce. The price to the Segovia mint was fixed at 5½ *reales* (1s 1¼d) a pound and to the Seville foundry at 6 *reales* (1s 3d), the cost of transport being included in both cases. For private purchasers the price was to be 5¼ *reales* (1s 1d) excluding transport charges.[16] These increases did not satisfy Angulo. As he pointed out, the most striking feature of the accounts was the continuous escalation in the cost of producing a pound of copper. The figures for 1789 to 1802 show that between those dates the actual cost of extraction, precipitation, calcination, smelting and refining for each pound of copper rose from 2 *reales* 19 *maravedises* (6½d) to 5 *reales* 28 *maravedises* (1s 2¼d), and amounted to a total of almost 21 million *reales* (£218,751). But the gross costs of running the mines, which included expenditure on 105 new houses to accommodate 500 people, the making and repairing of roads, purchasing tools and furnaces and maintaining the plantations, came to almost 27 million *reales* (£281,251).[17]

The financial situation of the mines was worsened by the policies in other fields pursued by the Spanish government from 1793 onwards. The French revolution and Charles the fourth's attempt to save the French king, Louis the sixteenth, from execution led Spain into a war with France. Godoy, a twenty-five year old army officer and the lover of the fifty year old queen, became chief minister. Under his incompetent rule there was a succession of defeats until the war ended in 1795. By the treaty of San Idelfonso the next year Godoy, by then decorated with the somewhat inapt title of 'the prince of the peace' committed Spain to join France in a war on Britain. Disaster struck from the start. The Spanish fleet was destroyed at cape Saint Vincent while Trinidad was lost to Britain and, for a time, Minorca. The country's economy suffered with the rising costs of war and the ever increasing incompetence of the government. As far as Rio Tinto was concerned there was little money available for capital investment. Expenditure on maintenance was cut back still more causing drops in production and thus even greater losses.

By the end of 1796 the mines' debts had escalated to 4 million *reales* (£41,666) as a result of rising costs, and the mines' officials had to be given authorization to draw on the public revenues of Seville province to obtain working capital. That December Angulo wrote a memorandum to the council of finance. He put the alternatives starkly. He could not agree to the sale of Rio Tinto copper at uneconomic prices to the royal mint and the foundry. There must be either a realistic price increase to prevent the mines closing down through lack of funds, or else the mines had to be let out on lease to a private contractor who would be able to manage them with greater economy and profit than any government department.[18] Angulo's advice went unheeded. What little energy the government had was expended in jumping to comply with orders emanating from Paris and in trying to stave off the French government's displeasure; a hard task, for even the excessively servile Godoy was considered not sufficiently compliant and at French insistence was temporarily removed from office in 1798.

In September of the latter year Jiménez, dogged by ill-health and dispirited at having to cope with the problems of the Rio Tinto mines on an ever decreasing supply of money, asked Angulo to relieve him of his post. A transfer was arranged for him to Seville as the superintendent of the royal mint and the accountant, Vicente de Letona, was promoted to be administrator of the royal mines of Rio Tinto in his place.[19]

Despairingly, Director General Francisco Angulo resigned himself to events. Having spent ten largely frustrated years trying to realize the vast profits which he knew could be obtained from the Rio Tinto mines for the benefit of the ungrateful and inefficient government of Charles the fourth, he gave up the struggle. As one of the most capable of all the royal servants, imbued with the ideals of his former master Charles the third, he eventually found himself welcoming the invasion of the Napoleonic army in 1808 as holding out some hope of efficient government and social reform. When the French emperor, Napoleon the first, contemptuously dismissed Charles the fourth from the Spanish throne and bestowed the kingdom on Joseph Bonaparte, Angulo agreed to become minister of finance. Doubtless he saw this recognition of his merits as an opportunity to apply himself more effectively to the interests of Spain, for it is impossible to doubt his dedication to his country's welfare.

But the course of events in Spain, the cynical disregard shown by the French régime for Spanish interests and the brutality meted out by the French army, must soon have caused him to have doubts about the way he had identified himself with the invaders. Finally and ironically, when his countrymen united with the British to chase the

French from Spain, he was declared a traitor to his own country and an enemy of his own people. In 1813, tacking himself on to the coat tails of the beaten Napoleonic army, he withdrew into France to live the life of an exile and an outcast.[20]

*The town of Castillo de las Guardas.*

# 6 *The devil of the mines*

## The Rodriguez family: 1798–1818

The twenty years from 1798 to 1818 which covered Vicente de Letona's administration of the mines coincided with some of the most disastrous years ever undergone by the Spanish economy. Much of the Rio Tinto story in that period reflects what was happening at a national level throughout Spain. Remote and inaccessible though the mines were, the people of Rio Tinto found their lives entwined with the political convulsions of Europe.

The problems facing the miners and their families during those years of national crisis were made worse by bitter disputes within the local community arising from Letona's inability to exert his authority. In this respect the mines' administrator was the victim of bad luck. Basically an honest and competent man at a time when life was so bad that most of his colleagues found it necessary to be dishonest and easier to be incompetent, he had the misfortune to marry a granddaughter of Francisco Sanz. Despite the purge of some of the Sanz family in 1779, and Francisco Sanz's own dismissal in 1783, a number of the family still occupied important posts in the mines' administration, and on Letona's appointment as chief administrator in 1798 he found himself under great pressure to adopt his in-laws' attitude that the mines' revenues existed primarily to meet their need for ready cash. In 1817 a government report reviewing the events of Letona's administration took the view that 'he is intelligent and capable but has not had sufficient force of character to withstand the domination of his father-in-law and his wife's other relatives.'[1]

His father-in-law was Atanasio José Rodriguez who had originally gone to Rio Tinto to work as a stone-mason and had married Francisco Sanz's only daughter at some date prior to 1779. Rodriguez was made keeper of the charcoal warehouse to mark his entry into the family and treasurer shortly afterwards. By 1779, too, Vicente de Letona had travelled to Rio Tinto to become a clerk in the counting house and in that year was promoted by the finance council to succeed Gabriel Sanz as accountant. Shortly before his appointment as head of the mines in 1798, the then middle-aged Letona married Rodriguez's young daughter and accepted her two brothers into the counting house as second and third clerks.

During the period of Director General Angulo's supervision there was little opportunity for Rodriguez senior to take advantage of his family connexions or his position as treasurer. But Angulo's withdrawal from active participation in the management of the mines as a result of his disillusionment with his political masters, which coincided with Letona's appointment as administrator, gave Rodriguez the opportunity he needed. When Letona and his wife moved into the large stone house reserved for the royal administrator they were accompanied by the entire Rodriguez family. From that moment started a long succession of complaints to Madrid about the way the treasurer treated not only the staff and workmen but also the inhabitants of the surrounding towns and villages. From the administrator's house Rodriguez ruled the mines and the neighbourhood around by corruption, threats and violence.

Amongst the country people Rodriguez became known as *el satanás de las minas*: the devil of the mines. His two young sons also acquired a reputation for imitating their father's tyrannical behaviour in the nearby villages, and his wife proved an insufferable meddler in the mines and local affairs, constantly playing the *grande dame* as the daughter of Francisco Sanz but equally prepared to get her own way by intimidation in her role as the treasurer's wife. What the Rodriguez family wanted they normally took, and those who opposed them found that they had to deal with the strong-arm thugs employed by the treasurer to help enforce his tyranny. Complaints to the council of finance were useless. The harassed and incompetent government of Charles the fourth was unable to provide any redress of the people's grievances against local despots such as Rodriguez, and the protesting letters were apparently merely filed without even acknowledgement or consideration.[2]

The general decline in the national economy at the end of the eighteenth century, as a result of government extravagance and war, had marked effects at the Rio Tinto mines. There was a considerable fall in the demand for copper within Spain and a steep decline in prices. The near exhaustion of the royal treasury led in 1799 to the closing of the Seville cannon foundry in an attempt to save money. For forty years or more the Seville works had been the principal purchasers of Rio Tinto copper and the closure led to a drastic curtailment of production and large-scale unemployment at the mines.[3]

It would have been possible forty years earlier in the face of such a calamity for the workmen to have quit Rio Tinto and looked for work elsewhere, for in those early days the mines were worked largely by young unmarried men who had travelled there from other parts of Spain. But the policies of Sanz and Angulo had led to the transformation of the mining camp into a settled community to which men took their wives on marriage and in which they raised their children. By the end of the eighteenth century many mining families had been there for two generations. At least one family, the Werts, had been there since the time of Wolters seventy years before, and three of its generations had already been born there and grown up knowing no other home. Such people had put down roots and when massive unemployment, and later famine, struck they could at first no more think of abandoning their homes than could the ordinary Andalusian villager. Having departed from the former tradition of mobility these miners became the victims of circumstances to a greater degree than their predecessors.

This situation was one which Rodriguez knew how to exploit. Any jobs that were available went to those who had the foresight and the means to offer him payment. In 1800 when Letona dismissed a number of men for stealing charcoal and fined them a total of 4,000 *reales* (£42) they were reinstated and their fines overlooked after each had paid Rodriguez 400 *reales* (£4 3s) as a bribe.[4] The government was informed but it was too preoccupied with other matters to take any interest.

In 1801 Godoy, restored as chief minister and as servile to France as ever, invaded Portugal when she refused to close her ports to British shipping. While 'the prince of peace' was leading his army on an unopposed route march through the Portuguese countryside, Napoleon Bonaparte sent 15,000 French troops into Spain. In face of protests from his Spanish ally, Napoleon enquired whether the king of Spain was tired of reigning and reinforced his expeditionary force with more men. Then, having established that the government of Charles the fourth was incapable of effective opposition, withdrew them all leaving the king and his ministers in a state of bewilderment and apprehension.

Letona did his best to overcome the effects of governmental paralysis on the mines and keep them in production, and as many men in employment, for as long as he could. But to succeed he needed money. Early in 1802 he appealed for the payment of promised but overdue funds and on 30 March he received the unhelpful reply: 'It not being possible to provide your excellency with the 300,000 *reales* (£3,125) which you say represent the outstanding debts, nor even the 200,000 *reales* (£2,083) due as the monthly contribution to allow you to continue in the present difficult circumstances . . . it is plain that you must carry out the necessary operations for the time being in the best way possible.'[5]

In 1803 Napoleon ordered Godoy to renew the war with Britain if he did not want Spain to be reoccupied. Godoy tried to buy neutrality by agreeing to give the French a monthly subsidy of 6 million francs.

In trying to raise the money each month the administration had little time to worry about complaints from the provinces, though its frantic endeavours were largely wasted since Britain was determined to bring Spain into the war as a means of attacking Napoleon on another and weaker front.* That year saw the beginning of the great famine which added further miseries to the affliction of unemployment at Rio Tinto, and which was cruelly exploited by Rodriguez with the aid of Juan Wert the chief supervisor of the mines' workmen.

In 1803 the shortage of corn caused the price in the vicinity of the mines to rise to 8 dollars (£3 4s) a bushel. There was a complete failure of the Andalusian harvest in 1804 and a great shortage in every subsequent year until 1812. From 1804–12 the corn price at Rio Tinto did not drop below 10 dollars (£4) a bushel. This higher price owed much to the activities of Rodriguez and Wert, for the governments of those years sent to Rodriguez as much money as they could spare, with the instruction to buy supplies in bulk and resell them at cost price to the people at the mines and the area around.[6] Wert was sent with this money throughout Andalusia to make purchases which, when they reached Rio Tinto, were resold by Rodriguez at a greatly enhanced price which the poor could not afford, and which yielded the treasurer and Wert a handsome profit. From 1803 to 1807 they were careful to refund the advanced money, but from 1808 to 1812 Wert allowed some of it to stick to his fingers, making 40,000 *reales* (£407) from this alone.[7]

There were other ways in which Rodriguez and his family swindled the crown with ever increasing impunity. They gave leases of mines' lands to friends at far less than their true values in return for bribes, and purchased equipment and supplies for the mines at higher prices than were necessary in return for part of the supplier's profits. Rodriguez even gave himself a lease of the gardens and orchards, and the right to collect the acorn and olive crops from the plantations, reselling the produce for high profits.[8]

The year 1808 opened with the arrival in Spain of a French army of occupation demanding the cession to France of all of Spain north of the river Ebro. Before General Murat's army took Madrid, Charles the fourth and Godoy fled to Aranjuez where the mob intimidated the king into abdicating in favour of his son Ferdinand the seventh. Both the old and the new king flew to Bayonne to see Napoleon; the former king asking to be restored, his son seeking confirmation.

Ferdinand was compelled to renounce the throne in favour of his father, and a renunciation was then forced from Charles in favour of the French emperor who proceeded to bestow the kingdom on his brother, Joseph Bonaparte. On 2 May (the great *dos de Mayo* of

*This was achieved by the British seizure of a Spanish treasure fleet in December 1804 which left Godoy no alternative but to fight. Ten months later the Spanish fleet sailed to destruction at Trafalgar, the sea battle fought in the bay of Cadiz not far from the confluence with the river Tinto estuary.

Spanish history) the people of Madrid rose heroically though impotently against the occupying army. In the brutality and carnage of the French retaliation a spark was lit which fired the country from end to end. A war of independence began which did not cease until five years later when the last French soldier was dead or driven back beyond the Pyrenees. It was a conflict which saw Spain ravaged in a singularly merciless and brutal fashion. Though the French managed to re-establish temporary control in Madrid, this resurgence of the national spirit led within a matter of weeks to the setting up throughout Spain of local *juntas* for organizing defence and raising militia, and the proclamation by the Spanish patriots of a regency to lead the fight against the French. The Spaniards needed artillery and reopened the Seville cannon foundry. The call went out from the regency council to Rio Tinto for copper to supply the Seville furnaces. Despite the poor conditions of the mines' equipment, Letona threw himself with energy into making the necessary preparations and in the years 1808–9 production once again soared.

Britain sent 20,000 troops under the command of Generals Moore and Wellesley* to supplement the Spanish resistance, but 1809 was a year of French military successes. By the start of 1810 a force of 300,000 French soldiers had moved rapidly into Andalusia and took Seville on 1 February. Work at Rio Tinto immediately stopped and Letona found he had to face a situation even worse than before, for all goods and foodstuffs for the mines were completely cut off by the enemy.[9] Such complete confusion reigned that for the first time since 1798 the flow of complaints about Rodriguez dried up.

During the next four weeks, however, the treasurer took the opportunity to extend his nefarious activities even while Letona was trying to restart work against the day when the country would need further supplies of copper for the war. In March, after a resumption of work that lasted only two weeks, the workmen downed tools in a protest against Rodriguez's tyranny and in particular the prices he was charging their families for food. They suspected, too, that the treasurer was using the excuse of bad communications with the regency council to keep back for his own use some of the money being sent to pay their wages. Rodriguez in a fury dismissed every one of them. He had not anticipated the men's response: they rioted for four days destroying mines' buildings, ransacking Rodriguez's orchards and gardens and looting freely. Their anger worked out, despair and hunger drove a great number of them to abandon their homes and to disperse throughout the province with their families, begging to keep themselves alive.[10]

Fear was expressed that these men might turn to banditry in

*Later first duke of Wellington.

despairing reaction against the barren hills and the poverty-stricken villages which had little food to spare them. It is possible that some did; at this time the *sierra* around the mines was infested with robbers who, because of the terrain and the feebleness of the law officers, were seldom brought to justice. For those workmen who stayed on at the mines life became even harder. A number of people died of starvation including four of the senior employees: the surgeon, the chief machinist and two supervisors.[11] The treasurer's brother-in-law Antonio Villapol (he had married Rodriguez's sister and so obtained the post of mines' legal adviser) was one of several senior members of the staff who decided to take to the road, though he was the only one who appears to have had the foresight to take with him some 63,000 *reales* (£656) of government money. The Rodriguez household alone lived in a state of plenty.[12]

This flight of most of the population turned the Rio Tinto *pueblo* of la Mina into virtually a ghost-town. The houses quickly became ruinous, grass grew in the streets and weeds choked the doorways of the houses which became infested with wild animals from the *sierra*.[13] The regency council heard the reports with alarm. Responsible for organizing the national opposition to the French, its members knew the importance of keeping the mines in a good state of maintenance if copper production was ever to be resumed. Henceforward, over some seven years, money was collected by great efforts and sent to Rodriguez to repair the underground works and the buildings, to buy food and to provide salaries and wages to keep the work-force together, but during all that time the funds were misappropriated. The cash for food purchases was largely expropriated by Juan Wert, though he seems to have taken care to give a share of the profits on the resale of foodstuffs to Rodriguez. The latter helped himself to the lion's share of the money provided for the wages and salaries of men who had either died or left the mines during the famine, and distributed the rest amongst the few who remained. His friend the apothecary was allowed five salaries: his own, that of the dead surgeon and those of the absent schoolmaster, notary and sacristan. The cash supplied for the upkeep of the equipment and the mines disappeared without trace, presumably into Rodriguez's coffers though later government enquiries were never able to prove the point. At this time, when men were dying for want of food, Rodriguez appropriated the mines' herds and added them to his own 200 head of cattle, protecting them with armed guards and feeding them not only on the common pasture but also over other men's land.[14]

During the summer of 1810 French troops in columns of a thousand or more frequently passed through the mines' area but surprisingly never made any inspection of the workings, perhaps assuming that because these were abandoned they were of no interest. In fact Letona had 2,461 *arrobas* (31 tons) of refined copper hidden there. Eventually the French activity around the mines ceased, and the invaders contented themselves with installing a garrison of 500 men at Castillo de las Guardas, repairing the ancient fortifications and sending out occasional patrols into the *sierra* to keep a watchful eye on the inhabitants. The castle was only twelve miles north-east of Rio Tinto and if work had been restarted at the mines the French troops were well placed to take them over. Thus Rio Tinto was spared any fighting. Once it looked as if there might be some action, for one evening a patrol of Wellington's men arrived at Zalamea la Real to stay the night. As the news spread the locals wondered if the visit was a prelude to an attack on Castillo de las Guardas, but with the dawn the British disappeared.[15]

In September of that same year of 1810, Letona made the journey to the beseiged city of Cadiz, the last stronghold of the patriots and the seat of their regency council. He told the five members of the council about the 2,461 *arrobas* of copper suitable for artillery manufacture which he held at Rio Tinto, and they decided to send men and mules from Cadiz to retrieve it before it was discovered by the French. Letona was also instructed to keep the mines in a state of readiness for resuming production at the earliest possible moment, but to employ the minimum number of men he could for the purpose. He was asked if any money could be raised by selling some of the lands or buildings attached to the mines but advised the regents that there was nothing for which buyers could be found. It was therefore decided to divert to mines' use as a matter of urgency 60,000 *reales* (£625) previously earmarked for army funds; an indication of the importance attached to Rio Tinto by the patriots.

At this meeting, too, there was a review of the long-term help which the mines needed from the regency. Despite the demands made on its meagre resources by defending Cadiz, provisioning the army and assisting the guerrillas in the hills, the council promised Letona a monthly supply of money: 3,750 *reales* (£39) for maintenance of equipment and buying fodder for horses and mules; 6,820 *reales* (£71) for workmen's wages; and 3,800 *reales* (£40) for salaries of the senior staff.[16]

Letona returned to Rio Tinto where doubtless Rodriguez was pleased to hear of the money which was to be diverted from the country's slender resources largely for his benefit. Soon the men and mules arrived from Cadiz to take the precious 31 tons of copper. It would have needed a train of 310 beasts and a sizeable escort of troops to take the metal over the hills to Huelva and from there along the coast to Cadiz, the route chosen as being the best to avoid encounters with the French. Rio Tinto tradition states that 446 *arrobas* ($5\frac{1}{2}$ tons) of the precious cargo fell into French hands, but in fact the receipt issued in Cadiz showed the arrival there of the full amount.[17] It was a fine gesture of defiance but it was the last contribution the mines were to make to the war effort. For the rest of hostilities the reduced population devoted its energies to hunting for food and to taking sides in the squabbles that were never long absent.

One of these involved the second chaplain of the mines' church of Santa Barbara, Francisco Lobo Parreño, who had been appointed by the finance council in 1806 but had been immediately suspended by the archbishop of Seville at Letona's request because of his alcoholism, after which he had disappeared without trace. Suddenly in March 1811 he reappeared and demanded his chaplaincy back. Since Letona had appointed Alberto González Calvo to the post, Lobo Parreño's demand was refused, whereupon he petitioned the Cadiz government for restitution, arrears of pay and damages against Letona. While awaiting the decision he set up house at the mines' *pueblo* of la Mina, pestering Letona and arguing his case in the villages around. While mass was being celebrated in Santa Barbara's on new year's day, 1812, he climbed into the pulpit while González was reading the gospel and started to deliver a prepared address recounting his disgruntlement. A shouting match resulted in which everyone joined and Letona and his supporters stalked out of the church.

The parish priest of Rio Tinto (modern Nerva) complained to the bishop of Orense, a member of the regency council and told Letona he should use force to make sure that Lobo Parreño was kept out of the church at all times. The priest of Zalamea la Real wrote a sharp note to the troublesome priest accusing him of embarrassing his clerical colleagues and telling him to refrain from his scandalous conduct. But Lobo was unrepentant and wrote another letter to Cadiz, telling the various stories he had heard at the mines about Rodriguez's fraudulent activities. The treasurer was doing his best to make the priest's life as uncomfortable as possible and the local people, who had little else to do otherwise, took sides in the dispute which, according to a civil servant visiting the place, 'turned it into a tower of Babel'. The regency council was eventually inundated with so many letters from the local clergy and others that in June 1812 it ordered Pedro Grimarest, the commander of the Niebla garrison, to investigate the matter.[18]

Grimarest was plainly a simple soldier, blunt and outspoken. To him the situation was quite simple. Since the government had originally appointed Lobo Parreño, then only the government could remove him. As it had not dismissed him he was, in Grimarest's view, still legally the chaplain. He accepted that the priest was a drunken nuisance but suggested that particular problem could be overcome by appointing him as a chaplain in the army where his alcoholism would be less obvious! As far as the accusations of dishonesty against Letona and Rodriguez were concerned, Grimarest dealt with these in an equally forthright manner: 'These things are very likely to happen in government departments.' Nonetheless, Rodriguez in his view was a troublesome man with an excess of pride, and the best thing to do was to send the entire family packing, with the exception of Letona and his wife, taking them under escort to a distance of thirty-five miles from Rio Tinto and telling them that if they ever went any nearer to the mines than that they would be arrested.[19]

Perhaps because of the tone in which the soldier couched his recommendations, perhaps because the Cadiz administration was too busy doing more important things, no action was taken on Grimarest's report. The siege of Cadiz was raised in August 1812 and the French were forced to evacuate Andalusia to counter Wellington's thrust to Madrid. The regency council then appointed Domingo Ibarrola, the accountant of the public revenues at Algeciras, to hold another enquiry in the following November into the reports emanating from Rio Tinto.

He took an opposite view to Grimarest. It was plain to him that since Lobo Parreño was still suspended from exercising his priestly functions, he was also disqualified from being a mines' chaplain. The priest had obtained a licence from the archbishop to appoint his own substitute while under suspension and, though he had not taken advantage of it, this represented a serious infringement of the crown's rights, in Ibarrola's view. For the licence presumed that the chaplaincy was a benefice with a freehold interest similar to a parish priest's, which would have meant that Lobo Parreño would have been entitled to draw the stipend and pay his substitute whatever he liked. Ibarrola had no doubt that Letona had acted throughout in the crown's best interest, and that the best thing was to do a deal with Lobo: the removal of the bishop's suspension in return for his agreeing to take an army appointment.

Ibarrola was convinced that the whole Rodriguez family were completely honest and he blamed the accountant Eugenio Serrano, whom he described as 'having very bad health, little knowledge, moderate education and slow intelligence', for being behind all the discontent, and proposed his dismissal. However, since Rodriguez and his sons were certainly disliked in the neighbourhood, Ibarrola thought it might be better to transfer them to other government posts elsewhere.

His report also estimated that it would need four years to get the mines in working order together with almost 6 million *reales* (£62,500). A sum of 60,000 *reales* (£625) was needed immediately to buy food for the workmen and the clerks who were begging for food and alms in the countryside around.[20]

As a result of this report a government decree was issued in March 1813. Lobo Parreño was declared to be properly dismissed but entitled to appoint his own successor and to make whatever financial agreements he wished with him. The accountant Serrano was dismissed and all allegations of dishonesty against other senior officials were declared to be without foundation. For the time being the mines would be kept purely on a maintenance basis.[21] A copy of the decree was sent by the general directors of revenue to Jiménez, the former administrator at Rio Tinto who then held the post of chief collector of customs at Cadiz, asking for his opinion. He replied indicating his approval of the action taken but must have been somewhat surprised when he received yet another letter, fulsome in the extreme, describing his reply as 'well-grounded, judicious, prudent and adequate, and most conclusive to a speedy resolution.'[22]

By June 1813 King Joseph Bonaparte had fled home to France. His army of occupation, weakened by the harassment of the guerrillas and the British army, and by the withdrawal of troops to fight on the Russian front, had been routed. Only a few pockets of resistance remained and for all practical purposes the war was over, though its effects were to continue to be felt for many years. For the administrative vacuum left by the collapse of the throne and its associated institutions, together with the ferment of political ideas which the war produced, were to set Spaniards in conflict among themselves and to provide the seeds for a harvest of bitter ideological discontent and civil disorder that lasted for more than a century.

The first constituent *cortes* in the country's history had devised at Cadiz the celebrated constitution of 1812. The *de facto* vacancy of the throne had permitted free discussion on the respective rights of crown and people, and the delegates had produced a formula for a restored but limited monarchy and a reinstated but less powerful church, alongside an elected single-chamber parliament with authority to pass laws, to control taxation and government expenditure, to administer the armed forces, to maintain the freedom of the press and to call ministers of the crown to account for their behaviour. And so at the end of 1813 it seemed as if the Spanish constitutionalists held the destiny of their country in their own hands.

By November of that year the regency council was planning the establishment of a peace-time administration on the lines envisaged in the 1812 constitution, and it turned its attention once more to the royal mines of Rio Tinto. Appreciating their importance in the country's economy it wanted them to start copper production as quickly as possible. With funds virtually non-existent, it decided the only way to raise the necessary capital was to sell off the non-mining parts of the property. But it sent the instructions for the preparation of a scheme of sales not to Letona or Rodriguez but to José Dominguez Moyano, the senior clerk in charge of the counting house since the removal of Serrano. Possible purchasers started to write in expressing interest, and discussions started with tenants to clarify their rights in the event of the proposed sales taking place.[23]

Then in August 1814 Ferdinand the seventh returned to Spain from his captivity in France. The enthusiasm of his reception and the popularity he so obviously possessed, convinced him that if he was ever to rule in the tradition of his ancestors this was the opportune time to shake off the fetters of constitutionalism. Before entering his capital he issued a royal decree rescinding the constitution and every decision of the *cortes*. The army arrested the regents, the ministers, the parliamentary deputies and all those who had sought to introduce limitations on the royal rights. Within a few weeks it became punishable by death to advocate the adoption of the constitution of 1812. The feudal seignorial jurisdiction, which had weighed so heavily on the countryside and its peasant population, was reintroduced. The holy inquisition was restored. Rigorous press censorship was instituted and all foreign books seized while Spanish society became honeycombed with government spies. Over 12,000 Spaniards were exiled by decree. At a time of such governmental reaction it was inevitable that the men appointed as ministers and senior crown officials would themselves be reactionaries, holding office because of their sympathy with royal policies rather than because of any inherent abilities they might possess. But even allowing for this, the six years following Ferdinand's return achieved remarkable standards of ministerial incompetence and a notable near-paralysis of governmental administration.

*Ferdinand the seventh.*

The inadequacy of the bureaucratic machine which was restored together with Ferdinand is amusingly illustrated by what happened when José Sanz, the son of the former administrator of the Rio Tinto mines, wrote from his home in la Palma to the king only three weeks after the triumphant royal entry into Madrid. His letter claimed that the Rio Tinto mines had belonged to Francisco Sanz who had been wrongfully deprived of them and whose claims for redress had never been finally settled. José therefore asked for a grant to restore the mines to him. The keeper of the records of the public revenue was asked to produce all the documents relevant to the application and to prepare a history of the Sanz claims. This official replied that the upheavals of war made it impossible to find the appropriate records and then filed the royal request with the Rio Tinto papers in the archives! José Sanz's letter was then laid before the council of finance for consideration. There it stayed for twenty years, appearing from time to time on the agenda of meetings, until in 1834 it was passed to the general director of the mines for comment and was apparently never seen again.[24]

It is true that on 23 December 1814 the minister of finance showed sufficient energy to send a strongly worded memorandum to the general directors of revenue demanding immediate resumption of production at Rio Tinto in order to provide copper for the new coinage to be struck at the Madrid mint, and on the following 6 January demanded a reply within seven days. The directors then told him it would be necessary to buy the copper from other sources but assured him that they were working strenuously to get the Rio Tinto mines on their feet again. In fact the directorate did nothing at all in the matter beyond sending the minister occasional memoranda assuring him that they were vigorously tackling the problem.[25]

By February 1815 the price of imported south American copper had risen to 20 dollars a *quintal* (£8 per 100 lb) and Jacobo Parga, the minister of finance, discussed with the king the possibility of letting the Rio Tinto mines to a private contractor for five or ten years as a method of getting production going again.[26] As a result it was decided to have a report made on the condition of the mines and Pio Laborda, the government subtreasurer in Seville, was sent a letter instructing him to go to Rio Tinto.

He protested at having to go to 'that desolate place' but set off a year later accompanied by Gregorio Azaola who was in charge of the recutting of the river Guadalquivir. His report, when it was written, confirmed all the allegations about the dishonesty of Rodriguez and by implication condemned the civil servants who had persistently failed to remedy the situation.[27] Laborda found that the treasurer was the real master of the mines, making appointments and granting contracts and leases all at a profit to himself, and that since at least 1810 not a single penny of the money provided for repairs and maintenance had been used for that purpose. Thousands of trees from the plantations had been felled and sold to timber-merchants in Zalamea la Real and the money for them misappropriated. Juan Wert was still giving contracts for supplies at inflated prices to people who made vast profits at government expense and gave him a percentage. Wages were being provided by the crown for men who had left the mines as far back as 1803 and the money was being drawn by various individuals associated with the Rodriguez family. The treasurer's elder son, Manuel Rodriguez, for example, had been committed to the Seville madhouse but his father was pocketing his annual salary of 4,000 *reales* (£42) as second clerk of the counting house along with the wages of a number of others. And Rodriguez senior was still terrifying the people in the area around. Laborda found him a cruel and vicious man.

The two main problems that Laborda saw in restarting work were that there was not one employee who was skilled in mining and metallurgy, and that the dishonest Rodriguez was all-powerful. The supervisors were illiterate and unintelligent labourers who had received promotion because of their servility to the Rodriguez family and ready compliance with its demands. Whatever else might be done, Laborda was insistent that it was essential to dismiss Atanasio Rodriguez and Juan Wert and prosecute them for fraud. He proposed that Letona, the mines administrator, should be transferred to the Cadiz saltworks at the same salary since he was basically honest and competent but unable to stand up to his father-in-law. Mad Manuel Rodriguez should be dismissed and allowed a pension. There were two other dishonest supervisors whom Laborda thought should also be discharged.

According to the mines' accounts, the twenty-eight years from 1783 to 1810 saw the production of 287,600 *arrobas* (3,595 tons) of Rio Tinto copper which was sold for a little over 41.3 million *reales* (£430,209). Production costs were 41.2 million *reales* (£429,167) and thus a profit of over 100,000 *reales* (£1,042) was claimed. But the costs did not include any figures for wages or capital expenditure and consequently the 'profit' really represented a substantial loss. The crux of the whole matter, according to Laborda, was that it was not the responsibility of the government to engage in manufactures: everybody knew that government officials were not interested in exercising economy, whereas a private individual could be expected to take care of his property because it was for his own benefit. He hesitated to recommend the freehold sale of the mines but felt they should certainly be let out to a contractor.

Laborda also observed that clergy problems were still causing controversy at Rio Tinto. The constantly drunk priest Lobo Parreño had refused to take an army appointment and had eventually been reappointed as a mines' chaplain, but his outrageous conduct meant that he was always at the centre of some tempestuous disagreement.

There were difficulties, too, with the parish priest of Rio Tinto (otherwise Nerva) who was claiming parochial rights over the mines and demanding clerical dues such as first-fruits, despite a royal decree of 1806 which declared that the mines were extra-parochial territory and which awarded the clerical perquisites to the mines' chaplains since the parish priest had no rights or duties at the establishment. Laborda therefore proposed that the mines should be constituted a separate ecclesiastical parish with its own priest fully independent of the nearby mother-church. This would have the additional benefit of abolishing the chaplaincies, thus solving the problem of the alcoholic Lobo Parreño.

But Lobo obligingly solved the problem himself. On 18 January 1817 he visited Juan de Bustamante, the priest of the nearby village of las Delgadas, to enlist his support in another of his interminable disputes with Letona and Rodriguez. He dined with Bustamante, went off for a while and returned with a friend called Francisco López, both of them being the worse for drink. When the time came for them to leave, the intoxicated chaplain declined an invitation to stay the night and made off with López for the mines. 'He went away at last', wrote the las Delgadas priest to Letona, 'to pay his tribute to mortality accompanied by Francisco López. Both fell into the river.' López managed to save himself but Lobo Parreño was drowned, his body being recovered from the flooded river Tinto the next day.[28]

At the end of 1817 little had changed at the mines. The treasurer was still hated by the people and defrauding the crown. The minister of finance, after considering Laborda's report for a year, noted that he agreed with its recommendations and that he was contemplating letting the mines on contract to Thomas Thomas, described as *un inglés* but obviously a native of Wales.

Thomas's application had all the attractiveness of originality. Claiming to be experienced in copper manufacturing in Wales, he had visited Rio Tinto on several occasions and believed 2 million *reales* (£20,833) would be enough to reopen the mines. This would be provided by a consortium of British merchants headed by himself, in return for the grant of a twenty-year contract which would stipulate that all production costs were to be deducted from sales receipts, and that the net profit was then to be divided between his consortium and the Spanish government in the ratio of three to one. Thomas also required clauses in the contract allowing him a free hand to dismiss any staff and workmen he wished, and to install twelve British master-workmen in positions of authority at the mines. He also wanted to be permitted to import machinery free of customs duties and to tour Spanish prisons selecting the fittest convicts to serve as mines' labourers, paying the government a rate for their services which he alone would decide. Furthermore, he required the grant of rent-free premises in Seville to serve as warehouses and a factory.

As an afterthought, he added that he would require a guaranteed price of 7 *reales* (1s 5½d) a pound for copper sold to the government and 9 *reales* (1s 10½d) for copper sold to merchants, and the imposition of an import duty of 8 *reales* (1s 8d) a pound on all foreign copper. The government was unhappy about the use of convict labour but remained open to persuasion on the point. It was prepared to concede all Thomas's other demands. However, the discussions were broken off by the Welshman when the government rejected another clause which he desired to add to the contract and which would have given him the right to mint his own coins for four years to a total of 5 million *maravedises* (£1,740).[29]

In January 1818 the government decided to act on Laborda's report. A royal decree declared that, once the future of the mines was decided, Letona would be replaced as administrator and transferred to other work. The treasurer Rodriguez, his sons and all his other relatives were dismissed from the royal service and would never again be employed by the crown.[30] Almost incredibly, it must have seemed to the long-suffering inhabitants of the Rio Tinto district, the government had at last taken a decisive action. The treasurer's twenty years of despotism were at an end.

*Open-air calcination: the system of burning off the sulphur content of the ores introduced by the Marqués de Remisa. (From a late-nineteenth-century photo.)*

7

# *And the birds ceased to sing*

## The return of private enterprise: 1818–54

One consequence of the disastrous effects of the peninsular war, which made it virtually impossible for the governments of both the regency and of Ferdinand the seventh to compel payment of the taxes demanded from the poverty-stricken Spanish people, was that the royal mines of Rio Tinto were starved of the capital investment essential for any large-scale production.

In the six years from 1811 to 1816 the government was providing 160,000 *reales* (£1,666) a year for wages and maintenance although no copper was produced apart from 148 *arrobas* (almost 2 tons) precipitated in 1815. When in June 1816 Letona wrote to the council of finance demanding money due as arrears of pay, the reply he received stated that payment was out of the question: the total cash in the royal treasury amounted to only 1,394 *reales* (£14 10s 5d)! He was given authority to draw 14,365 *reales* (£150) monthly from the public revenues of the towns of Aracena and Huelva. But both towns were having difficulties collecting their taxes, and the following December the Huelva officials refused to hand over another penny to the mines. In September 1817 those of Aracena followed suit. By June 1818 wages at Rio Tinto were in arrears, according to Letona, as from August 1814 and no bills had been paid since April 1815.[1]

In July 1818 the supreme council considered the possibility of letting the Rio Tinto mines to another contractor, but rejected the proposal and ordered steps to be taken to raise such capital as the crown needed to resume operations. The next month responsibility for the mines' administration was transferred to the board of public credit which in October relieved Letona of his post of administrator and designated him 'director' of the mines, a specially created post which the board intended would restrict him to a consultative and advisory role. His replacement was the manager of the lead factories of Presidio and Canjayar in Almería province, José de Miaja y Pingarron, who was given instructions that his first task was to collect the arrears of money due from the public taxes of Aracena and Huelva. Despite considerable effort he met with no success and soon he was affected by the contagious apathy of the central government. A request for a report on the state of the mines, made by the board of public credit in December 1818, was not answered by Pingarron until July

1820. His account of the attempt to collect the money owing from the two towns was begun in February 1819 but laid aside and not completed until January 1822. Meanwhile all maintenance work continued to be neglected and the condition of the machinery, buildings and underground works deteriorated still further.[2]

The future for the mines looked very black in the view of those working there. Attempts were made to obtain iron for producing copper by the immersion process. But though in 1818 and 1819 respectively, 460 and 80 *arrobas* (5½ tons and 1 ton) of copper were manufactured in this way, even these small quantities could not be sold, so depressed was the state of Spanish trade and industry.

The revolt in January 1820 of 22,000 troops in Cadiz both provided the country with a short spell of constitutional rule and raised hopes of resumed working at Rio Tinto. Riego, the demagogic officer who led the rising, compelled Ferdinand to dismiss his ministers and to replace them with liberals dependent on military backing, and they reopened the Seville foundry to provide the army with desperately needed artillery. Pingarron made plans to restart mining but lack of funds proved as much a problem for the new régime as for the old. Eventually the most that the government could offer was a small but regular monthly supply of iron for precipitation. With this it was possible to produce 200 *arrobas* (2½ tons) of copper every month for almost four years until, in October 1823, a French invasion restored Ferdinand to absolute rule. And, with 35,000 French troops left to garrison the country, the king was able to economize on his own army and the Seville foundry was again closed and the monthly iron deliveries to Rio Tinto discontinued. According to Pingarron, unrelieved gloom descended on Rio Tinto as hope subsided, a gloom that was only slightly alleviated by the resumption of small deliveries of iron in the following year.

In December 1823, two months after Ferdinand's triumph, two French brothers called Rollac offered to take a lease on the mines together with those of Guadalcanal for twenty years subject to the government first undertaking any necessary repairs, but the ministry of finance rejected their proposal with the declaration that the Rio Tinto mines were one of the greatest possessions of the Spanish crown and would be worked by the government for the country's advantage.[3]

Following this, the board of public credit busied itself for a while with reports on the hardships being undergone by the 500 inhabitants then living at Rio Tinto, who had been 'reduced to the greatest want and misery', and on the economic problems caused for the towns and villages up to thirty miles around by the paralysis of the mines.

Studies were made of the possible profits that could be obtained by canalizing the drainage waters to increase the amount of copper produced by the immersion of iron, and by introducing English machinery to manufacture sheet copper. But no one tackled the basic problem of how to provide the capital which was needed to finance the work envisaged in these studies, and the men at the mines realized that the board's discussions were of no more than academic interest. But even if the money had been available, it was unlikely that the mines could have been restored to life under the then mines' administration, for the whole of the management had succumbed to the apathy, the inefficiency and the corruption which characterized the central government of the period. Between 1820 and 1825 an average of over 190,000 *reales* (£1,979) a year was provided from the public funds for wages and repairs at Rio Tinto, over half of which, some 102,000 *reales* (£1,062) a year, was siphoned into the pockets of the senior officers at the mines under the pretence of payment of wages to a number of non-existent employees.[4]

This dishonesty and indolence were only to be expected. With the government of the country so incompetent it must have seemed to Pingarron and his colleagues that their own conduct was justified merely as a form of self-preservation. Wages were always in arrears; there was the ever present threat of a complete closure of the mines and their consequent unemployment; and any attempt at starting work in earnest was always frustrated by the inadequacies of the bureaucrats in Madrid. In March 1824 Pingarron complained that six years of dealing with the board of public credit had convinced him that there was no possibility of that department's officials ever coming to any decisions or taking any action. He had negotiated the sale to the Segovia mint of 2,000 *arrobas* (25 tons) of copper held in stock at Rio Tinto, but his requests for permission to proceed with the transaction were completely ignored by the board. In annoyance he wrote to the directors of revenue, but it took them seven months to reply authorizing him to go ahead. By then the mint could raise the cash to pay for only 500 *arrobas* (6¼ tons) at the uneconomic price of 4½ *reales* (11¼d) a pound. But at least with the 56,000 *reales* (£583) he obtained he was able to pay some arrears of wages, buy some charcoal and start repairs on the ruined houses in the mines' village. Pingarron had hopes of continuing to supply the mint, the director of which was paying as much as 5¾ *reales* (1s 2¼d) a pound for imported copper and was prepared to take 4,000 *arrobas* (50 tons) annually from Rio Tinto, paying 4½ *reales* (11¼d) a pound at the mines or 5¼ *reales* (1s 1d) for copper delivered at the mint. However, Pingarron's letters to Madrid on the subject went unanswered.[5]

In January 1825 a royal decree was issued declaring that although the mines were to continue to be worked by the crown, they were to become independent of the central government for the purpose of day to day administration. The mines' management was to be free to raise loans against the security of the copper production and to be allowed to buy machinery to make sheet copper. Simplified regulations for running the establishment were to be prepared, and the mines' officers were to have freedom in arranging contracts and fixing prices in accordance with their knowledge of the markets. These radical changes were the proposals of Fausto Elhuyar who, as a newly appointed inspector of mines, had visited Rio Tinto in 1823 and seen the needs of the mines quite clearly. His ideas, however, ran completely contrary to the growing tendency to centralization which characterized Ferdinand the seventh's reign, and this decree appears to have been a temporary aberration on the part of López Ballesteros,

*The Marqués de Remisa.*

the minister of finance who was responsible for it. The senior officials of his ministry objected strongly to its terms and managed to prevent it from ever being put into effect.

In May the minister announced the dismissal of Letona from his advisory post as director of the Rio Tinto mines and his replacement by José Martinez Marcos. The intention was to subject Pingarron to greater control, and with this in view Martinez was given instructions to take over most of the administrative work. His letter of authority, however, did him little good since Letona refused to step down for him or hand over his office and house until such time as he was assigned a pension. With Pingarron's sympathetic assistance, Letona carried on and drew his salary for another three years until in 1828 the ministry at last decided on the amount he was to receive.[6]

Between September 1825 and December 1827 the government was considering an application for a lease of the mines. Santiago García had petitioned for a twenty-year lease, offering to pay 200,000 *reales* (£2,083) a year for the first ten years and thereafter 300,000 (£3,125), and undertaking to supply the government with copper at 9 to 10 per cent below the market price in Cadiz for the time being. However, as the discussions went on it became plain that García's intention was to work the mines and their associated properties without any regard for preservation or maintenance in order to make as much profit in return for as little expenditure as possible. His proposals were ultimately rejected but in the course of the talks on them the finance ministry officials had to consider the hard facts relating to Rio Tinto. It was at last accepted that in the state of the country's economy the crown was not likely to be able to work the mines efficiently and that the only way to make any money from them at all was to let them to a contractor. With this in mind regulations were prepared to control the conduct of any lessee and to provide for government inspection of the mines at four-monthly intervals.[7] A notice advertising the sale of a twenty-year lease was then sent to the managers of all mines in Spain, the viceroy of Navarre and the Spanish consuls general in Paris and London, and published on 21 May 1828 in the Paris newspapers, *Du Commerce* and *Le Constitutionel* and the London *Times* and *Morning Chronicle*.

Despite the government's attempt to arouse the interest of possible foreign applicants, the three bids which were received all came from Spaniards. Two of them offered such derisory sums that the ministry of finance gave serious consideration only to the third, the offer of Gaspar, *Marqués* de Remisa. Even then the negotiations occupied almost two years. Under the contract, which was finally signed in March 1830, Remisa undertook to pay a rent of 215,000 *reales*

(£2,239) a year for the first ten years and 300,000 *reales* (£3,125) for each of the final ten.[8]

Remisa had already had considerable experience of administering the mines of Guadalcanal and if anyone was likely to be successful at Rio Tinto it was he. He was an extremely successful businessman, efficient and somewhat ruthless. The fine portrait of him by Vincente López is indicative of the unusual nature of the man. For at a time when few of the Spanish nobility would have been prepared to pursue a business career, and when most middle-class businessmen preferred to have their portraits painted to convey the impression that they were gentlemen, Remisa chose to be shown in his office wearing his black business clothes, with an open cupboard, crammed with bundles of papers, against the wall, and on the desk account ledgers and letter-books.

When the mines were handed over to the new contractor there were 152 occupiable houses, fifteen warehouses and workshops, six smelting and refining houses in working order and sixteen in ruinous condition, plus the administrator's house, the guards' barracks, the chemist's shop and the school-house. In the inventory the value of the whole establishment was given as 3 million *reales* (£31,250), only a little more than the valuation made forty-three years earlier at the time of the 1787 inventory. Allowing for inflation, and the fact that the value of the tree plantations had increased to 1½ million *reales* (£15,625) by 1830, the run-down state of the works themselves in the latter year is evident.[9]

Remisa was soon complaining. During his negotiations for the lease, twelve assays of the ore body made by government analysts had shown variations in the copper content of between 1.3 and 4 per cent, giving an average of 2.6. But he found when he started his operations that the copper never exceeded 2.25 per cent and was often less than 0.5. Consequently he was forced to work on a much larger scale than he had anticipated and with far greater costs. He was helped by the fact that the Rio Tinto ores existed in great abundance and that access to them was easy. But the need to mine bigger tonnages affected every aspect of his operations. He needed to expand the work-force and by the end of his first year could not find enough men in the district to meet his requirements. Consequently he had to increase wages and pay between 12 and 18 *reales* (2s 6d and 3s 9d) a day, which he claimed was twice the amount paid in any other Spanish mine, to attract skilled and unskilled workers from other parts of the country. In addition, Remisa needed vast and expensive quantities of wood for reduction purposes, and in his first twelve months he had to rebuild completely the smelting houses and to install nineteen new furnaces.

He therefore protested to the finance ministry that unless he was freed from the contractual obligations which regulated the extraction of minerals and the conservation of the properties, and was allowed to run the mines in his own way, he would have to close down. This elicited no reply from Madrid but Remisa quickly found that the details of the contract could be ignored with impunity, for the government's attitude to Rio Tinto was still one of apathy. Moreover Remisa had as his secretary José García Rodriguez, a former clerk of the finance ministry who had been sent to the mines to make formal delivery to the new contractor and stayed on in his employ. At their first meeting Remisa had offered García employment and the young man quickly proved his value. Knowing the inadequacies of his former colleagues, and the ineffectiveness of the system they were required to operate, he was able to advise Remisa precisely what he could expect to get away with.

The two men decided to make the first breach in the contract by carrying out a wholesale destruction of the forestry plantations to obtain the wood that was needed for fuel. The contract restricted the amount of timber that could be cut annually, regulated the method of felling and stipulated the numbers of new trees to be planted each year as replacements for those which Remisa was authorized to chop down; but these provisions were completely ignored without even a protest from the inspectorate of mines or the finance ministry.

To some extent Remisa was helped in his enterprise by the death of Ferdinand the seventh in 1833. The king was succeeded by his infant daughter, Isabella the second, under the regency of her mother Queen Maria Christina. This precipitated a civil war between the supporters of Ferdinand's disgruntled brother, Don Carlos, and the government, and gave civil servants other things to worry about than the Rio Tinto mines. Thus, free from interference, Remisa was able to increase production in his own way and make the mines pay well. During the twenty years he held his contract there was an average annual production of 21,000 *arrobas* (262½ tons) of manufactured copper, most of which was taken by the Segovia mint and the Seville artillery foundry.[10]

A visitor to Rio Tinto at that time was the English writer Richard Ford, who was living in Spain between 1830 and 1833, and recorded his impressions of the mines in *A handbook for travellers in Spain*:

'The red naked sides of the copper mountain, la Cabeza Colorada, with clouds of smoke curling over dark pine woods, announce from afar these celebrated mines. The immediate approach to the hamlet is like that to a minor infernal region; the road is made of burnt ashes and *escoria* [slag], the walls are composed of lava-like dross, while haggard miners, with sallow faces and darkened dress, creep about, fit denizens of the place. . . . The village is principally occupied by the miners; there is, however, a decent *posada* [inn]; the *empleados* [members of the staff] and the official people have a street to themselves. The view from above the church is striking: the town lies below with its stream and orange groves; to the left rises the ragged copper hill, wrapt in sulphureous wreaths of smoke; while to the right the magnificent flat fir bank, which supplies fuel to the furnaces, *la mesa de los pinos*, is backed by a boundless extent of cistus-clad hills, rising one over another.

'A proper officer will conduct the traveller over the mines. . . . Entering the shaft you soon descend by a well, or *pozo*, down a ladder to an under-gallery. The heat increases with the depth, as there is no ventilation; at the bottom the thermometer stands at 80°Fahr., and the miners, who drive iron wedges into the rock previously to blasting, work almost naked, and what few clothes they have on are perfectly drenched with perspiration; the scene is gloomy, the air close and poisonous, the twinkling flicker of the miners' tapers blue and unearthy; here and there figures, with lamps at their breasts, flit about like the tenants of the halls of Eblis, and disappear by ladders into the deeper depths. Melancholy is the sound of the pick of the solitary workman, who alone in his stone niche is hammering at his rocky prison like some confined demon, endeavouring to force his way to light and liberty. . . The stalactites are very beautiful; for wherever the water trickles through the roof of the gallery, it forms icicles, as it were of emeralds and amethysts, but these bright colours oxidize in the open air, and are soon changed into a dun brown.'

But the Rio Tinto landscape which Ford recorded, and which was still in essence the scenery and setting known to Francisco Sanz, was soon to change as a result of the effects of advancing technology. Because of the low copper content of the ores being mined, Remisa could improve the economic viability of his operations only by ordering the excavation of greater quantities of mineral. This involved both recruiting greater numbers of workmen and finding a new way of treating the ores; for the system used up till then, by which the lumps of minerals were burnt in kilns and the metal thus extracted then smelted, was no longer capable of coping with the vast amounts being mined. A new method of dealing with them had to be devised. The solution to the problem was found in a process known as open-air calcination, a system that was to provide great profits but was accompanied by lamentable environmental consequences. It also represented another breach of the contract since no provision for it was contained in the agreement.

Open-air calcination, devised by Ignacio Goyanes, became the first stage in the smelting process and involved roasting the poor quality ores in pyramids on the large area of flat land known as los Planes. The piles, known as *teleras*, measured approximately 12 by 7 yards at the base and stood about a yard high, each containing some 4,000 quintals (200 tons) of ore. One *telera* needed 1,000 *arrobas* (12½ tons) of brushwood and 125 *arrobas* (just over 1½ tons) of roots to set it alight. Once started, the roasting lasted until the fire died out five or six months later. The high temperature reached during this process, and the contact with the oxygen in the air, had the effect of burning off most of the sulphur in the ores and (more important from Remisa's point of view) changed the copper, iron and sulphur parts to chemical forms more soluble in water.

When the fires were out, the calcined ores were carried on mule-back to the lixiviation tanks, constructed of rough masonry and lined with asphalt, 7 yards long, 4 yards across and 1 yard deep, each holding 2,000 *arrobas* (25 tons) of ore when two-thirds filled. During the next nine days each tank was filled with drainage water from the mines, then drained and recharged a number of times. The water drawn off contained salts of iron and copper in solution, and went into other tanks of the same size in which were placed piles of pig-iron for the process known as precipitation. The water was periodically drained from these tanks in turn, and the scales which had formed on the pig-iron, and contained about 55 per cent copper, broken off. Every ton of copper produced in this way needed 2.7 tons of iron.

When the water was drawn off from the tanks containing the pig-iron it was no longer wasted as had happened previously during the precipitation process, for it went into a third tank where the sub-salts of iron in the water were allowed to settle on the bottom. The resulting mud-like deposit contained about 10 per cent copper, and was scooped out of the tanks and rolled into balls about 5 inches in diameter which in their turn were calcinated in blast furnaces, 28 *arrobas* (5¼ cwt) of them at a time. In twenty-four hours 176 *arrobas* (a little over 2 tons) could be treated in each furnace manned by a smelter, his assistant, a labourer and two men pumping the bellows.

The *cobre negro* produced in this way was then refined in Remisa's reverberatory furnaces (the first of which started work on 3 August 1831). In every twenty-four hour period one of these furnaces consumed 220 *arrobas* (2¾ tons) of pinewood and treated 220 *arrobas* (2¾ tons) of crude copper, the refined metal which was produced being run off into basins and then ladled into moulds. Even the slags from the blast and reverberatory furnaces was not wasted by Remisa, for 90 *arrobas* (rather more than 1 ton) of slag were mixed with 120 (1½ tons) of flux and reduced in the furnaces to yield a further 20 *arrobas* (5 cwt) of black copper assaying at 85 per cent.[12]

Producing copper in this way proved highly successful and provided the metal more cheaply and quickly than before. But, as the tonnages mined and taken to the calcination ground at los Planes increased, life started to become highly uncomfortable for the local people. For when the *teleras* were lit in the open air, clouds of stinking white sulphur-dioxide fumes rose upwards, killing trees and vegetation in the vicinity or, if there was no wind, hung at ground level choking human beings and animals alike. Remisa was correct in thinking that this was the most economic way to work the mines of Rio Tinto, and his processes were continued when later the mines reverted to royal control and later still when they were sold to a British-based company; but the whole quality of life at Rio Tinto began to deteriorate and 'the smoke problem' became an important factor in social and industrial unrest during the next seventy years.

The mining *marqués* was not only head of the mines and of the manufactory. He was the head of the local community, too, and like his predecessors was troubled with clergy problems. The abolition of the chaplaincies, and the creation of the new ecclesiastical parish at the mines, meant that the priests possessed a greater degree of independence than before and that any attempts to control them had to be carried out through the archbishop of Seville, who could be as slow to take action as the finance ministry officials had been. In 1836 a new priest was appointed called Carlos María González, and soon Remisa and his staff were bombarding the archbishop with complaints about the priest's conduct while Alejandro Vicente Ezpeleta, the civil service director residing at the mines and charged with keeping a watch on the royal interests, was sending long accounts to the civil governor of Huelva province of the interminable disputes in which the priest had become involved. According to Ezpeleta, González's behaviour during services and the bluntness of his speech caused people to walk out of church either grumbling or laughing while many people stopped attending church completely. González was accused, too, of having an unusually strong inclination for caning the boys of the *pueblo*, while it was said that the coarseness of his language made the miners' wives blush with embarrassment![13]

On 3 September 1837 the priest caused a near riot on the feast day of the patron saint of the mines' village, San Roque. Since 1800, when Rio Tinto had escaped a plague visitation which hit the surrounding *pueblos*, the people had taken his life-size image from the church, together with that of Our Lady of the rosary, on the evening of the first Sunday in September after the evening service of the rosary; and carried them through the streets in solemn procession. This was in fulfilment of a vow made by the villagers at the time of the pestilence when San Roque had answered their prayers for deliverance.*

In 1837, however, the interfering González expressed his disapproval of the festivities and games with which the miners and their families celebrated the day but, failing to get any alterations, he announced that for that year the procession would be held in the afternoon. His idea was to disrupt the day's merry-making and he was left in no doubt as to what his parishioners thought about him. But then on the Sunday, with no prior announcement, González brought out the statues for procession after the scarcely-attended morning mass. In the afternoon most of the families left their feasting to attend the sermon in the church under the impression that the saint was to be processed afterwards. By the time that the priest disappeared after the service they must have discovered that the ceremony had already taken place, but they returned that evening for the rosary determined on having a procession. The crowd was immense. Most people could not get into the church but overflowed into the *plaza* outside. The service over, the congregation tried to prevent the priest from leaving the church and accused him of prejudice against the saint. Tempers were quickly lost. González and the sacristan managed to make their escape and run to their houses, locking their doors after them.

A cry went up for the images and they were taken from their niches. Canopies were erected over them and they were carried through the streets to receive the devotions of the people. Ezpeleta, the government director, was brought to the *plaza* by the shouting, and on his arrival found the square packed with kneeling figures reciting the rosary around the images. He stayed to ensure the safe return of the statues to the church and then left. Within a few days González had bombarded the archbishop of Seville, the ministry of finance, the civil governor and the mines' management with complaints against his parishioners. The archbishop protested to the governor of Huelva province about the violence the people had shown towards the saint and Our Lady. The governor told Ezpeleta that the demonstration should have been prevented by force. And Ezpeleta and the mines' staff complained at length about the enormities of the priest's behaviour.[14] The angry exchanges continued for another two years until González quitted Rio Tinto, but the civil authorities soon decided that they did not want to get involved in the altercations. They were too concerned with fighting the Carlist faction in the civil war which was then being waged, and with solving the problem of the national debt by confiscating much ecclesiastical and all monastic property, to be over-concerned with one awkward priest among so many.

*The feast day of the saint is now celebrated on 15 August.

In the autumn of 1837 Ezpeleta retired and Vicente López Preve was appointed to watch the government's interests at Rio Tinto in his place. The new director had the idea that Remisa should start making copperas by adding vitriolic earths to the natural drainage waters used in the cementation tanks. This was done without the knowledge of anyone in the government and was a further breach of the contract, for the crown was suffering a loss for which no allowance had been made. But Remisa was delighted with the results. It greatly increased the amount of copper obtained by cementation, and from 1841 onwards provided over 1,000 *arrobas* (12½ tons) of copperas each month. By 1845 seventy-seven cementation tanks had been constructed, though Remisa was then faced with the prospect of losing this easily-made income in 1849 when his contract ran out. However, López the director proved still more helpful and wrote to the government stating that the manufacture of copper and copperas in this way was absolutely new and had been discovered by an Englishman living in Seville called Felipe Prieto.* As a result of this falsehood Prieto was granted a patent for the 'new invention', and he entered into a contract by which he licensed its use at Rio Tinto to a newly formed company, la Compañía de los Planes, for fifteen years. The partners in the company consisted of Prieto, Remisa and some of the latter's financial backers.[15] By this means Remisa was able to ensure a further ten years' interest for himself in the mines even if the government refused to renew his main agreement when it came to expire.

The expulsion of the Carlist army from Spain was followed by a period of political instability at the centre of government. In 1840 a new law, which had tried to alter the elective basis of government in the municipalities, had provoked a rebellion in the towns throughout Spain. Queen Maria Christina had resigned the regency and been succeeded by General Espartero, the hero of the war against Don Carlos. His lack of political ability had produced a further series of risings backed by the Vatican which had been outraged by the treatment afforded the Spanish church. Though Espartero had subdued them all quite ruthlessly by 1842, he was a poor administrator and by the time of his own downfall in 1843 he had lost what little government control had previously existed over local authorities. Remisa had seen in the disturbances of this time an opportunity to acquire local political influence at Rio Tinto which was denied to him while the mines were part of the *ayuntamiento* of Zalamea la Real, for that council was largely controlled by the Lorenzo family who were large local landowners.

On his plea, not without merit in itself, that the growing settlement

*The name was presumably a Spanicisation of Philip Black.

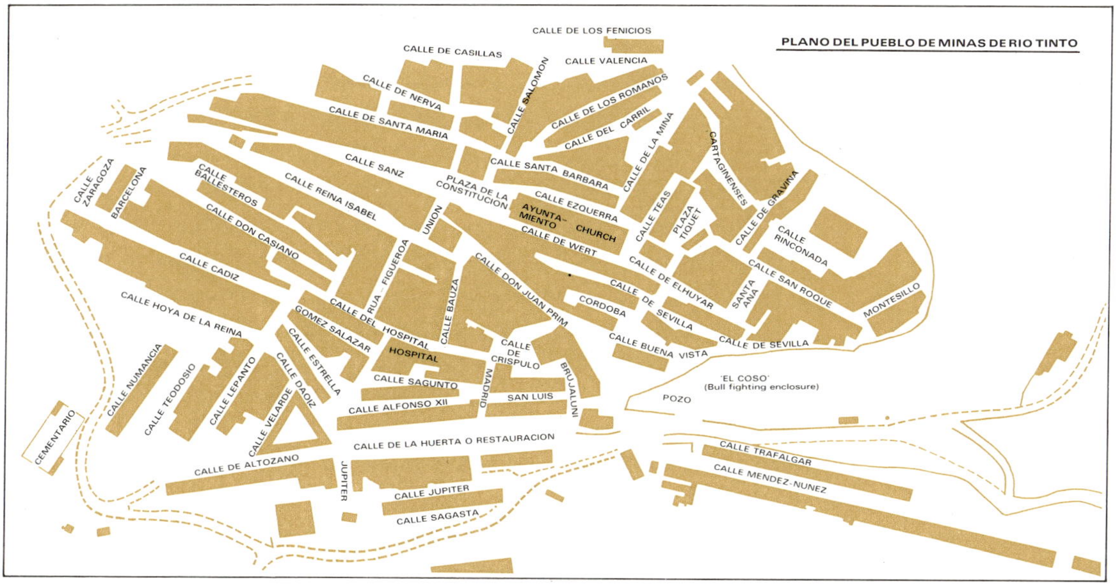


*Map of la Mina (Rio Tinto) as it was in the mid-nineteenth century. 'Pozo' is a public fountain, 'cementario' is the cemetery, 'el coso' the bull-fighting enclosure. (Reconstructed from a map of the 1880s.)*

at the mines needed a proper municipal administration, the government agreed in 1841 to sever the area within the mines' boundary from the Zalamea municipality and to make it an *ayuntamiento* in its own right under the name of las Minas de Río-Tinto,* with la Mina as its principal *pueblo*. But Remisa was to be disappointed. The officials of the new town proved to be fairly irascible and determined to be independent of his control in civic matters even though many of them were mines' employees.

The new council claimed the freehold ownership of all public buildings such as the church, the gaol and the slaughterhouse, and also the streets. Remisa had conveyed control of these to the new *ayuntamiento* but officially they were crown properties on lease to the mining contractor and the government refused to part with them. Similarly, requests from the council that the waste lands be placed under its control were rejected, although the finance ministry invited it to supervise the building of new houses which Remisa had been given permission to erect on the waste.

The local officials were disappointed at not being allowed to issue the regulations for the new housing, but Remisa was equally frustrated with the difficulties they made for him in supervising the erection of the buildings. In 1846 the *ayuntamiento* put pressure on Remisa to rebuild 170 old houses that formed the heart of the old mines' *pueblo* and which housed over 1,000 people, but Remisa agreed only to provide the materials which were to include 400 pine trees for the roofs. The work had got as far as rebuilding the walls when a party of drunken hotheads set fire to the mines' plantations. The woodland

*The hyphen has now disappeared and the *ayuntamiento* has become las Minas de Ríotinto. The *pueblo* of Río-Tinto continued as part of Zalamea la Real until 1885 when it was separated and made into the *ayuntamiento* of Nerva.

was saved from complete destruction by the prompt action of the townspeople, but Remisa declared that there was now a shortage of timber and that the mines' needs must take priority over the town's. He therefore refused to deliver the promised trees to complete the roofs of the houses. In his view, since all the workmen indulged in drunken orgies, all were as bad as one another, and as far as he was concerned the familes affected could roof their own houses. Faced with the prospect of a thousand people living in roofless accommodation and a rainy season ahead, the *ayuntamiento* appealed angrily to the government and Remisa was finally forced to give way.[16]

The heavy drinking of which Remisa complained was common not only to Spain but to other European countries of the time, but at Rio Tinto it was possibly made worse than normal by the living conditions in which people had to exist. Until the rebuilding of the centre of the mines' *pueblo* in 1846, little had been done since the time of Angulo to improve the workmen's housing. Life was made even more unbearable by the smoke from the calcination ground at los Planes which was increasing rapidly in quantity with the dramatically higher output which Remisa was getting. The *teleras*, the piles of burning ores, reached to within 140 yards of the *pueblo*.

As Remisa's profits grew so did the sulphur fumes, pervading the men's houses and choking their lungs. The trees and the vegetation gradually withered under a permanent coating of white powder. The workers' livestock was unable to graze on the common pasture. There were days when even the hot summer's sun could not pierce the sulphurous clouds hovering above the mines. Even the birds ceased to sing and disappeared from the stunted woodlands.[17] The *ayuntamiento* protested vigorously but ineffectively to the government about the desolation of the Rio Tinto countryside and the growing unpleasantness of life there.

In 1847 the government sent an inspector to report on Remisa's activities. He found that although the *marqués* had produced over 21,000 *arrobas* (262½ tons) of copper a year it had been done only by persistently breaking the terms of the contract. The plantations had shrunk to one-eighth of the size that they had been when Remisa took over, and the careless extracting of ores had caused losses equivalent to 2,000 *arrobas* (25 tons) of copper annually between 1838 and 1846.[18] This unrestricted exploitation of the mines produced a display of government anger. Even though it had received its annual rent regularly and thus at last made some money from the mines, it nonetheless declared that when the main contract expired at the end of 1849 Remisa would not get the extension for which he had applied. The inventory made in the latter year justified the accusations of spoliation. The total value of the mines and their possessions had declined to a little over one million *reales* (£10,417), while the woodlands, which in 1829 had been valued at $1\frac{1}{2}$ million *reales* (£15,625), were estimated to be worth only 287,000 (£2,990).[19]

But although Remisa left Rio Tinto at the end of his contract it was not the end of his interest there since the agreement with him and Prieto allowed their Compañía de los Planes alone to use the cementation process at the mines until 1859. Under the grant the government undertook, as the inheritors of Remisa's former obligations, to provide the company with 30,000 *quintals* of ore per month, in return for which the company promised the crown 1,500 *arrobas* ($1\frac{3}{4}$ tons) of refined copper at a price of 56 *reales* an *arroba* (equivalent to $5\frac{1}{4}$d per pound).

At the end of 1849 the government took over from Remisa a mine with a proven record of profitability. True, Remisa had spent little on maintenance but he had demonstrated that there were large profits to be had. If the crown had been able to maintain output it would have been able to put repairs in hand and still have shown long-term gains if reasonable economy had been practised. But just as the government had handed the most profitable part of the business on a plate to the Compañía de los Planes, so in 1849 it gave yet another slice of its own profits away in the form of a contract to Mariano de la Cerda, the parish priest of Rio Tinto mines since 1839.

Cerda had patented what he described as a 'new electro-chemical process' of artificial cementation and its was agreed that he should receive one half of all the ores remaining after each month's deliveries to the Compañía de los Planes. Under the agreement with him the crown also undertook to supply him with precipitation troughs, three factory buildings close to the mines and supplies of fuel. Prieto objected strongly that the priest was getting an unfair advantage, since the cost of transporting ores to los Planes was considerably higher than those which the priest would incur. Cerda thereupon agreed to set up his establishment further off and was given an area of land which has ever since been known by the name of Cerda. In return for his concession he was granted the large mansion formerly occupied by Remisa.

At the same time as the crown gave away the best chance it had of making money from the mines, there was a reorganization within the finance ministry which resulted in a heavy-handed bureaucratic machine being set up to control the government's part of the mines in place of the flexible system of administration established by Remisa. Despite the examples available from Rio Tinto's past the civil servants failed to appreciate that flexibility is an essential for any mining enterprise. Thus responsibility for Rio Tinto was transferred in 1849 to another division of the ministry, the directorate of state property (*la dirrección general de fincas*), whose officials did not include anyone with a specialist knowledge of mining. The results are evident from the statistics. Between May and December 1849, for example, the los Planes company had produced 4,747 *arrobas* (59 tons 6 cwt) of refined copper and the crown manufactory the considerable amount of 15,431 *arrobas* (193 tons). For the whole of that year the government sales had shown a profit of over 922,000 *reales* (£9,604) a return of almost 100 per cent on the operating costs of 924,000 *reales*. But in 1850, with Cerda taking his share of the ores, government production for the whole twelve months was only 17,352 *arrobas* (217 tons) instead of an anticipated 30,000 (378 tons). For the year 1852 the figures were: crown manufactory 18,600 *arrobas* (233 tons); Compañía de los Planes 24,379 (305 tons); and Cerda 16,074 (201 tons).

During 1850 to 1852 Cerda wrote virtually every day to the directorate of state property requesting such things as timber supplies, additional land for burning his *teleras* and extra factory and warehouse space. As a result the directorate devoted more time to his establishment than had been expended by the civil service on the whole of Rio Tinto for many years past. The unusual degree of consideration which Cerda received was due in great part to his friendship with senior government officials and influential politicians. Consequently the officers of the directorate found it politic to express themselves as being well satisfied with Cerda's results and to reject Prieto's accusations that Cerda had not invented any new process but was merely using the same techniques as the Compañía de los Planes and was thus violating Prieto's patent.

In 1852 the directorate issued new regulations for the government's establishment at Rio Tinto, providing intricate controls which caused the mines' officials considerable administrative difficulties, and took almost all the decision making away from the men on the spot and gave it to its own officials in Madrid. This was the time of the premiership of Bravo Murillo which aimed at restoring the power of the church, destroying the representative nature of the *cortes* and legislating by ministerial decree. Absolutism was re-established and open criticism of government policy forbidden by Murillo who threatened to hang dissident army officers 'in their own sashes'.

But even Murillo's policy of civil service centralization and repression of criticism could not disguise the fact that the crown's interests were being harmed by the preference shown by the government to the two private establishments. In 1853 the storm broke: it began with a protest from the Seville foundry at having to buy copper from the private contractors. According to the foundry, Cerda's refined copper assayed at 98 per cent and Prieta's at 92 per cent, compared with the crown manufactory's copper at 100 per cent.[20]

The figures were leaked to the Spanish mining magazine *Revista Minera* which mounted a campaign against Cerda in particular. In a scathing article published in July 1853 it ridiculed his 'electro-chemical' process and referred to his friends in high places who had made life difficult for those civil servants who had tried to expose him as a fraud. It demanded that his company should be kicked out of Rio Tinto and his influential allies from their government posts. Cerda began a libel action which he quickly dropped. The magazine followed up with another critical article and this time Cerda decided to go through the courts.[21] The hearing of the action took place in May 1854 when the verdict was in favour of the magazine.[22] Cerda lodged an appeal but this was dismissed the following November.[23]

Meanwhile Murillo's administration had fallen and been succeeded by that of General O'Donnell which was pledged to return to constitutional forms of government. But O'Donnell's assumption of power in 1854 did not provide any answer to the political and economic problems facing the country. Despite his early reputation as a liberal, he had done nothing to reform the bureaucratic apparatus of the state by the time of his own fall in 1863. He found it to be to his own advantage to leave the civil administration unreformed, and much of the responsibility for the failure to make Rio Tinto a successful crown enterprise in those years must be attributed to the nature of the central government and to the inadequacies of those men whose duty it was to lead the nation.

# 8 *That noble corpse*

## The dead hand of bureaucracy: 1853–73

The twenty years leading up to the sale of the Rio Tinto mines in 1873 to a foreign consortium were a time of intense frustration for the crown's employees at Rio Tinto. The majority of the men on the spot were well-qualified mining and metallurgical specialists but the incompetence of the civil servants and the politicians at the centre of national life made it impossible for the officers in charge of the mines to discharge even their most rudimentary duties efficiently. Life became a tedious struggle against bureaucracy. In 1853, for example, there was a severe drop in the amount of copper produced because under new regulations all stores which it was desired to purchase had to be authorized in advance by ministry officials, who completely failed to reply to continuous requests for permission to buy gunpowder for blasting. Consequently work was halted and production by the crown manufactory dropped to 7,324 *arrobas* (91½ tons) for the year.

Agustin Martinez Alcibar, then manager of the mines, was unable even to maintain order amongst his men since most of his disciplinary powers were taken from him to be exercised in Madrid. There were frequent fights between his workmen and those of Cerda's because the smoke from the priest's *teleras*, which were situated close to some of the entrance shafts, often prevented the horse-whims from raising ores to the surface which led to stoppages of work and reduced pay. When, as a result of complaints by Cerda, the manager was told to control his men he replied to the officials in the ministry that since he no longer had the power to prevent insubordination they would have to deal with it themselves.[1]

His letter resulted in his instantaneous removal and his replacement by a Madrid civil servant, Ramón Luz, who was sent to the mines and given the lofty title of 'royal commissary'. Despite the style he was just as restricted in his powers as Martinez Alcibar had been, save that he was under explicit instructions to dismiss all mines' employees who might make criticisms of ministry officials or claim that the latter were assisting Cerda at the expense of the crown: an interesting demonstration of Cerda's special relationship with the finance ministry. The faults, of course, did not all lie on one side. As Luz discovered, his employees were quite capable of exploiting the chaotic state of the mines' administration and of indulging in a little peculation. On his arrival to take up his duties in February 1854 he personally made a stock-taking of the crown warehouse which revealed a large surplus of copper bars in store above the amount recorded in the books. The discovery coincided with an outbreak of cholera at Rio Tinto and the surgeon ordered a thorough disinfecting of a number of buildings including the warehouse. The copper bars had therefore to be taken to Seville on muleback for storage, but only the amount shown in the books arrived there. The commissary accused his staff of removing the surplus quantity on the way and disposing of it, but his men blandly assured him that he must have made a mistake in doing the stocktaking.[2]

In May Director General Aribau, of the department of mints, mines and state property, made a visit of inspection and told Luz to substitute cast iron for wrought iron in the precipitation process on grounds of economy. Because of the impurities in the cast iron the result was a drop in the amount of copper produced but, rather than report this to the director general, the commissary tried to remedy the situation by enriching the drainage waters with vitriolic earths just as Remisa had done. This made it necessary to take men off underground work in order to stand at the tanks and continuously stir the water to make the earth dissolve as quickly as possible. Despite this the copper obtained by the precipitation method fell by half during the year. The director general also decided on the installation of a new type of smelting furnace, the first of which was lit in his presence and that of the inventor amid scenes of festivity. Difficulties with these led to yet another drop in production which the commissary failed to notify to the directorate. Such was the nature of the régime that Luz knew his superiors would not welcome the truth if it was disagreeable, and he replied to an enquiry from the director general about how the new furnaces were doing with the assurance that excellent results were being obtained.[3]

Because of its inefficiency the government went to absurd lengths to protect itself from criticism. When Luz made arrangements for a supply of cast iron to be paid for in kind with timber from the rapidly diminishing Rio Tinto plantations, on highly disadvantageous terms and in contravention of a recent royal decree on government contracts, the director general sent him a severe but private reprimand adding, however, that he would try to get approval from the minister for this breach of the regulations in such a way that there would be no criticism of the department. This was achieved by the publication of a royal edict thanking Luz in profuse terms for obtaining so quickly the cast iron he had needed and on terms of great profit to the crown.[4]

In an attempt to win approval in Madrid, Luz sent the directorate an idea to reduce costs. He proposed that the system of carrying out underground mining by giving piece-work contracts to senior men, who in turn selected their own teams of workmen, should be replaced by one in which the contracts would be awarded to fathers of families who would employ their own wives and children. This would enable the total number of workmen to be cut and make it possible to have a general reduction in wages. The directorate, however, refrained from commenting on his suggestion.

After only one year's administration by the royal commissary, copper production had been reduced by half and costs had increased to such an extent that the forecast of 2 million *reales* (£20,833) profit on the year had become a considerable loss. Each pound of copper made in the crown manufactory had cost 5 *reales* 5 *maravedises* (1s 1d) to produce and had been sold at 3 *reales* 8 *maravedises* (8¼d). The percentage of copper being recovered from the ores had fallen from over 2 per cent to less than 1 per cent as a result of both inadequate smelting and a fall in the copper content of the ores. On the other hand, for that same year of 1854 the profit made by the Compañía de los Planes came to 750,000 *reales* (£7,288) and by Cerda to 120,000 (£1,250).[5]

That was the year of Cerda's disastrous libel actions against the *Revista Minera*, but despite these the priest continued to enjoy his government contract and was even permitted to increase his prices to the royal mint and the foundry. His friends in the directorate passed on to him all the technical reports on his operations submitted by critical civil servants together with a great deal of confidential information about the los Planes company.[6]

In February 1855 the crown's senior staff at the mines wrote to the directorate about the commissary's incompetence, backing up their complaints with carefully compiled statistics from the mines' offices. Aribau, the director general, responded by instructing Luz that in future the commissary alone was to have the keeping of the statistics in order to prevent their being used to criticize the department. Shortly after this the director general was dismissed and the senior officials of his office hurriedly submitted strongly-worded reports about the mismanagement of the mines. The end of the year stock-taking at Rio Tinto showed a deficit of 6,000 *arrobas* (78 tons) of copper compared with the book entries, and a minute scrutiny of the accounts showed that there was a loss of almost 376,000 *reales* (£3,916) on the year although Commissary Luz had sent figures to Madrid showing a profit of 315,000 *reales* (£3,281). Some bitter comments were passed on Luz's abilities and then very quietly he was transferred back to his former job at the finance ministry.

The government of O'Donnell continued to be completely out of touch with the needs of the situation at Rio Tinto. Convinced that the mines were always going to be run at a loss, the council of ministers gave instructions to concentrate on reducing the amount of that loss. Consequently there were drastic cuts in the money allocated for essential operations. Thus in 1856, when there was a rapid increase in the demand for copper and the price in the Seville market rose to 4 *reales* 4 *maravedíses* (10¼d) a pound, the staff at the mines prepared to take advantage of it by increasing production and thus cutting losses. But orders arrived from Madrid to reduce production, and despite appeals from the mines these were not rescinded until it was too late to take advantage of the temporarily increased prices. Even then the supply of cast iron for the precipitation process was completely cut off for some months by the action of officials in the directorate of state property, who told the newly appointed manager that they had done this to teach him a lesson because he had written directly to the minister of finance instead of following the proper procedures and going through them! Lucas de Aldana, the chief engineer at the mines, estimated that this piece of petulant stupidity cost the crown half a million *reales* (£5,208) in lost sales.[8]

The problem caused by the workmen leaving the mines at harvest time to work in the nearby villages for the high wages paid to temporary harvesters, which Director General Angulo had resolved in his own time, had re-emerged with the decline in real terms of the mineworker's wages and the loss of the benefits Angulo had arranged. By 1856 the miners were only too pleased for a chance to get away from the smoke-polluted atmosphere of Rio Tinto and from their overcrowded housing to earn some extra money. To prevent the mines from closing down completely, the manager wrote to Madrid asking for permission to pay the essential workers, such as the whim-drivers, an extra *real* (2½d) a day during the harvest season to keep them at their jobs. The directorate refused and forbade any increases of wages at Rio Tinto by an order printed in the official *Gaceta*, an order which Aldana described as 'a decree worthy of a lollipop maker'.[9]

In November of that year the newspapers returned to the attack on the government for neglecting Rio Tinto and for favouring the Compañía de los Planes and Cerda. The priest had somewhat pompously announced a war to the death between himself and the los Planes company and it was obvious from past experience that in any fight between the two private enterprise establishments the most probable loser would be the public purse. The newspaper *el Criterio* had been primed with facts and figures by an anonymous member of the mines' staff and declared that the government's losses on the year would be over a million *reales* (£10,417). It demanded the publication of full accounts and action to terminate the maladministration.[10] At the beginning of 1857 the government decided to write off 700,000 (£7,292) of the 1,700,000 *reales* (£17,708) which the official accounts showed as having been lost at Rio Tinto in the four years 1853–6.[11]

In the face of the mounting accusations of fraud and corruption being made in the press against Cerda, the priest's protestations of injured innocence began to sound somewhat hollow. The press campaign and the complaints by the government departments that were forced to buy his copper eventually produced action. In 1858 the finance ministry ordered that no more purchases were to be made from Cerda and he agreed to surrender his contract in return for compensation. His friends managed to obtain for him a handsome payment of almost half a million *reales* (£5,208) for renouncing his rights under the agreement which then had only five years to run.[12]

When Prieto and Remisa's contract for the los Planes company expired the next year, the whole Rio Tinto mines once more came under the control of the crown and the mines' officials looked forward to a bright future. The railway era in Spain was just beginning and it looked as if the country was about to enter on a period of considerable economic activity. Between 1850 and 1860, for example, imports and exports doubled and living standards began to improve. In 1859 a Carlist invasion to dethrone Isabella collapsed and the Spanish army achieved victory in its north African campaign against the Moors. The momentary glory and successes of O'Donnell's administration produced a temporary atmosphere of self-confidence which was mirrored in a government promise to make the royal mines of Rio Tinto a success and 'an ornament to the crown'. Unfortunately, despite this rush of heady enthusiasm, the government of O'Donnell was no more capable than its predecessors in its handling of the national finances or in curtailing inefficiency and curbing corruption.

Consequently opposition against the régime began to mount, much of it being channelled into a movement to establish a republican form of government. In 1863 the opposition groups rejected invitations to participate in party politics within the framework of the *cortes*, and in a series of manifestos poured scorn on the self-styled constitutional government of O'Donnell with its rigged elections and backstairs' intrigue. O'Donnell's government collapsed. But before its fall it had started toying with the idea of selling off crown property to raise some much needed money and a British civil engineer called Joseph Lee Thomas was invited to the Rio Tinto mines to inspect them. His observations, which were published in London, were daunting in the extreme.[13] He took the view that the only way to work the Rio Tinto mines effectively was to double the labour force from 2,000 to 4,000 men and to double the yield of ore to 10,000 tons a month, though he had no doubt that anyone trying to do this would have a difficult job in view of the hostility of Rio Tinto workmen to new arrivals and their belief that the mines were their own special perquisite.

He saw the future of Rio Tinto as being in the provision of large quantities of mineral for exportation to England for use in sulphuric acid making. But this would require both the adoption of open-cast mining, an expensive operation since it would require a vast work-force and machinery was then very primitive, in place of the existing system of underground galleries, and also the construction of a railway to the fishing port of Huelva. He described three lodes of mineral at Rio Tinto of which only the southern was then being worked, and this only within a restricted area of some 500 yards in length and 75 yards in width. In Thomas's view, any purchaser should begin by using an old Roman adit known as los Molinos on the south side of *cerro* Salomón to provide an easy entrance to open up the south lode to its full extent, but the cost would be considerable.

The railway he proposed was a concomitant of his suggestions, for to transport the ores to Huelva by any other means was not feasible. If the figure of 120,000 tons, which he calculated as the minimum required to work the mines economically, was to be exported each year, the cost of carrying them to Huelva by road (and at that time no road existed) would be 23 million *reales* (£239,584) a year, compared with 8 million (£83,334) by railway. He estimated, too, that the total cost of mining, transporting by rail and shipping 120,000 tons annually would amount to £267,000 by the time of delivery in Liverpool. At the current United Kingdom price of £3 18s a ton this would yield an annual operating profit of £162,000. Without a railway to carry the ores to Huelva the transaction would show no profit at all.

The results of government assays provided for Thomas showed the ores to have an average copper content of 4.8 per cent which he refused to believe. (Of the forty-one samples no fewer than nine assayed at 9.4 per cent copper indicating that the specimens had not been properly selected.) He calculated 3 per cent as being a more likely figure. About one-fifth of all the ores being mined were in extremely small pieces and these were left underground until they had undergone partial decomposition through exposure to the air and the damp, and were then used in the cementation process. This was still the principal method used at Rio Tinto for making copper, since by that time not a tree remained for miles around as a result of the smoke damage from the *teleras*, and thus fuel was unobtainable save at an enormous price. Charcoal for the furnaces, for example, cost 3 *reales* for a *marcilla* of 29 pounds, or £2 6s a long ton.

Comparing these mines with those of Agordo, belonging to the Austrian government near Venice, he was critical of the results being obtained. The Rio Tinto ores, although probably containing 3 per cent copper, yielded only 1.5 per cent; a loss of half the copper they contained. The ores at Agordo assayed at only 1.5 per cent copper but their more careful treatment led to a loss of only 0.1 per cent. The Venetian system involved three separate calcinations and lixiviations compared with only one of each at Rio Tinto. Thomas doubted whether the high cost of wood would permit the adoption of the Venetian method, but if any purchaser decided to continue with copper production at Rio Tinto then it would be better for him to reduce the calcined ore in a blast furnace using coke rather than in a reverberatory furnace using wood. The coke would have to be imported, probably from England; but being more economical, and needing less skilled labour, it would do the job more efficiently and cheaply, even allowing for extra transportation costs.

The decision not to offer the mines for sale after all meant that Thomas's report had little immediate effect; but when at a later date another government's financial problems led to yet another change of mind, and the mines were purchased by Hugh Matheson's consortium, the main proposals made by Thomas provided the foundations of the expansionist programme then adopted.

The year 1864 saw the first of several unsuccessful uprisings and the spectacle of yet another panic-stricken and incompetent government endeavouring by legislation to stave off its overthrow, though the more it adopted repressive measures the more did educated Spanish opinion come to think of revolution as inevitable. By 1865 the administration was in desperate financial straits. Queen Isabella offered to present to the nation two-thirds of all crown property, in return for a cash payment of a quarter of its value, in order to help the government out of its difficulties. Her proposal misfired when Emilio Castelar, a republican orator and professor of history in Madrid, replied to it with a newspaper article demonstrating that the crown estates were not part of the sovereign's private property but merely adjuncts of the crown. Thus they already belonged to the nation and the effect of Isabella's proposal, if carried out, would simply be to transfer a vast sum of public money to the queen's own pocket. In retaliation for his attack the government ordered the rector of the university to dismiss him, thus triggering off a wave of student riots in the course of which the ministry fell from office.

On 18 September 1868 the Spanish navy mutinied at Cadiz and was soon joined by the army. Their demand for the queen's overthrow led to Isabella fleeing to France and the establishment of a regency under Generals Serrano and Prim. They began by promising universal suffrage, freedom of the press and of conscience, the removal of church control over education and the right of workmen to form associations. But the revolutionary *juntas* which were rapidly establishing themselves throughout the country made the government's task much harder and carried the prospect of large-scale civil war. The country threatened to fall apart and the growth of armed disorder forced the government to abandon its plans to end conscription for military service.

The London *Times* was expressing the common view of the situation when it compared the events in Spain with the French revolution:

'It is not difficult to foresee that it will stand out in future days as one of the most terrible judgements and warnings of history.... Since the first fall of the French monarchy in 1792 there has been no revolution so indicative of a transformation of the popular character of the nation that has effected it.'

This opinion was proved by events to be over-pessimistic but, as the country drifted towards disorder, the situation seemed desperate enough; and the management of the Rio Tinto mines, dispirited and unable to obtain any co-operation from Madrid, lost all interest in efficiency and in maintaining production. A government commission, consisting of the engineers actually working at Rio Tinto under the chairmanship of the inspector general of mines, Ignacio Gomez de Salazar, was asked to prepare a report on the future of Rio Tinto. The commission recommended similar measures to those proposed by Thomas: open-cast working and the construction of a railway to Huelva; but it must have known there was little chance of their implementation in the foreseeable future under any Spanish government, for the capital outlay required was too great for the nation's available resources.[15]

In 1869 Serrano the regent and Prim the prime minister provided the country with a charter prescribing a limited monarchy working in conjunction with an elected *cortes*. The two generals spent a year looking for a new king while Prim complained, 'Finding a democratic king in Europe is like looking for an atheist in heaven.' Meanwhile the high cost of trying to restore order throughout the country led Prim to the decision to sell the Rio Tinto mines as a way of raising money. In 1870 four events of importance in Spain's future took place within four days. On 23 June, Leopold of Hohenzollern, a prince of the Prussian royal house which within a couple of years was to become the dominant ruling family in Europe, rejected Prim's offer of the Spanish throne. The same day an amnesty was signed permitting the return to Spain of extremist left-wing exiles. On 25 June, Serrano signed a law authorizing the sale of the Rio Tinto mines. And on 26 June, Isabella formally abdicated the Spanish throne in favour of her son Alphonso and thus provided the formula for the eventual restoration of the house of Bourbon.

On 16 November, however, Amadeo of Savoy the second son of King Victor Emmanuel of Italy, was elected king of Spain by a large majority in the *cortes*. But in an atmosphere of mounting violence, with extremist newspapers and politicians calling for armed revolution to overthrow the conservative-liberal government, the prospect of good order looked remote. On 30 December Prim died, having been shot three days earlier when leaving the *cortes*, and it must have seemed extremely unlikely that any foreigner would be prepared to bid for the Rio Tinto mines. Nonetheless the government proceeded with the project and in December 1870 and January 1871 a commission set to work evaluating the freehold value of the mines themselves, all the land within the Rio Tinto mines' boundaries and all the buildings owned by the crown together with the mines' equipment. They arrived at a total value of 104,375,769 *pesetas*, or £4,160,000.[16]

During 1871 the Carlists and the republicans formed an alliance to make parliamentary government unworkable and to try to bring down the régime. Socialism grew rapidly, receiving an impetus by the declaration of the third republic in France and the heroic struggles of the Paris *communards*. The anarchist movement gradually spread its hold over the working-classes of Barcelona and Andalusia, becoming particularly strong at this time in Rio Tinto. The prospects for any foreign businessmen who might have been considering the purchase of the Rio Tinto mines were extremely discouraging. In fact when the sale was advertised on 11 May 1871 in all the principal European newspapers only four offers were received, all for sums below the government valuation. The best offer was from a group of businessmen headed by Hugh Matheson, the Scottish head of a London banking house.

The Matheson offer was not an ill-considered gamble, though it seemed such to public opinion throughout Europe. Heinrich (otherwise Enrique) Doetsch, a German businessman trading in Huelva, and his partner Sundheim, knew the mines and the area around. They were also good judges of Spanish politics. Travelling to London, Doetsch and Sundheim met Matheson and persuaded him that he should organize a bid. Matheson undertook to form a consortium of international financiers, men operating on a large scale and on world-wide basis. In Matheson's mind was the thought that if they bought the Rio Tinto mines and formed a company to work them on an unprecedented scale, extracting not 120,000 tons of ore a year as envisaged by Thomas in his report but 700,000 tons a year, the profits would be colossal.

Matheson knew that the political situation in Spain was bad but accepted Doetsch and Sundheim's view that within a short while stability would be restored either by the existing government or by a strong-handed successor in the Spanish tradition. He appreciated that there would always be difficulties of varying complexity, but the company he proposed to erect would be the most important industrial concern in Spain. Its immense size would give it a dominant role in the Spanish economy and a voice to which any Spanish government would be compelled to listen. Thus Hugh Matheson saw the upheavals in Spain not as a deterrent to purchasing the mines but as an opportunity of which to take advantage. He authorized Doetsch to begin negotiations with the Spanish government on behalf of Matheson and Company while he himself set to work to form his alliance of financiers.

The negotiations were complicated and delicate. They took over a year to complete and involved Matheson himself in a visit to Spain with a mining specialist, David Forbes, FRS, to evaluate the potential of the mines and with an engineer, George Bruce, to consider the construction of a railway to Huelva. During that time the country continued in a state of upheaval. In January 1872 the government outlawed the first socialist international, and in April Don Carlos launched the second Carlist war in an attempt to seize the throne while a large section of the Spanish army was in Cuba trying to suppress a rebellion there. But at the height of all the turmoil, on 26 December 1872, another law was authorized permitting the government to dispose of the mines for a smaller sum than that specified in the earlier valuation.[17]

Meanwhile, King Amadeo was involved in increasing conflicts with his government. Following Prim's assassination the coalition had broken up and a more radical administration had been formed. The king wanted a ministry of conciliation and the liberals, the party that claimed to be his only real supporters, refused to consider this. Torn by jealousy and rivalries within their own party, and failing to make Amadeo into the puppet they wanted, they turned against him, and on 11 February 1873 he abdicated after a reign of only two years. The same day the first republic was declared. Technically the declaration was an illegal act for under the constitution the *cortes* should have been dissolved on the abdication and elections held for a constituent parliament. But the liberal party was anxious not to be excluded completely from power, while some of its leading members wanted to use the crisis to oust their leader, Ruiz Zorilla, from his party leadership. A coalition was immediately formed between the liberals and the federal party.

Estanislao Figueras, the federalist leader, took command as president of the republic with Emilio Castelar as the minister of state. These two appointments were significant, for Figueras was considered nearer to the English conception of a parliamentarian than any of his contemporaries, while Castelar enjoyed a reputation abroad as a man of moderation. José Echegaray, the liberal finance minister who had been negotiating the sale of the mines to Matheson's consortium, remained at the same ministry. Putting the sale into effect was considered a matter of the utmost importance by the new government and had top priority in its programme. The value attached to it by the republican administration is seen in the fact that on 14 February, within three days of the declaration of the republic, Echegaray issued the draft of the law designed to complete the sale and give the new owners of the mines the right to build the railway they would need.[18] On the same day a decree was issued signed by president Figueras declaring the government's intention to sell the freehold of the mines to Matheson's consortium for 92,756,592 *pesetas*, equivalent to £3,500,000 sterling. [19] The fear that the political upheaval might make Matheson withdraw his offer doubtless lay behind the speed with which the government acted.

Up till the last moment the senior officials at the mines protested at the proposed sale and the manager, Juan Diego López Quintana, promised that if it could be averted for only a year there was an excellent chance of increased production and profits. The minister of finance retorted that these assurances were utterly absurd in the light of the previous history of Rio Tinto and the views of mining experts. Gloomily the manager and Aldana recalled a previous correspondence on the subject when the promises of Luz the royal commissary had been dismissed as 'poetry' by bureaucrats who were 'experts and not visionaries,' 'men whose records breathed cold reason and scientific syllogism. . . These men were not poets and for that reason they slept calmly over the remains of that noble corpse. For that reason the words "it is impossible" were the convenient solution of all the problems of life at Rio Tinto. In my heart there is poetry. You minister understood it and sent me to Rio Tinto. I, a poet, felt in Madrid the necessity to sacrifice my rest and my health. . . I, a poet, felt afflicted at the sight of Rio Tinto and guided by the sorrow in my heart I wrote over the word "impossible" this note: *Fiat lux et facta est.*'[20]

But the protests could not deflect the government from its course. Matheson and Company, quietly and with no fuss, collected together the money required as the initial deposit for the purchase: £422,680 in gold. In great secrecy it was transported by indirect routes, at times by train, at other times in carts, through France, over the Pyrenees and across Spain to Madrid under the supervision of William Macfarlane, one of the Matheson's employees, completely undetected by the forces of the various factions involved in the civil war then in progress.[21] Thus by a deal unique in mining history the cartel acquired not only the mineral rights of the Rio Tinto mines but the very ground itself and all subsoil within the mines' boundaries.

The Spanish mining engineers packed their bags and departed as a new phase in the history of Rio Tinto began.

# Part Two:

THE RIO TINTO COMPANY LIMITED

# THE MEN WHO DID THINGS

*Hugh Matheson, founder and first chairman of the Rio Tinto Company.*

# 9 *Don Hugo and friends*

## International capital and the Rio Tinto Company: 1873–98

The consortium that Hugh Matheson brought together to purchase the Rio Tinto mines from the Spanish government was an interesting group of men with close ties with the city of London and European commerce: G. A. Schroeder and L. G. Dyes, the president and deputy president respectively of the Deutsche National Bank of Bremen; Edwin Clark, of Clark, Punchard and Company, a leading British railway constructor; Heinrich Doetsch and W. Sundheim of the Huelva firm of Sundheim and Doetsch, businessmen of German origins with financial interests in various Spanish mines; Martin Ridley Smith, a partner in the London underwriters of Smith, Payne and Smith; William Macandrew, director of the Glasgow-based Union Bank of Scotland and Peter Denny, a director of the merchant bank of Arthur Heywood, Sons and Company of Liverpool. But of them all the most interesting was undoubtedly Hugh Matheson.

His was a personality of great complexity and depth. On the surface he appeared a typical Victorian banker: one of the best-known men in the commercial life of the city of London, whose considerable contributions were to that great expansion and development of the city's financial institutions which gave Britain a notable role in nineteenth-century world trade.[1] But his family origins were in the western highlands of Scotland, in those wild parts between the Five Sisters of Kintail and the narrow sea which separates the mainland from the isle of Skye, where for centuries his ancestors, the Mathesons of the Lews, had lived on their lands as minor chieftains. Early in the nineteenth century two Matheson brothers left home, The younger, Duncan, went to Edinburgh, became prominent in the public life of Scotland's capital and sheriff (or presiding magistrate) of the city. There his son, Hugh McKay Matheson, was born in 1822. The elder of the two brothers, James (later Sir James Matheson of the Lews), travelled to London where as a partner he shared in the excitement of establishing the important firm of China merchants known as Jardine, Matheson and Company.

As a young man Hugh McKay Matheson went into business in Glasgow and then decided to follow his uncle to London. In 1848, when only twenty-six, he established the banking house of Matheson and Company in Lombard Street, London, and began a close association with his uncle James, becoming in the course of time a director of Jardine Matheson. It is difficult to trace the extent to which James Matheson influenced and moulded Hugh's character. Probably the similarities between them owed much to their Scottish background. But in all events, young Hugh Matheson developed many characteristics akin to those of his uncle. As a businessman he was prepared to gamble for great stakes. Shrewd at appraising risks, he was capable of seeing possibilities in situations where most men saw only dangers. Like his uncle, too, he became a staunch liberal in politics. In time he was to embrace fervently such causes as Irish home rule and the extension of the suffrage, and to oppose with passion the imperial policies of the conservatives. Contradictory features, it may be thought, in a man who in his business life was a complete (even if benevolent) despot, and whose immense fortune was derived in part from the opium trade with China and from providing the railways, the equipment and the guns for the different factions engaged in the Chinese civil wars.

The most surprising aspect of Hugh Matheson's life was his religion. An intense and devoted presbyterian, he was alarmed at the state in which he found the English presbyterian church when he came south of the border. He dedicated himself to its revival, poured money into its coffers and became acknowledged as its leading layman. From his own pocket he financed the building of Holy Trinity church at Hampstead, close to his London home, and gave generously of his time and money to the presbyterian mission in China. Nor was that all, for his religion was not superficial. He was a passionate Christian, who, when in London, left his office on five nights a week and went straight to the east-end to conduct missions of rescue for the prostitutes and the alcoholics who dragged out their existence in the filth and squalor of the slums of Spitalfields and its neighbourhood. And amongst these dregs of humanity he came alive in a way that would have startled his city colleagues. For his kindness and the warmth of his personality won him much affection, just as it did with many of the labouring people at Rio Tinto.

Yet with all this he could still be narrow-minded and bigoted. Perhaps not surprisingly, given his nonconformist background, he loathed Roman catholicism, but his real venom was reserved for the ritualist wing of the church of England of which his friend William Ewart Gladstone, the liberal leader, was a prominent member. These inconsistencies cannot be explained by assuming that he was one of these men who hated in the abstract but loved individuals. For, even in the funeral address preached over Matheson's coffin, the minister of his own church at Hampstead felt obliged to record that some of the congregation considered that Matheson's righteous mien indicated a lack of charity apparently incompatible with his impulsive generosity towards the poorer members of the congregation and children.[2] Certainly the Rio Tinto Company's postal clerk in its London office received no mercy when a deficiency of 5s was discovered in his accounts. He was dismissed on the spot and the money deducted from the pay owing to him.[3] Matheson was equally an enigma to most of the British staff at Rio Tinto and to his fellow directors on the board of the Rio Tinto Company which he formed following the purchase of the mines. The German directors thought him particularly idiosyncratic and on occasions expressed surprise at his concern for the spiritual and social welfare of the Spanish employees when at the same time he was prepared to justify the use of troops to suppress their political aspirations. (Not that they themselves had any objections to the latter.)

Hugh Matheson, too, had a strong sense of his family history and a pride in his descent. He was certainly entitled to be proud of his own achievements, for he advanced the family name and fortune of the Mathesons of the Lews even further than had his uncle, Sir James. Rather oddly, in view of his standing in the city of London, he did not attempt to win admission to the inner ranks of its establishment, or to hold those offices which would have led him to the lord mayor's chair and its accompanying baronetcy. He was made one of Queen Victoria's deputy lieutenants for the county and a justice of the peace, but these were prizes awarded for his charitable work and were, in any case what any gentleman of his social standing might have expected. The highest public honour he ever received came not from his own country but from Spain. In 1884 he was made a *caballero gran cruz* (a knight grand cross) of the order of Isabella the catholic. He was one of the few foreigners, apart from heads of governments or ambassadors, to be awarded this distinction and its associated title of *excelentisimo* ('most excellent'). The broad white sash, with its wide yellow border-stripes, and the gold and red-enamelled star gave a deep satisfaction to the austere teetotaller, for the award in his case represented no diplomatic exchange of courtesies but was an acknowledgement of the fact that by that date he had become one of the most powerful men in the economic life of Spain. And the exercise of power was what Hugh Matheson enjoyed more than most other things in life.

Such was the man who brought together the combination of capitalists who purchased the Rio Tinto mines and then created the Rio Tinto Company which began life with its registration on 29 March 1873. At its first meeting, held at Matheson and Company's office at

3 Lombard street at noon that same day, Matheson was appointed chairman and Doetsch, Dyes and Edwin Clark became directors. The new board quickly got down to work. Until September it met every second day, appointing staff, arranging for public participation in the company, raising funds and sending orders to the as yet small staff at the mines on the basis of the reports and letters that came pouring in as the enterprise got under way. (Not until the following October was it felt that the project was sufficiently far advanced for the directors to be able to relax and to hold board meetings only once a week.)*

At the beginning of April two clerks were appointed and sent to Rio Tinto to establish an office. A German engineer, Theodor Blum, was appointed as head of the mining and extraction department and as acting mines' manager a few days later. David Forbes was confirmed as the company's consultant mining engineer and George G. Bruce as consultant railway engineer at the end of the month.[4] Forbes submitted his first report on the reorganization of mining at Rio Tinto on 25 May and Bruce his plans for the proposed company railway on 11 June.[5]

Much of the month of June was devoted to writing and redrafting the prospectus which was to be issued on 8 July inviting public participating in the company and to deciding the beneficial terms which the original members of the consortium would receive. Under the agreement with the Spanish government, Matheson and Company acting on the group's behalf, had paid the first instalment of £422,680 towards the purchase price of the mines and given nine *pagares*, or promissory notes, the last of which was due for payment in 1883. The consortium proposed transferring its rights in the mines, and its obligations to pay the *pagares*, to the new company for £150,000 in cash, £600,000 in debentures bearing interest at 8 per cent per annum (which in May 1875 were converted into ordinary shares) and £250,000 in ordinary shares; a total of £1 million in return for the cost of the first payment. In addition there were other gains which the members of the group were to get and which were considerable in themselves. Sundheim and Doetsch of Huelva were appointed as the Rio Tinto Company's agents in Spain and Portugal; Matheson and Company as agents in Great Britain, Ireland, France and all countries outside Europe; and the Deutsche National Bank for the remainder of the European continent. By their contracts the agents were to receive a commission of 1 per cent on sales and 2½ per cent on purchases. Clark, Punchard and Company received a contract to build the railway between Huelva and the mines and also to construct the pier at Huelva which would be necessary for the deep-water ships arriving at the port to take on ores for export. Smith, Payne and Smith, together

*The weekly meeting lasted until the first world war when a fortnightly meeting was instituted.

with Arthur Heywood, Sons and Company and the Union Bank of Scotland handled the public share issue.[6]

The board estimated that the cost of constructing the railway and the pier, and of equipping them with the necessary locomotives and waggons, would come to £1 million; and that another £1 million would be required as working capital to convert to an open-cast system of working and to provide the necessary plant and buildings. Since the purchase price of the mines themselves came to almost £4 million, the new company's total cash requirements were estimated at £6 million, but the directors decided that if they could raise £2 million as capital the remainder of this sum could be amortized over ten years by using profits from sales. Consequently they registered the company's capital as £2,250,000 made up of 225,000 shares of £10 each, and in their prospectus offered 200,000 of these to the public to

*Rio Tinto in 1973: The old railway does not meet modern requirements and is gradually falling into disuse.*

obtain the necessary £2 million in cash. The Deutsche National Bank subscribed £560,000 towards the public issue (for 56,000 shares), Hugh Matheson £228,000 and Edwin Clark £200,000 while the other members of the consortium purchased additional shares for £29,000. Thus of the 225,000 voting shares with which the new company began its existence the consortium itself held 126,000; more than sufficient to ensure that control would stay in its members' hands.*[7]

The claims made for the company's prospects by its directors were considerable. Their optimism to a great extent reflected the current prosperous state of business and in particular the buoyancy of mining shares. The achievements of the Tharsis Sulphur and Copper Company of Glasgow, which had acquired the Tharsis mines near Rio Tinto in 1866, had encouraged a great deal of speculation. By 1873 its £10 shares were being quoted above £40. In that year, too, it extracted 374,000 tons of ore and declared a dividend of 40 per cent.[8] The directors of the Rio Tinto Company claimed in their prospectus that they would surpass this performance, mining a minimum of 700,000 tons of ore annually of which 200,000 tons would be calcined at Rio Tinto and 500,000 tons sold in the world's markets. They estimated that the ores contained 48 per cent sulphur and 2½ to 3½ per cent copper. Those earmarked for export, the prospectus declared, should yield over 100 per cent profit. The average selling price in England for ten years past had been £3 3s 3d a ton and, since it would cost the company only £1 7s 0d to mine and to transport each ton to Liverpool, there should be a profit of £1 16s 3d a ton. Once the instalments due to the Spanish government were paid, the directors envisaged dividends of at least 35 per cent of the share capital each year.[9]

This prospectus appeared to many to be almost fraudulently optimistic. Its claims and promises were challenged in the press by those who disbelieved that the new company could extract 700,000 tons of ore annually, and doubted that the markets could absorb this enormous quantity even if it did. In any case, the critics argued, such large-scale production was likely to lead to a marked decline in prices which would make the profits anticipated in the prospectus unrealizable. The Tharsis directors, speaking from a position of authority in view of their own success, would not believe that their new rival could be financed on a bare £2 million. In a printed broadsheet they declared that the proposals of the Rio Tinto Company's directors for capitalizing the company were inadequate, and that the provision which Hugh Matheson and his colleagues had made for themselves was over generous and to the detriment of other shareholders.[10] But the confidence of the Rio Tinto directors, reflected in their own heavy shareholdings, lent credence to their statements, and the

*These were bearer shares. In 1881 the share capital was increased to £3,250,000 by the issue of a further 100,000 shares with a nominal value of £10 each, sold at £22. In 1897 at the request of the larger shareholders the shares were split into £5 issues and into 325,000 preferred and 325,000 ordinary, the rate of dividend on the preference shares being 5 per cent. In November 1905 a further issue of 50,000 £5 ordinary shares was made, and in May 1929 a similar issue. Thus by the latter date the share capital had become £3,750,000 consisting of 325,000 preference shares with a face value of £1,625,000 and 425,000 ordinary shares with a face value of £2,125,000.

mining boom in progress was a sufficient inducement to the share-buying public. And so the company was successfully launched. In the first few years the forecast of the amount of pyrites to be mined was fully justified but, as many had anticipated, world demand was not sufficient to take up this extra amount available, and a price war developed between the mining companies.

Buyers of pyrites, the ores which combine sulphur with copper and iron, were principally the sulphuric acid manufacturers on whom the world's chemical industries depended. At the annual meeting of the Rio Tinto Company in 1876 Hugh Matheson made a typical Victorian businessman's assertion when he declared, 'The civilization of a country may be estimated by the extent to which it uses sulphuric acid.' (His statement brought from one shareholder the perhaps not unexpected response, 'I am sure we shall promote the civilization of the world to the utmost of our power.') But with production greater than demand the sulphuric acid makers were able to exploit the situation and to force price reductions on the pyrites' producers. Nonetheless the Rio Tinto Company was soon in a position to compensate for this by increasing the quantities of pyrites mined each year and before long was able to deliver to its shareholders the promised profits.

The original shareholders, however unhappy they may have felt in the first few years about future prospects, if they kept their shares certainly found that the promises of the prospectus were honoured. By 1884 Rio Tinto had become the greatest mining centre in the world: in that year the company produced 1,370,000 tons of ore, of which over a million tons was treated at Rio Tinto and the rest exported. Copper manufactured at the mines alone reached 12,668 tons in that year compared with 946 tons made in the first year (1876) of the company's manufactory at the mines.[11] In 1905 the £5 ordinary shares were changing hands at £65 each. In 1912, 2,407,000 tons of ore were extracted, 652,000 tons being exported and the rest being used at the mines to produce sulphuric acid and 21,062 tons of copper.[12] Even as late as 1929 the £5 ordinary shares were quoted at £50. The fact that the company was so sensationally successful, in a period when business was never easy, is its own tribute to the vision and industry of the men who created the company and ran it for the first twenty years or so of its life.

But in 1873 the great exploits still lay in the future. There was much to be done before and after the first meeting of shareholders at the Cannon Street Hotel. A London office had to be set up, and premises were acquired at 17 Cornhill in the city of London in August. Staff had to be recruited for London and Spain. Rodney Fennessy was recruited as company secretary, from a similar job with the Continental Union

Gas Company, at a salary of £600 (to rise to £1,000 over three years) with his annual income tax of £5 12s 6d to be paid by the company.[13] A number of clerks were engaged at £120 per annum and a German translator appointed at £300 a year.[14] In those halcyon days the company paid its employees income tax as a matter of course and office hours were 9.30 a.m. to 4.30 p.m. daily from Monday to Friday and 9.30 a.m. to 1 p.m. on Saturday.

Although it was intended to send at least one deputation of directors to Rio Tinto each year to inspect progress, discuss problems and issue detailed instructions, it was essential to ensure that the mines' staff responsible for the day to day management was the best obtainable. The head of the company's activities in Rio Tinto and Huelva was the general manager who was expected to take responsibility for all aspects of the company's work. The first of these men was Mark Carr, who administered the Spanish affairs of the company from 1873 to 1879 for the high salary of £2,500 a year, plus ½ per cent of net profits up to an additional £3,000 a year above his basic salary.[15] His assistant, who also served as manager until Carr's arrival, was the German Theodor Blum who was paid £1,500 a year.[16]

Such men, highly experienced and skilled, were exceptionally valuable to the company but, because of their standing in their own profession, they could be highly opinionated and were often involved in personality clashes with one another and even, on occasions, with the directors. Between 1873 and 1876 Theodor Blum was often in conflict with both his chief, Mark Carr, and with David Forbes the mining consultant, refusing to obey technical orders with which he disagreed on such matters as dam building, the utilization of machinery and the best methods of digging the open-cast pit then being made. Yet such was the board's opinion of this short stocky German, who stumped heavily around on his walking stick, that they were unwilling to dismiss him or accept any of his frequent letters of resignation. His refusal to bother with paper-work infuriated them. Once, when he persistently failed to submit his monthly statistics to London, Matheson cabled to him demanding them immediately. Blum sent a cable in reply: 'What you want with figures? You send me money. I send you copper. What more you want?'

During his third summer at the mines Blum took umbrage at a comment by one of a visiting party of directors who suggested that the extraction of ores was behind schedule. He lost his temper, harangued the board deputation on the difficulties of the work and declared that if they did not like the way he was doing the job then he would resign. To his great surprise the ruffled directors told him that his resignation was accepted and that he must leave the mines on the next train. He barged into the laboratory to tell his young technical assistant Alexander Hill, declaiming against Matheson and the board. 'Look you, Hill, never believe in directors of big companies. They have not one soul to be saved nor one backside to be kicked!' He swung his walking-stick, sweeping beakers, phials and flasks to the floor, and stalked out to pack his bags.[17]

Of the many problems with which Matheson and his directors had to cope, the one that undoubtedly loomed largest was the financial one. In the very first year, Matheson and Company had to raise a loan of £1 million for the Rio Tinto Company as a temporary relief until the shares were taken up by the public. Between 1873 and 1876 the Spanish government was constantly worrying the company to redeem the promissory notes early, because of the pressure being applied by the London Association of Foreign Bondholders and by the British government to get the Spaniards to pay the state's overdue debts. In 1876 the Spanish government offered some of its British creditors the Rio Tinto *pagares* at a discount in order to be discharged from its obligations, but found that the promissory notes were not acceptable. One reason for the rejection of this offer was the doubts about the future of the company. The more respectable credit institutions considered mining companies in general to be a risky investment, and the board of the Rio Tinto Company in particular to be gamblers. The prospectus had set 1881 as the date when the company would be a profit-making concern, but there was considerable doubt whether the company could survive until that time without incurring a considerable burden of debt.

At annual meetings of the company there was always a critical element among the shareholders. The reason lay largely in the fact that each year, to help maintain share prices, Matheson gave glowing reports of progress at Rio Tinto and how good the prospects were, although he was never able to declare even a modest dividend as the costs of the work were always higher than foreseen and all profits had to be ploughed back into the mines in an effort to meet by 1881 the targets set in the prospectus. Time after time Matheson had to inform meetings both that lack of rain at the mines had hindered the production of copper and that world copper prices had continued to fall. It was only to be expected that on one occasion a shareholder, demanding to know when he and his fellows could expect some return for their outlay, should comment bitingly that he knew that 'the directors cannot control prices any more than they can control the weather' but that he had been induced to subscribe only because of the extravagant promises of the prospectus.[18] Not only the stock-markets but some of the shareholders were losing confidence.

In 1875 another request from the Spanish government for the company to anticipate the *pagares*, together with the increasing public doubts about the company's future, led Matheson to propose the redemption of the promissory notes as a way of proving the soundness of the enterprise. Consequently the company issued in 1875 its 'Spanish coupon' 5 per cent mortgage bonds and a further issue of 7 per cent mortgage bonds, both issues backed by Matheson and Company; and by the end of 1876 the instalments for the mines had been completely paid.[19] The 1881 increase in the share capital by 100,000 shares, with a face value of £10, which were sold at a premium setting the price at £22 a share, gave the company £2,200,000 which enabled it to pay off the debts associated with purchasing the mines.[20] But before the company reached the haven of that year the continual financial problems must have been a sore trial to all save the most stout-hearted members of the board.

The price obtainable for the sulphur in the pyrites was always a matter for concern. In 1873, after a sales' war between the Tharsis company and the Mason and Barry company (who were working the San Domingo mine in Portugal), the price per unit of sulphur that the pyrites' producers were able to obtain stood at 7d. The beginning of operations by the Rio Tinto Company coincided with a drop in price to 5d. The three companies eventually came together in 1878 and agreed to fix the price to buyers at 6d a unit, though relationships between the three were so strained that an independent negotiator had to be brought in to smooth the way, receiving a fee of £6,000 from the Rio Tinto Company, £1,000 from Tharsis and a lesser sum from Mason and Barry.[21]

Shortage of funds lost the company a glittering prize in 1876 when a group of German chemical manufacturers asked Rio Tinto and Tharsis if one of them would set up a plant there to process pyrites' residues. Tharsis refused as they were not prepared to help make the German chemical industry independent of the British soda manufacturers. The Rio Tinto directors were attracted by the idea and considered ways of raising the necessary capital, but had to withdraw from the discussions when it became obvious that participation in such a scheme would so strain their resources as to endanger their Spanish operation. In consequence the Germans themselves founded the Duisburger Kupferhütte and the Rio Tinto Company missed what proved to be a profitable opportunity for diversifying its activities.[22]

Other reasons for the difficulties of the 1870s lay in industrial and agricultural change. A new soda-manufacturing process was invented by the Belgian, Ernest Solvay, and in 1874 he licensed Ludwig Mond and J. T. Brunner to use it in England to make ammonia-soda. Soon they were making 52,000 tons a year and outselling the soda manufacturers who were using the older Leblanc process which relied on sulphuric acid, and who started to decline in numbers and importance. Thus a major sales outlet for pyrites began to disappear. The other principal British sulphur users of the time were the fertilizer manufacturers, whose sales were cut severely by an agricultural depression as large imports of American wheat flooded into Britain, thus further reducing demand for pyrites.

For a while it became necessary for Matheson to apply severe controls to all expenditure and to look for ways of reducing costs. A constant source of irritation to him was the behaviour of Sundheim and Doetsch in their role as company agents in Spain and Portugal. Their contract had, in Matheson's view, been drawn too generously. They were earning large sums as commission for performing simple actions which the staff at the mines could have done as well, and often more competently. He tried to get Sundheim and Doetsch to

*Guillermo (or Wilhelm) Sundheim.*

renegotiate their contract but they refused. The 'great deliverance robbery', as it became known in the Rio Tinto Company, in July 1876 gave Matheson the stick he wanted with which to beat them.

In that month a delivery of bullion for the mines was despatched by deliverance (that is, a stagecoach) from the agents' premises in Huelva. In the hills between the coastal plain and Rio Tinto a gang of masked bandits held up the coach at rifle point and stole the gold consignment. The loss was the considerable one of £8,200 and Matheson demanded that the agents reimburse the company since they were responsible for the bullion's safe delivery at the mines. Over the Cornhill board-room table Doetsch indignantly rejected the claim. As far as he was concerned his company's duty to protect the bullion ceased when the boxes were loaded onto the deliverance in Huelva. For five months the matter was raised at every weekly board meeting and finally counsel were instructed to begin legal proceedings on behalf of the Rio Tinto Company.

Sundheim and Doetsch were in a weak position and anxious not to run the risks of a court action. Apart from the probability of having to pay compensation and costs, Doetsch would not be able to stay as a director of the Rio Tinto Company, and an adverse verdict could lead to yet another action to put aside the contract under which they held their agency. Doetsch started to make concessions, agreeing first of all to the company moving its bullion under its own escort of armed guards, and finally accepting Matheson's offer to withdraw the action and to let him keep his directorship in return for a payment of £4,100 representing half the value of the lost bullion, and a new agency contract which transferred much of the work which Sundheim and Doetsch had previously done to the staff at the mines.*[23]

But economies and the exercise of financial caution could not provide the complete answer to the poor performance of the Rio Tinto shares in the stock-markets. By 1878 the Rio Tinto £10 shares were changing hands for as little as £2 10s. Then there occurred an affair of which only the bare details are recorded and of which many points are still obscure. Dyes and Schroeder of the Deutsche National Bank, who were also members of the board of the Rio Tinto Company and apparently its largest single shareholders at that time, found themselves in financial straits. They were threatened with a German government inspection of the books of their bank and, if they were to emerge unscathed, something would have to be done to increase dramatically the value of their Rio Tinto shareholdings.

At this stage Julien Deby, a Belgian mining engineer, who had never visited Rio Tinto, was commissioned to write a report on the mines. When this was published in London it contained an estimate

*By 1904 the other agency agreements had been terminated by negotiation.

that the ore reserves of the mines were many millions of tons more than had previously been claimed. Clearly, the report was written to mislead its readers into thinking that the author had both inspected the mines and based his conclusions on technical information supplied by the company. As a result of this publication share prices rose steeply, increasing by over £10, and the Deutsche National Bank passed satisfactorily through its inspection. Deby was rewarded with a life appointment as a consulting engineer to the company and the annual salary of £5,000, though he still never visited the mines nor submitted any reports thereafter.

Mark Carr, the general manager in Spain, resigned in protest giving as his reason that Deby's report was pure fabrication. Peter Denny and Martin Ridley-Smith both resigned from the board and David Forbes surrendered his consultancy. None of them made any public protest, perhaps through fear of causing a panic-selling of shares with a consequent loss to investors, and the affair was kept a company secret. The resignations were not even noted in the subsequent annual report.[24]

The gamble, morally dishonest though it may have been, did the trick. But it was a near thing. Operating profits on the sales of produce had risen from £3,200 in 1873 to £236,000 in 1878, but these had been wholly absorbed in the capitalization of the mines until the latter year when for the first time there was a net distributable profit for the business, some £4,000. In 1879 the net profit on the year's operations soared to £114,000 which made it possible for Matheson to announce at the 1880 annual meeting the first dividend for the ordinary shareholders which, though it was a modest 5 per cent, provided an important psychological boost. The 1879 profits from sales were almost £444,000 and from then until 1895 they were normally above £600,000 a year, with peaks in 1888 of over £1,076,000 and in 1890 of over £842,000. Net profits available for distribution to shareholders during the same period fluctuated considerably with the needs for capital investment from year to year, although in ten of the sixteen years they were still above £300,000 and the dividends to shareholders averaged 10 per cent. On a long term view, the forecast in the prospectus of an eventual 35 per cent dividend when the mines were in full production was justified by the results for the years 1879 to 1913, for the average dividend on ordinary shares in those years was slightly over 38 per cent.

In 1896 profits from sales rose to £828,000 and the net profit available for distribution to shareholders to £639,000, and in 1897 the figures were £987,000 and £752,000 respectively. From then on until the outbreak of the first world war the yearly average sales' profit was

£1,700,000 and the net profit in those sixteen years fell only twice below the £1 million mark. The peak was in 1906, the figures being £2,414,000 and £2,171,000 respectively. From 1898 to 1913 dividends declared on ordinary shares were never less than 50 per cent, and for ten of the sixteen were 70 per cent or more. Dividend peaks were: 80 per cent in 1899; 85 per cent in 1900; 80 per cent in 1905; 110 per cent in 1906; 87½ per cent in 1911; and 90 per cent in 1912.

For Matheson godliness and profits went hand in hand. In 1880, when he announced the first dividend to prolonged cheers, he was the subject of laudatory speeches from the floor. In replying to the thanks of the shareholders he told them that, although his exertions had been at a cost to his health of which they had not been aware, he had been successful because he had behind him 'the skill and diligence of an excellent staff and the blessing of Divine Providence.' (the minute book records shouts of 'Hear! Hear') However, he had reaped his reward, he declared, because of the benefit which the company was able to confer on the Spanish people. He then revealed what it had been necessary to keep a secret during the lean years when the shareholders might not have welcomed the information: the existence of the company's schools in Spain, the only elementary schools in the country in which the holy Bible was taught. He added, to applause, that this would 'have a most excellent effect upon the conduct and character of the workpeople.'

Even though Matheson believed the worst years were past it was not possible for him to sit back and expect the company to run itself. Lacking the connexions with the British alkali industry possessed by the Tharsis company, its smaller but extremely active rival, the Rio Tinto board was forced to become cosmopolitan in outlook. The board members, with their wide international interests, themselves constituted a team of aggressive worldwide salesmen. The sulphur prices of the period though low were fairly stable, and the basis of the company's prosperity was guaranteed by its large-scale exportation of pyrites to the world's sulphuric acid makers. However, there were difficulties even here. In 1882 the British alkali manufacturers put pressure on the three main pyrites' producers to lower their fixed prices. The Tharsis Sulphur and Copper Company, more dependent on the trade than either Rio Tinto or Mason and Barry, and fearing the development of an improved sulphur recovery process which might enable the manufacturers to cut their consumption of pyrites, urged the other two members of the price-fixing agreement to accede. But they were not willing to lower prices and the Tharsis company therefore withdrew from its co-operation with them.

The next year the alkali industry's experiments led to the adoption of the Claus kiln which made it possible to obtain sulphur at 3d per unit, half the price of the sulphur in the pyrites being provided by Rio Tinto and Mason and Barry. The Tharsis company price was readjusted to 3d and the next year Rio Tinto and Mason and Barry were forced to follow suit when many of their customers refused to renew contracts. At the 1885 annual meeting of shareholders Matheson promised that there would be no drop in income from pyrites despite the 50 per cent price reduction as output would be expanded to compensate for the lower price. In fact it proved impossible to sell the extra pyrites which were mined and between 1884 and 1887 the dividend fell from 14 per cent to 3 per cent.

However, as each year went by demand gradually increased and, provided it did not lose too many customers, the Rio Tinto Company knew that it had little to fear. A few customers were lost in 1890, when forty-five alkali makers combined in the United Alkali Company, the world's largest chemical concern formed in an attempt to improve efficiency, to rationalize production and to beat off the challenge from Brunner Mond. But the majority of the firms involved in the merger had been buying their pyrites from Tharsis which therefore suffered a greater drop in sales than Rio Tinto.

In November 1895 a delegation from the Superphosphate Producers' Association called on Matheson with a proposal that, in return for a price reduction, the association would buy all its members' pyrites requirements from the Rio Tinto Company. As this association represented all the manufacturers engaged in the superphosphate trade in the United Kingdom it was a tempting offer and one which, if accepted, would have hit Tharsis hard. However, Matheson decided to reject the chance to acquire a supply monopoly in this field on the grounds that he might have been leading the company into a trap. For once the association had become the bulk buyer for its members it would have been in a position to exert pressure for even lower prices by threatening to transfer its business to Tharsis or Mason and Barry. Paradoxically, in this case competition between the pyrites' producers for the orders of the individual superphosphate manufacturers seemed to Matheson the best way of maintaining prices and profits.[25]

In reaching this decision Hugh Matheson had an eye turned towards the United States' markets. The sulphuric acid manufacturers there had for long been using Sicilian brimstone to produce their product. But in 1894 there had been changes in the United States' tariff structure which made it possible for the Iberian pyrites to compete with the Sicilian product on comparable terms. The Rio Tinto Company directors were anxious to unload their excess production in this new potential market and, as the American sulphuric acid makers' contracts with the brimstone suppliers expired, the Rio Tinto Company was successful in gathering a good crop of new customers. Tharsis also went into the new market but Rio Tinto, with its aggressive sales' policy, benefited most.*[26]

Throughout this period the price of copper, the company's other great money-earner along with sulphur, fluctuated greatly due to a series of trade depressions and increasing production on the American continent. In 1879 the world's main suppliers of copper were Chile which produced 49,000 tons, Spain and Portugal which together produced 32,000 tons (of which the Rio Tinto Company manufactured 7,179 tons at the mines), followed by the United States which produced 23,000 tons. The total world production for that year was 151,000 tons. But the expansion of the United States' copper mines in the early 1880s totally changed the picture within nine years. In 1887 world production was up by 50 per cent on 1879 and the United States was the largest single producer providing 79,000 tons, while Spain and Portugal increased their figure to 47,000 tons (of which almost 18,000 were made at Rio Tinto) and Chilean production decreased to 29,000 tons.†

However, the growth of the United States' mines helped to make copper prices less stable, as did the periodic spells of prosperity and depression in European business, and in particular in the new electrical engineering industry which was a principal user of copper.

The price of Chile bars varied between £108 per ton in June 1872 and £38 per ton in December 1886. As the large copper companies were inflicting severe wounds on one another's profits by price-cutting sales' policies, the Rio Tinto Company began secret discussions with Tharsis and the major American producers with a view to reaching agreement on limiting production to bring an end to the acrimonious sales' war and to help keep up profits. These talks were hurriedly abandoned by Matheson following an approach by Hyacinthe Secrétan, head of the Société Industrielle et Commerciale des Métaux, the largest manufacturer of brass and copper products in France and probably the world's biggest buyer of ingot copper. In October 1887 he formed an informal consortium backed by the Comptoir d'Escompte (the second largest of the French banks), the Paris Rothschilds, the Crédit Lyonnais and the Banque de Paris et des Pays Bas which undertook to provide advances and guarantees of up to £2,800,000 to allow him to try and corner the world market in copper in a classical operation of its kind.

Secrétan began by entering into a secret agreement with the Rio Tinto Company, by which the company agreed to restrict output and Secrétan undertook to buy all its copper at fixed prices, for a

*By 1909, 84 per cent of American sulphuric acid production was based on pyrites, most of them supplied by the Rio Tinto Company. Moreover, profiting from the lesson learnt when it had failed to build a recovery plant in Germany in 1876, the Rio Tinto directors had a metal extraction plant built at Wilmington in Delaware for the recovery of the copper in the cinders returned by their customers.

†The Rio Tinto contribution of 7,179 tons in 1879 then represented less than a quarter of the Iberian copper being made and 4 per cent of the world's production. The 18,000 tons in 1887 meant that Rio Tinto by then supplied a third of the copper made in the peninsula and 8 per cent of the total world supply.

minimum period of three years. In the following four weeks he signed similar agreements with all the world's major producers in thirty-seven separate countries which gave him control of over 80 per cent of world output. Then in November and December he bought all the available copper remaining in the markets, causing a spectacular rise in price. His idea was to dictate the market price of copper by these methods, keeping it higher than the price at which he was buying and giving his own Société des Métaux the advantage of cheaper prices than he gave its competitors.

Matheson had known Secrétan since the Rio Tinto Company had been established in 1873 and had a great respect for his abilities. He had no hesitation in joining him. And Matheson's judgement was right, for Rio Tinto, like all the participating companies, benefited greatly. As the world price rose so did the amount which Secrétan paid Rio Tinto under the contract. In 1888, despite a major dislocation caused at the Rio Tinto mines by rioting workmen, profits on sales of copper jumped by 40 per cent, and it was possible to increase the dividend for that year from 10 to 17 per cent. By September of that year the market price of copper stood at £115 per ton, and though the company was receiving only £72 a ton under its agreement with Secrétan, this was almost 100 per cent more than it was getting a year earlier in the free market.[27]

The French syndicate was making vast profits as a result of Secrétan's monopoly and its members used much of this money to purchase shares in the mining companies whose copper they were buying, thus making additional profits by sharing in the copper producers' gains. By the beginning of 1888 the French group had acquired a controlling interest in a number of companies including Rio Tinto. With the increased dealings on the stock exchanges of Europe, share prices rocketted and a copper boom began. Though the annual output of the larger companies was limited by their contracts with Secrétan, the upper limit had been set high enough to allow many of them to increase production.

As the price of copper rose the smaller companies, outside the control of the syndicate, also responded by producing more copper. At the same time many of the industrialists buying the metal allowed their stocks to decrease and held off buying as long as they could in the hope of the price falling once again. Secrétan was forced to increase his stockpile by mopping up the extra production, and to do this he was compelled by the end of 1888 to borrow £6,880,000, half of which was in the form of unsecured loans. In that year world copper production exceeded sales by one third, and some 260,000 tons were manufactured of which Secrétan could afford to buy only 160,000.

By March 1889 the whole of the resources of the Comptoir d'Escompte were tied up in copper purchases. Secrétan was desperately trying to raise further loans but failed. The Société des Métaux suspended payments to its suppliers and, when on 1 April it refused to take delivery of its monthly supply from Rio Tinto, Matheson gave notice to the Comptoir that it might be called upon to honour its guarantee given in support of the Société. Similar notices were served by other suppliers and the Comptoir d'Escompte, unable to meet its obligations, announced it would go into liquidation. Denfert-Rochereay, the Comptoir director who had persuaded his colleagues to finance Secrétan, committed suicide. In the ensuing panic the market price of copper fell to £35 a ton. The Comptoir was eventually rescued by the Bank of France and reorganized, though it proved impossible to pay its debts. The Tharsis Sulphur and Copper Company and the Rio Tinto Company both made claims for the difference in price between the one guaranteed under their contracts and the lower market price at which they had been forced to sell. Rio Tinto even obtained judgement in its favour in the English courts in 1891 but this did little good as the French court of appeal refused to allow the judgement to be received in France.[28]

After the initial confusion it was seen that the failure of Secrétan's breath-taking attempt to establish a world monopoly was not as bad in its effects as many had feared. Matheson, in giving his account to the Rio Tinto shareholders of what had happened, pointed out that Secrétan until March 1889 had paid punctually for all copper deliveries Consequently his suppliers had enjoyed sixteen months higher prices than would otherwise have been possible. Moreover, the spectacular collapse pushed many of the smaller producers out of existence, to the Rio Tinto Company's eventual advantage.[29]

It was presumably during the period of the French syndicate's activities in 1887–9, and the period of panic immediately afterwards, that the French branch of the Rothschild family acquired its controlling interest in the Rio Tinto Company. The company's voting stock then consisted entirely of bearer shares and it is therefore impossible to trace the process by which the family, acting as always in great secrecy, was able to become the dominant shareholder. The Rothschild passion for remaining behind the scenes, however, meant that the extent of the family's voting control remained unsuspected by the public for many years, while its dislike of putting too much down on paper makes it difficult to gauge the extent to which the Rothschilds influenced Rio Tinto policy. From the early 1890s onward they started appointing their own nominees to the board although, since there was never a contested election, even this

PICTURE OPPOSITE
*Baron Alphonse de Rothschild.*

was not evident to the majority of other shareholders.*

Following the failure of Secrétan, an attempt was made by French copper buyers and banks to inject some order into the market by establishing a new quota system. This was supported by Matheson but the negotiations broke down with disagreement between the European and the American mining companies. In 1891 Baron Alphonse de Rothschild acted as intermediary in a fresh round of talks which brought agreement between the Rio Tinto Company and the American producers, but failed to satisfy the Tharsis company which decided that, after all, a free market would be more advantageous to its own activities.[30]

The next year a European copper producers' committee and an American producers' association were set up in an attempt to bring agreement between the two sides. The negotiations were of necessity in secret, since any rumours of a possible compact sent share and copper prices rising only for them to fall sharply when the stories were denied. In 1895, for example, a rumour circulated that the Americans were going to cut exports to Europe by 60,000 tons in return for the European producers' cutting back their output by 7½ per cent. Shares changed hands rapidly at increasing prices but at last came down with a bump when an official denial was issued. The Tharsis company's general manager commented: 'It seems to have been a stock exchange movement all the way through, with the object of advancing Rio Tinto shares and disposing of a large lot.'[31] But without the co-operation of Tharsis, no combination was likely to succeed; and Rio Tinto was sufficiently progressive, and its mineral deposits so much greater than any other European producer's, that the restrictions of a cartel were not likely to be to its long term advantage. Certainly, the large profits made in the years between the end of the nineteenth century and the first world war seem to lend support to this viewpoint.

It was inevitable in the years from 1873 to 1898 that the company would be the subject of many attacks and criticisms. Normally they were ignored. Only occasionally would Matheson comment on them, and then merely to say that they were based on inaccurate information. Even the most violent of all the press attacks, made by the journalist-politician Henry Labouchere in *the World*, brought no reaction from him. In so far as these criticisms accused the directors of financial juggling and fast dealing, Matheson would doubtless have replied, if he had spoken his mind, that it was these characteristics which had contributed so largely to the company's success. And the company was a great success. Long before his death in 1898 the old man had the satisfaction of knowing that it was due principally to his vision, to his talents in the fields of finance and administration, and

*The French Rothschilds would seem to have kept their controlling interest in the company not only through the good years but through the bad and at least up till 1954.

above all to his readiness to gamble for heroic stakes, that the company which he had founded, and the certain failure of which had been so loudly proclaimed by disbelievers in its early years, had survived, and survived so gloriously.

*Opening of the railway: Rio Tinto, 28 July 1875.*

# 10 *A vision realized*

## The resurrection of the mines: 1873–1914

In the three centuries or so following the rediscovery of the mines in 1556, the thought of them had often inspired dreams of riches in the minds of many individuals. But for the greater part those who tried to translate fancy into fact found the dream unattainable. For them the name of Rio Tinto became synonymous with frustration and disillusionment, and perhaps penury or even a grave in the Nerva churchyard. Francisco Sanz and Gaspar Remisa alone had successfully made the mines a profitable enterprise, though in doing so they had resorted to questionable activities, but even they had not visualized the full extent of the riches which could be won. It needed Hugh Matheson with his unique combination of talents to convert the vision into reality. There was sufficient of the Scottish highlander in him to make him a romantic and a dreamer; from his uncle James he had learnt how to play for big stakes; while his experience in the city of London money-market had taught him how to temper enthusiasm with hard practicality at every step. And his own personality, no doubt moulded by the hard uncompromising presbyterianism that dominated his life, gave him the stern self-discipline and application to work that was necessary for anyone wanting to embark on the task of creating at Rio Tinto the largest and most profitable mining enterprise in Europe.

What the Rio Tinto mines had needed during all the time that they had been subject to the stifling control of Madrid civil servants had been a large-scale injection of capital and well-organized market outlets, together with administrators capable of making imaginative and daring decisions and possessing sufficient skills to carry these through to successful conclusions. The government, desperately anxious to obtain quick relief from its financial problems, was willing to sell both the freehold of the mines, and the land within the Rio Tinto boundaries, to the Matheson group which had access to capital in large amounts. Thus were made possible the indispensable conditions for success at Rio Tinto. For though the known ore bodies there were the largest in Europe and well known to everyone with an interest in mining, the costs of acquiring the freehold and of developing the area were so expensive that few individuals, or for that matter companies, could have afforded them. It was no accident that the purchase was finally made by an international consortium with close ties with the city of London and European commerce.

The achievements and the successes that followed the purchase of the mines and the creation of the Rio Tinto Company were due to a combination of expertise allied with a great expenditure of money. The basic factors in the prosperous progress and development of the Rio Tinto workings were the construction of a railway to Huelva, the adoption of open-cast mining, the production of copper at the mines by large-scale techniques and an aggressive sales' policy in the world's sulphur markets.

The route which the railway was to follow from the coastal town of Huelva to the mines was dictated largely by the terrain. The first twenty miles across the coastal plain offered little difficulty to George Bruce, the consulting engineer who planned the railway and supervised its construction, but the thirty miles between the plain and the mines provided many problems. The hills were such that it was necessary to follow the course of the river Tinto, blasting and cutting into the steeply descending and thickly wooded slopes to provide a level platform a few feet above the river for the 3 ft 6 in gauge.

Work was first started on the easier part of the route, between Huelva and the hills. At the beginning of June 1873, only six weeks after the registration of the Rio Tinto Company, Clark Punchard and

*Starting to construct the Rio Tinto railway, June 1873.*

*One of the first Rio Tinto railway engines: sketch from the notebook of Manuel Iglesias, a visitor to the mines in 1875.*

Company, the contractors, started to construct their trial sections. As a result Bruce had to report to the board that stronger foundations would be needed than he had anticipated and that steel rails of 50 lb to the yard, in place of 45 lb rails, would be required. The sum of £500,000 allowed for building the line would therefore be insufficient and Bruce submitted a new estimate of approximately £788,000. By simultaneously laying track at several places, building bridges over rivers and tunnelling through hills the railway was completed on 28 July 1875, three months ahead of schedule, at a cost of £767,190.

The amount of effort required to complete the task in just over two years was considerable. Vast quantities of ancient slag had to be moved from the mines to provide the ballast for the rails. There were 51.9 miles of track to be laid, rising from sea level to a height of 1051 feet at Rio Tinto station. The bricks for the embankments and the linings of the five tunnels had to be made and fired on the spot. Eight bridges had to be built to carry the railway across ravines or over the river. There were frequent difficulties caused by rock falls, and problems in obtaining possession of much of the land over which the railway had to pass. Under the decree authorizing the sale, the buyers of the mines had the right to acquire the land which they needed for a railway by compulsory purchase. But the invoking of the procedures before the Spanish courts was a long process and the company preferred to pay more generous compensation than might otherwise have been necessary to keep delays to a minimum. When agreement with a landowner proved completely impossible, Bruce anticipated the court orders, for which the company's legal representatives were applying, by going straight ahead and building the line despite the outraged protests of the proprietor of the piece of land concerned.

The necessary water for the locomotives that were being brought from England had to be provided by building a dam (known as 'the railway dam' to distinguish it from the south dam immediately above it) and by sinking wells at a number of the stations that were being erected along the route. These stations were designed not only to serve the small communities of railway workers and their families, which had to be created along the track to take care of its maintenance, but also the fare-paying passengers. The government appreciated that the line would be of considerable importance for the economic and social life of the region and insisted on the Rio Tinto railway providing for passengers by adding first and second-class carriages to a number of its daily trains.*[1]

The construction of a pier at Huelva was essential. Before the British company started its work there had been little in the way of ore exports from Rio Tinto and what there had been had consisted of

*This obligation lasted until 1953.

rich cupriferous pyrites. It was such a load that the good ship *Ada* of Portmadoc was carrying from Huelva to Tynemouth on her maiden voyage in July 1865 when she disappeared with all hands and leaving no trace.[2] In the days before the pier was built the Rio Tinto mineral had to be ferried from the Huelva shore in barges across the shallow water to where the seagoing ships waited mid-stream in the estuary. This system of loading vessels was not appropriate, however, to the vast quantities of pyrites that would be streaming down from the hills once the railway was completed.

George Bruce's design for the pier was simple but brilliant. He proposed that from the Huelva marshalling yard of the Rio Tinto railway the track should lead on to the pier by way of a bridge constructed over the marshy ground at the water's edge. The 633 yard long pier was to be built on cast iron screw piles erected on top of a timber platform sunk into the river bed. Laden railway waggons were to be set moving six at a time along the track by a shunting-engine in their rear, being propelled up the incline of the pier's upper deck to its topmost point. From there they were to coast down individually to halt over the hoppers through which their ores were to be discharged, out of doors in the undersides of the waggons, into the holds of the ships moored below. From there the waggons, under their own momentum, were to run back to the end of the pier to where locomotives would collect them and return them to the marshalling yard.

At first, however, there was a considerable delay in getting the

*The pier at Huelva, August 1875.*

Spanish government's approval for the plans. Once consent was obtained and the work started there was another hold up when the government chief engineer objected to the design of the bridge crossing the marshy ground and also declared that the cast iron screw-piles of the pier would not be going to a sufficient depth to ensure the stability of the structure. The protracted discussions had to be started once again and ministerial changes delayed them still further, but finally, as a result of demonstrations with models and weights, Bruce was able to convince the Spanish officials that the design was safe. Even then it was necessary for the company to give an undertaking that, if the government ever wished to put into effect an earlier proposal of its own to build a stone quay from the shore parallel to the company structure, then the Rio Tinto Company would reconstruct the end of its pier at its own cost to give easier access for shipping to the government jetty.[3] Since Doetsch doubted that the government was ever likely to act on this proposal, and Bruce declared that in any case it was impossible to build a stone quay of the sort envisaged, Matheson accepted this condition without any argument.

While the discussions were in progress Clark Punchard and Company gave up their contract for the pier, explaining that the engineering skills required were beyond their experience, and were replaced as contractors by John Dixon and Company. The estimate of £80,000 was turned by the delays into a final cost of £101,505 and since the work was not finished until 23 March 1876, eight months after the completion of the railway, it was necessary for the Rio Tinto Company to buy seven barges to carry ores out to the cargo boats during the interim.

The company foresaw that in time Huelva would become a substantial port. The Tharsis company had already built a railway and erected a pier of its own there; and as the mining industry of the region expanded so the level of imports and exports from all sources could be expected to grow. Consequently the lower deck of the Rio Tinto pier was designed to accommodate smaller cargo-ships and also the traffic of the Seville branch-line of the Madrid, Saragossa and Alicante railway. This line eventually opened in 1880 though, following frequent disagreements between the railway and the Rio Tinto Company, the contract was terminated in 1896. (The railway company demanded £2,300 compensation but settled for £1,000 and permission to use the Rio Tinto pier at a charge of 20 per cent more than the rate for the town pier if this was ever necessary, though in fact the right was never exercised.)

The Rio Tinto Company was right in its assessment of the changes

*The lower deck of the pier, circa 1885.*

that were to come. In the period before the railway and the pier had been built, the company had to send most goods destined for Rio Tinto either on mule back or by stage coach from Huelva to the mines. Larger consignments had to be unloaded from the seagoing ships into barges and taken up the river Tinto as far as the town of San Juan del Puerto. There they were loaded onto Buitron railway trains and carried by rail to a railway siding six-miles beyond Valverde del Camino before making the final sixteen miles of the journey to the mines in bullock-carts or, broken down into smaller quantities, on mule back.* This way of carrying goods from Huelva to Rio Tinto had cost 25s a ton.

The Rio Tinto railway and pier reduced carrying costs to 2s a ton, which benefited not only the company, importing heavy machinery from England and exporting vast quantities of pyrites, but all local purchasers of consumer goods. Consequently the mines' population, which expanded so rapidly in the years after 1873, was guaranteed its food and other supplies at far lower prices than would otherwise have been possible, and a real improvement in living standards took

*The Buitron railway opened in 1869 and closed in 1968. Until 1954 the Rio Tinto Company still owned the triangular piece of ground near Valverde which had served as its railway depot there prior to 1875.

place in which cheaper prices were as important as higher wages.

The whole history of the Rio Tinto mines between 1725 and 1872 had centred around the production of copper. The British company, however, was interested in making not only copper but also in selling pyrites to the world's sulphur users.* The deposists at Rio Tinto were part of a series of continuous masses of ore stretching for seventy miles from near Seville westwards to San Domingo in Portugal. Of all the deposits in this series those at Rio Tinto were unquestionably the largest and, in the opinion of David Forbes, the company's first mining consultant, probably equal to all the others put together.

Forbes reported to the board that the lodes were made up of a compact admixture of iron pyrites with a smaller quantity of copper pyrites, both with a high sulphur content. As a result of the action of air and water, oxidization had caused the break-up of the pillars and walls of the underground workings which thus became subject to frequent and dangerous rock falls. The oxidization process was accompanied by the generation of heat so that in some parts of the mines temperatures reached as much as 98 or 100 degrees Fahrenheit and the atmosphere was almost suffocating. Because of these factors he disapproved of any suggestions to continue mining by underground methods and advocated the introduction of a system of open-cast quarrying. Mining economics also provided arguments in favour of this.[5]

The three lodes at Rio Tinto (Forbes mistakenly believed, as had the Spanish mining engineers, that there were four) all lay close to the surface and were of considerable width. Open-cast mining would require the removal of a large amount of overburden between them and the surface, which would be an expensive business, but the vast amount of pyrites that could be extracted once this was done would justify the heavy expenditure. In 1873, the system was already being employed at the Tharsis and San Domingo mines on a larger scale than previously known in Europe. At Rio Tinto it was to be used even more extensively, not only yielding great profits but attracting to the company's service some of the most intelligent and ambitious young British mining engineers.

It was decided to make the first of the open-cast pits in the south lode, on the lower slopes of Salomón between the pueblos of Rio Tinto and Nerva. Getting the plant and machinery and recruiting the workmen took almost a year. Since Mark Carr, the general manager, was busy in Huelva overlooking the construction of the railway and pier, the work on the pit became one of the responsibilities of Theodor Blum, assisted by Neil Kennedy a young Englishman. A length of some 400 yards was marked out by the two men and then subdivided

Between the purchase of the nines in 1873 and their sale in 954 the Rio Tinto Company nade 1,257,541 tons of copper, ither by cementation at Rio 'into or by smelting at the mines r in Britain. It mined 07,983,705 tons of pyrites, of hich 40,386,532 tons of ordinary yrites and 19,638,635 tons of ashed pyrites went for export or ere delivered to Spanish pyrites' asters. (The remaining ,958,488 tons were treated at the ines to recover the copper.)

by the making of six cross-cuttings. Work then started on removing the overburden (the cap of earth lying over the mineral and on average 30 yards in depth) from these subdivisions. Blum could find only 837 labourers to begin the work and he blamed the shortage of men on the political state of Spain and the confusion resulting from the republican government's inability to maintain order. But in fact the paucity of labour was largely due to the high wages being offered by Clark Punchard and Company which allowed them to acquire most of the surplus labour in the area for constructing the railway.

The removal of the overburden was a costly operation. The first open-cast cutting involved removing well over four million cubic yards of waste and transporting it to the dumping grounds. The cost was 2s a cubic yard, a total of £300,000 and a considerable sum for the time. However, the cost spread over the next twenty years added less than 6d a ton to all the pyrites extracted and represented a considerable saving on the expenses of underground mining.

An open-cast pit of this type required direct access to the railway and to provide this it was necessary to drive a tunnel through the lower slopes of *cerro* Salomón. Thus the railway was planned to come

*Huelva: The railway machine shop.*

along the Tinto valley, passing the present Rio Tinto station and then cutting into the hill (and through the south lode open-cast pit when the latter was dug as far down as the eleventh floor) emerging on the north side of Salomón to communicate with the open-cast pits which it was intended to excavate there. But digging through the hill to make the 500 yard long central tunnel was more difficult than George Bruce had anticipated. Even though work started simultaneously on both sides it took two years to complete. There were falls of rock and sudden inundations of flood-water from the caves in the hill above which caused loss of life and held up work. The cupreous acid water also caused delays because of its rapid destruction of the metal tools used by the men and of the water-pumps. A Cornish pump was installed, made of copper and bronze, which proved resistant to corrosion but it became plain that by the time that the railway line between Rio Tinto station and Huelva was finished (which happened in July 1875) the tunnel would be nowhere near complete.[7]

Thus Bruce had to adapt his plans to provide for the continuation of the railway track straight along the south slopes of the ridge to the present Coso station. From there was built an inclined plane, the

*Open-cast pit in the north lode, 1900.*

contractors being Head, Wrightson and Company, to the top of the exposed mineral of the south lode. Up this was laid a double track in parallel of 3 ft 6 in. gauge. Loaded waggons coming down one line, from the level at which the mineral was being worked, pulled up empty waggons on the other line, providing a continuous stream of carriages to take the ores for export down onto the main railway until the tunnel was ready in November 1876. At the same time another track was laid to convey the pyrites intended for use in making copper at Rio Tinto to the calcination ground at Cerda.[8]

The ores which were being mined varied greatly in their contents. Along the north wall of the lode there were some that reached a copper content as high as 10 per cent but the greater part of the ores held as little as $1\frac{1}{2}$ per cent. At first it was intended to select those pyrites with a high copper content for export, separating them from the others which would be retained for calcination. But using large-scale open-cast working it proved impossible in practice to distinguish between the richer and the poorer ores, for the whole mass of the lode had to be simultaneously quarried and taken away in the waggons either to Huelva or to Cerda.[9] But since greater profits could be obtained from pyrites with a rich copper content it was decided, despite Forbes's earlier views, to mine these selectively by means of underground mining in the south and north lodes, and this work started in 1878.[10]

To open up the north lode a heading was driven in from the valley close to the source of the river Tinto. It took three years to reach the mineral, and to widen the heading (which entered the ore body 40 yards below the top of the pyrites) sufficiently to take the necessary railway track.[11] However in 1889 it was decided by the board that both the difficulties in mining this lode by underground methods (for it had been extensively worked in the pre-Roman and Roman periods) and the need to extract as much as possible from it, made it desirable to switch to open-cast methods and the next year work began on removing the overburden.[12] In 1882 the San Dionisio mass in the north-west of the area was explored and five shafts sunk along its length. By 1885 the eleventh-floor tunnel had been extended to enter the mass so that the Alfredo underground mine could be started,[13] and in 1907 the Atalaya open-cast pit was begun thus making it possible to attack both the narrower and the wider sections of this mass more effectively.[14]

The Rio Tinto landscape was altered out of all recognition by these great open pits and their attendant waste dumps which filled in valleys and created hills where previously none existed. Both sides of *cerro* Salomón were gradually torn away leaving deep scars and exposing the red ironstone capping and the brilliant blues and greens of the copper-staining. At first great armies of men and women hand-loaded the vast tonnages of waste and mineral into the waiting railway waggons. Then at the end of the nineteenth century Bucyrus-Erie steam shovels were introduced from the United States, operating on railway lines and moving the blasted rock ten times faster than could be done by the hands of men. Their success was so phenomenal that later the Rio Tinto Company added to them by purchasing the steam shovels which had been used in the construction of the Panama canal and shipping them to Spain. The great open-cast pits of Rio Tinto provide impressive and easily appreciated testimony to the skills of the men who created them.

Less obvious, because they are buried deep in the earth, are the achievements of the engineers responsible for the underground workings. When the Rio Tinto Company took over the mines it inherited the shafts and galleries that survived from the period when the Spanish crown had owned Rio Tinto and worked the mines on what was known as the pillar-and-stall method. Inside *cerro* Salomón the Spanish mining-men had excavated seven different levels, with

*Bucyrus-Eyrie steam shovel used in making the open-cast pits at the beginning of the twentieth century.*

*This photograph of circa 1880, taken during the making of the south lode open-cast pit, shows the large quantity of mineral left untouched between galleries when using the pillar-and-stall method of underground mining.*

*Diagram showing cut-and-fill system.*

floors of 33 ft 10 in. between each. The cross cuts and the drifts at each level were laid out to the same measurement, and the ores extracted in sections of 13 ft by 13 ft, leaving pillars 20 ft square to support the floor above. The company continued the system, though to provide greater safety it established intervals between the floors of new galleries of 41 ft and ensured that the pillars on the different levels were above and below each other in a vertical line. And the company mined to a greater depth: the seven floors existing in the south lode in 1873 had by 1900 been extended to twenty-two.

The great drawback with this pillar-and-stall method was that it permitted the extraction of between only 20 to 30 per cent of the ore body. The south lode underground, by 1900, had become a mere shell and the men in charge were in a state of continual despair, wondering from where they could get the 1,000 tons of mineral a day which they were required to produce to reach the target of 300,000 tons per annum. In the August of that year Robert Palmer joined the company as its chief mining engineer and, in the report he wrote for the board on the subject, declared that all the excavated apertures would have to be filled in to enable the remaining ores to be mined.[15] His proposal was for a formidable project which would involve taking down from the surface many millions of tons of rock, not only to plug the gaps and galleries which already existed but also to replace the pillars and floors which would have to be cut out. This system, known as cut-and-fill, seems to have originated in Germany in 1868 and to have been in use in the nearby mines of Tharsis by the 1880s.[16] Though costly, it enabled all the ores to be removed and the Tharsis company had found it an economic proposition when the pyrites contained more than 1.5 per cent copper.[17] The adoption of Palmer's proposals for the Rio Tinto south lode underground saved it from an early shut down.

Once the old workings had been supported by the introduction of waste rock and thus rendered stable, the pillars were cut out and the floors between the levels gradually removed. This was done working from underneath and by removing strips of about four yards by two yards at a time. The filling material was dropped down shafts from above, its level rising as the ore above it was removed. The excavated ore itself was passed down through shafts until it reached the bottom level of the mine where it was loaded into waggons and from there hauled through the tunnels to the open air. Eventually the decision was taken to cut out strips of rock of nine yards or so in length though this caused a number of difficulties for, when the width of the transverse cut was increased, subsidences sometimes occurred causing the collapse of both the access galleries and the filled-in galleries on the higher floors, and starting fires.[18] The latter developed very rapidly once they began and the only way of controlling them was by cutting off the stope concerned by building a wall in front of it.

The raw pyrites that were removed so laboriously from the underground workings and the open-cast pits and then shipped to Britain, France and Germany, were sold for their sulphur and their copper. Chemical manufacturers removed the sulphur by roasting the ores in furnaces and, having extracted it, sent the cinders on to other manufacturers for the removal of the base metals they contained, of which the copper was the most valuable. Originally the company planned to export the greater part of its output and to retain only a small amount for treatment at the mines. However, the amount of pyrites that it proved possible to extract so exceeded expectations, and were so much greater than needed for sales to the chemical industry, that it was necessary almost from the start of operations for the company itself to make use of the major portion of the ores mined. The average annual tonnages for every five-year period from 1876 to 1910 are given on the next page and illustrate this trend.

| Period | Pyrites for shipment | Ores for local treatment | Total mined | Copper made at mines |
|---|---|---|---|---|
| 1876–80 | 236,194 | 526,560 | 762,754 | 4,673 |
| 1881–85 | 308,223 | 844,305 | 1,152,528 | 11,752 |
| 1886–90 | 383,900 | 904,115 | 1,288,105 | 18,018 |
| 1891–95 | 474,466 | 911,459 | 1,385,925 | 20,700 |
| 1896–1900 | 632,215 | 934,801 | 1,567,016 | 20,684 |
| 1901–05 | 650,103 | 1,248,202 | 1,898,305 | 21,014 |
| 1906–10 | 628,656 | 1,268,604 | 1,897,260 | 22,789 |

In the last year for which the accounts of the royal mines of Rio Tinto are available (that is, 1871) some 65,000 tons of ore were mined and 974 tons of refined copper were produced (by cementation, followed by precipitation in vitriolic waters, then smelting and finally refining in a reverberatory furnace). This copper cost on average £62 8s per ton to manufacture, more than the price obtained for it in Seville.[19] By improvements in the various processes David Forbes, by the beginning of 1875 and even before the railway had been fully brought into operation, was able to reduce the cost per ton of Rio Tinto copper to £25 free on board at Huelva.[20]

One element in the struggle to reduce the expenses involved in making the metal was the cost of extracting the ores. Here the introduction of open-cast working helped to reduce the average cost of mining each ton of ore from 6s 8d in 1871 to 2s 3d by the beginning of 1876. Another important saving was made by altering the scale of the cementation process. At the calcining ground Forbes had the soil removed down to the bare rock, ready to receive the minerals being conveyed there by the tramway which he had installed. On arrival they were broken into even smaller pieces and then burnt in heaps of 100 tons, being covered with ore dust to moderate the draught and to prevent too great a heat, which otherwise would have made the central portion melt and fuse together. Later the amount in individual heaps was increased to 800, then 1,000 and finally 1,500 tons. Those of 800 tons were found to be the most efficient from the point of view of thoroughly calcining all the pyrites. Larger *teleras* always contained a proportion of unburnt ore which had to be returned to another *telera* for further calcination. However, the heaps of 1,000 and 1,500 tons had the important advantage that they required far less wood per ton of pyrites to start them burning. Once the fires were out the calcined ores were taken to the tanks for washing which enabled 60 per cent of their copper content to be recovered immediately by the precipitation process.

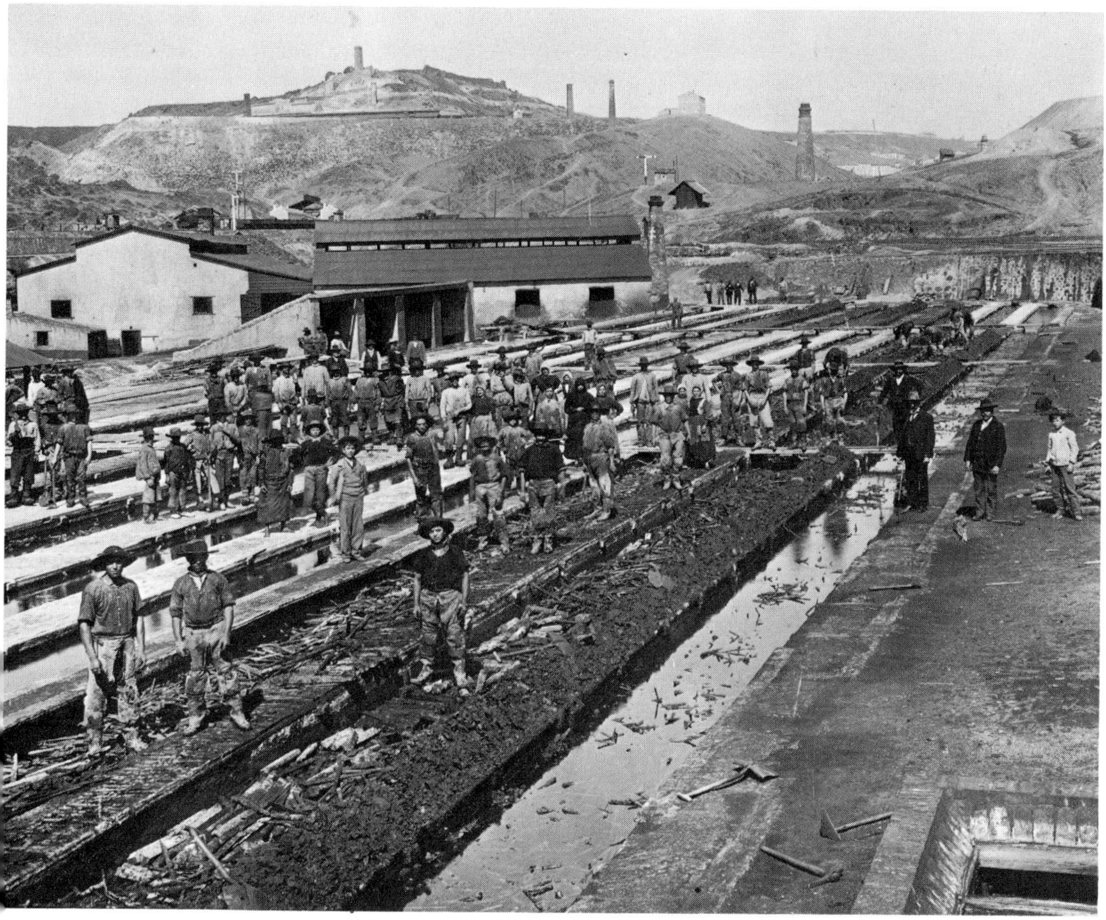

*Cleaning out the cementation ·anks at Cerda: circa 1890.*

The water for the leaching was brought down from the source of the river Tinto itself, the springs in the mines, via a series of conduits.* The pig-iron in the tanks was scoured by workmen using stiff brooms to ensure that new surfaces were constantly being presented to the action of the water, two tons of pig-iron being required to produce one ton of copper.[22] The cost of this pig-iron when it was delivered at the mines by the Seville merchants was £11 a ton in 1875; but once the railway was opened it was imported from England and cost little more than £5 a ton including import duty, yet another important economy.[23]

The burning of the piles of ores, the *teleras*, made life periodically unbearable for the inhabitants of the mines' area and Remisa had found considerable discontent among local people in connexion with this earlier in the century. However, the increased tonnages being mined by the Rio Tinto Company from 1873 onwards led to a great extension of the numbers of *teleras* burning at any one time. Moreover,

In 1883 there was a flow of 960 ubic metres of water a day from ne north lode containing 720 rams of copper per cubic metre. rom other sources there was a ow of 2,900 cubic metres a day ntaining 370 grams of copper for ery cubic metre.

as the clouds of sulphur grew larger and thicker, the smoke damage to vegetation and crops spread even wider and there were bitter protests from local farmers and landowners, though it would have been impossible for the company to have abandoned the process in those early years without effectively going out of business.

There is no doubt, though, of the devastation to the landscape and the harm to health caused by these *teleras*. John Allan, a young man who went to work at Rio Tinto in 1886, was appalled at the effects of the smoke-pollution which he saw. He had gone to Spain with his head full of the conventional Victorian ideas about the blessings which nineteenth-century science had bestowed on mankind but was quickly disillusioned. His despondency was expressed in a poem of admittedly uneven quality, but containing some striking metaphors, which he called 'A modern inferno':

The classical talk of the south, and poets
    ne'er tire to tell
Of the lands where the gods of old, the
    gods of the beautiful, dwell;
Of olive groves and orange trees, and maids
    with lustrous eyes,
Where the turtle-dove says 'Coo', and his
    mate 'Coo, coo' replies.
The science men talk of the benefits science
    has given our time,
Of engineering, chemistry, steam, and poverty
    left behind;
Well, I'd heard both sides of old and 'twas
    put into my mind
If I went to some works in the south I should
    see both combined.
I went, God help me, I went expecting to see
    such bliss
As never before was known in the old world or
    in this;
'Twas this I was told to expect. Let me tell
    you what I find,
I'll give you just the picture eternally
    graved on my mind.

The earth is red, the sky bright blue,
No tree or green thing breaks the view;
On every side death reigns supreme,
For six long miles no life is seen,
But barren ground and charréd stumps,
With here and there some ruined humps
Telling where once a winepress stood,
With vineyards and a little wood
Of olive or of orange trees,
Ere science o'er this land did breathe
Her poisoned breath, polluted air;
Which, withering every blossom fair,
Has left instead of nature's plenty
A howling desert bare and empty.

For science, like some dragon old
Whose breath was poison while his scales were gold,
Attracts the dazed spectator with the glare
Of her bright scales; and then, ere he's aware,
She sucks him in, extracts his richest blood,
And cries, 'All this I do for your great good.
Your mind will grow so great, no more you'll need
On beer and flesh of beasts to drink and feed.
I'll take the meat, teach you to eat the bone,
And of each cherry, you shall have the stone.
So that, in short, you'll always bless the day
When friendly science lighted your dark way
Her noble ends will feed the enfranchised mind,
And all bucolic tastes be left behind.
The science man but little needeth her
And gets that little for his screw a year.'

But e'en as science here doth nature damn,
Thrice doubly doth she curse the health of man.
With haggard cheeks and sunken eye
And palsied hand, the wrecks of men go by
Collected from the refuse of the earth;
A cesspool of creation 'tis and worse.
Perhaps kind heaven deemed this fumigation
Might show them hell, and so work their salvation.
I know not, but indeed I do infer
That hell is empty and her people here.

From one point of view Allan's pessimism was not justified, for the 'science men' were eventually able to solve the problem created by their predecessors; and in the meantime the company was able to argue that the material gains from the process of open-air calcination were so great that they more than offset the damage and discomfort it caused. The mere fact that men continually flocked to Rio Tinto from all over Spain to seek work at the mines was proof, it was maintained, that the smoke problem did not make life intolerable to the degree claimed by the company's critics. This argument lost much of its force, however, as a result of the 1888 riots at Rio Tinto which were in part a protest against the pollution caused by the *teleras*. Shortly after the riots a royal decree severely limited for a time the amount of ores the company was allowed to calcine in the open air. The decree was reversed following some adroit political manoeuvring by the company but it was realized that the system could not be retained indefinitely.

The company was calcining over 1,000 tons of pyrites daily sending about 300 tons of sulphur into the atmosphere as sulphur dioxide. The mines were capable of providing 4,000 tons of pyrites every day and the company was anxious to extract this amount to increase its production of copper. But to calcine this amount in the *teleras*, and to quadruple the quantity of sulphur dioxide polluting the atmosphere, would both have annoyed the government and made life at Rio Tinto completely unbearable. Consequently experiments were begun, even before the 1888 riots, to try and solve the problem and to find a way of making the copper amenable to extraction without

*The treeless Rio Tinto landscape: this photo of 1897 shows the British village of Bella Vista.*

discharging the sulphur into the air.

In 1879 Heinrich Doetsch, the Huelva-based director, got the board to adopt his ferric chloride process for experimental use at the mines. This was a process devised in 1873 by the French scientist Joly, for which Doetsch obtained the patent, which involved washing pyrites, crushed to an eighth of an inch in size, in tanks with a ferric sulphate solution containing chlorine. It was, however, very expensive as the chlorine had to be made by heating ferric sulphate crystals and salt in a reverberatory furnace. In 1884 Brenton Symon read a paper to the Mining Institution of Cornwall in which he suggested that pyrites of about one inch in size should be mixed with calcines to give a more suitable copper-bearing liquor using less iron in the precipitating tanks. By this means the ferric sulphate was reduced to ferrous sulphate.

Doetsch was influenced by this paper and took out a patent on what was known as the second Doetsch process in 1885 and persuaded his fellow directors to introduce it at Rio Tinto. This new method consisted of mixing three parts of small pyrites with one part of calcines adding 1 per cent of salt and 0.25 per cent of manganese ore to the heap which was then washed with the acid water from the mines. H. F. Collins, who was in charge of the process at los Planes and Naya, reported in 1887 that the addition of the salt and manganese in no way increased the amount of copper recovered and that the same result would be achieved by using the natural cementation method, that is, simply washing the heaps with the mine water.

Doetsch, however, dismissed this report and was not prepared to accept the cancellation of his contract and the process continued to be used until his death (by which time he was deputy chairman of the company) in 1894. His insistence on the importance of his patented method seems to have sprung from his financial interest in its use. For not only did he receive a royalty for allowing it to be employed, but he also sold the company the salt and the manganese which it required. Doetsch and his partner Sundheim had been pioneers in the mining of manganese in Huelva district and also had financial interests in salt production. When there was a slump in the manganese trade they found that they could move their surplus stock only by selling it to the Rio Tinto Company, and it would seem that this was their primary consideration in promoting the process. Indeed, such a large quantity of salt had been purchased from Sundheim and Doetsch that when the method was abandoned the company had 3,635 tons of salt and 1,092 tons of manganese ore on its hands, all mixed together and impossible even to give away.[24]

When the Doetsch method was discontinued there were some

20 million tons of pyrites and calcines mixed together in heaps. In 1908 it was decided to recover the pyrites from these and in 1912 the first *lavadora* plant was erected. Between 1912 and 1918 alone almost 3 million tons from the heaps had been treated, and almost 1,040,000 tons of high grade pyrites recovered. But there was a great quantity which had undergone decomposition and a flotation plant had to be installed to recover this.

The experiments with the Doetsch process delayed the introduction of natural cementation which was to provide a simple solution to the smoke problem. Meanwhile, the company felt obliged to mitigate the effects of the smoke damage caused by the *teleras* by paying compensation to local farmers. But they soon found that this opened up a new and complicated field for disputes. Claims came in from people living in a wide area around the mines, and the company each year regularly paid compensation in varying amounts to over 1,600 persons owning some 11,000 separate properties in an area of 300 square miles around Rio Tinto. As a result the company soon decided that it would be cheaper, where it was possible, to purchase as much of the affected land as it could in order to reduce the amount it would be liable to pay out over the years.

The system of compensation was beset with difficulties. An accurate assessment of each farmer's or landowner's losses was impossible and few claimants received as much as they thought they should get. Yet none of them was prepared to sue the company in the courts since it would have been an expensive business and the company was rich enough, if it wished, to exploit the complicated and slow-moving legal processes to its own advantage. On the other hand, there were plenty of 'smoke farmers' who appeared to rely more on the compensation payments than on their farming to get their livelihoods.

In 1895, following the scrapping of the Doetsch method, it was decided to experiment with the natural cementation process and in 1896 this was gradually brought into operation. It was in fact an improved method of cementation that soon proved it was unnecessary to roast the ores in *teleras*. It involved a careful reconstruction of the washing ground, which was provided with a new drainage system, underground air flues and tall chimneys. Ores brought straight from the pits were built up in heaps of 100,000 tons made up of alternating layers of large pieces and small pieces mixed with dust. Water was run over the heaps, its flow to various parts being controlled by the ridges of the heaps themselves. As oxidization of the ores took place, the temperature within each pile rose, and the air vents in the chimneys were gradually closed to accelerate the spread of the oxidization throughout the entire heap. Great care was required at this stage of the process as in hot weather it was easy for the great piles to ignite and once alight a fire could not be put out. When the oxidization was completed, water was again run over the heaps to leach out the soluble copper through the drains.

Once the copper residue in the heap was reduced to 0.3 per cent the ores were then exported under the name of washed ores to the chemical trade for the extraction of the sulphur content. The Tharsis company had been using this method since 1889. But so bad was the relationship between the two companies that it was not until 1899, when the chairman of the Rio Tinto Company, J. J. Keswick, who had succeeded Matheson the year before, visited Tharsis at the invitation of its chairman, William Rutherford, that it was realized that the rival company was using the process and had developed it to a high degree. After that the Rio Tinto staff were able to draw on Tharsis experience to improve their own system. The open-air *teleras* were gradually done away with and in 1907 the last of the great heaps of burning ore, which for so long had disfigured the Rio Tinto landscape, was roasted. The new system, apart from leading to better living conditions at Rio Tinto and the eventual end of compensation payments, also gave the company enhanced profits, for the sulphur in the pyrites, instead of going up in smoke as part of the copper-producing process, was saved for the sulphuric acid makers. In 1903 the sale of these washed ores was 118,000 tons. By 1913 it had risen to 825,000 tons and provided a considerable income for the company.[25]

It was in 1898, with the first major reduction in the quantity of pyrites being roasted, that William Nash, the head of the company's lands department, noticed that some land which had been abandoned for years was once again being brought back into cultivation.[26] Local farmers, however, continued to demand compensation on the grounds that land once rendered sterile by the sulphur fumes was not likely to be suitable for many years, and indeed the last payment was made as late as 1919. It was with the intention of establishing how far the agricultural land around the mines had been rendered unfit for farming that the company decided to begin an experiment in afforestation. In the autumn of 1908 Nash planted the first eucalyptus trees on mines' property near the Zumajo dam, the seedlings for the experiment being brought from Granada.

They seemed to take well and Walter Browning, then general manager of the mines, called in Spanish government forestry experts who recommended planting pine (*pinus pinea*), oak and cork trees. The success of these led Browning to recommend a policy of large scale afforestation and in 1916, on the advice of the university of

*Present day Bella Vista: it is now virtually impossible to photograph the village because of the dense woodland surrounding it.*

Copenhagen, he employed as the company forester Kai Hase, a Dane. Hase decided to plant over 22,000 acres of barren land with pine trees and great numbers of women and boys were employed to walk over the bare hills, driving one inch holes into the soil with spiked hammers and putting a seed in each. Slowly the hillsides became covered with forests of aromatic young pines, providing the mines with an abundant supply of fuel and doing much to improve the surroundings in which the people of Rio Tinto lived and worked. Pasture farming, too, was introduced and herds of sheep, goats and cows once again began to wander over the slopes of the valleys as the grass sprang up. In the nature of things, the change could not happen overnight but by 1954, when the British Rio Tinto Company handed over the mines' property to its Spanish successor, there were 33,000 acres of forest and farm land: a new green landscape where only forty years before there had been nothing but parched soil and brown dust.

The boldness, the exuberance and the contempt for difficulties which characterized the operations of the Rio Tinto Company in its formative years were typical of Victorian and Edwardian Britain, but this fact does not detract in any way from the impressive technical achievements of the men in Spain and in England who realized Matheson's visionary project. Their skills, their imagination and their self-confidence served the company well, for practice and policy do not always proceed hand in hand, even the most imaginative technologists are at the mercy of their administrative superiors and self-confidence is not always a sure foundation for success. Theirs was a working together in which luck had no place. By 1914, as the war clouds gathered and the lights began to go out all over Europe, it must have seemed to anyone who had known the mines during the past forty years of their history that, whatever the future might hold, he had seen Rio Tinto's golden age.

*A group of Rio Tinto workmen, circa 1890.*

11

# *A time to weep*

## The riots of 1888

Tragedy and tears have never been long absent from Rio Tinto, though for the greater part the bitterness has soon subsided and the recollection of them eventually dimmed. But the calamity of February 1888 has etched itself into the minds of the people there like acid into a metal plate. The bloodshed and the sorrow of that time are still remembered vividly and described graphically by old and young alike. The intense shock left its imprint on the communal memory and its mark even on common speech. For when Rio Tinto people wish to say 'a long time ago' they are likely as not to use the expression '*en el tiempo de los tiros*': 'in the time of the shootings'. So catastrophic seemed the tragedy, so great was its impact on men's imaginations, that it came to be regarded as a demarcation line between the past and modernity; synonymous with some distant time, presaging the present and expiring on a cold February day among the bodies strewn over the cobbles and between the ornamented cast-iron benches of the main square.

But the story of that tragedy is not so simple as people now tell it to the interested stranger. The causes were more complex than they now remember. And the step by step progression to those agonising moments, when men of the *regimiento de* Pavia fired volley after volley into the throng of panic-stricken demonstrators in front of the town hall, is not to be found in the folklore told in the taverns, but is to be traced in the reports and the telegrams written and sent by the long-dead participants in that sad event.[1]

The disaster had its roots in the most difficult of the many problems facing the Rio Tinto Company's British staff in Spain in the last quarter of the nineteenth century: the recruitment and control of labour. Between 1874 when workmen were first taken on in large numbers and 1888 when industrial unrest first erupted into violence, the work-force of the mines was expanded from 1,000 to 9,000 men, the additional 8,000 originating in many different parts of Spain and Portugal. The company appreciated the need to house these new employees and their families, and to provide various welfare facilities for them. It failed to realise that such a large influx of people would not only destroy much of the social fabric of the old mining community, arousing fierce hostility among the people already there, but would also produce strong social tensions among the mass of rootless newcomers.

This failure to anticipate the long-term social consequences of the rapid population increase at Rio Tinto was not the result of stupidity or wilful ignorance on the part of the company. There was no framework of reference, no comparable situation elsewhere, from which the directors could have foreseen the probable future and in the light of which they could have taken remedial action. For the period, they were reasonably humane men and their chairman, Hugh Matheson was a Christian of high ideals with a well-developed social conscience, anxious to provide the mineworkers and their families with decent housing standards, an exceptionally good medical service and better than average educational facilities. At Matheson's suggestion one of the first decisions made by the board of the new company was that, despite the need for rapid progress with the open-cast programme, there would be no Sunday working in order that the men could have a day of rest and recreation; a resolution made in the knowledge that it would involve both recruiting even more labourers to complete the work according to schedule and offering still higher wage rates to get them.[2]

The directors could never understand the animosity that was shown towards them by their Spanish workmen in the early years and attributed it solely to the traditional Spanish suspicion of foreigners. Undoubtedly this was an element in the situation but the principal reasons for the hostility lay deeper and were to be found in the stresses of life at Rio Tinto. It seems probable that much of the industrial conflict that arose at the mines was inevitable and, though poor labour management exacerbated it at times, there was little the directors could have done about it even if they had diagnosed the root causes. For to give the men the living and work conditions demanded by some of the politically more extreme employees was economically impossible. The mines had to be worked on an unprecedented scale to make them a profitable business proposition and, though by one set of standards the company soon became profitable, there was still a financial tight-rope which had to be walked to enable it to fulfil its contracts and pay for the further expansion of the mine-workings. When trouble arose over wages the directors had to remember that labour costs were one of the heaviest of all charges, and thus had to subject all claims for increases to the closest scrutiny. When complaints were made about the welfare services and social amenities, they had to remember that the provision of these must not become disproportionately expensive.

There were, too, other complicating factors beyond the company's control and these determined to a great extent the form that the industrial unrest was to take. The most important were the political divisions and disturbances which dated back to the post-Napoleonic period of Spanish history. From 1814 onwards Spain was in an almost continuous state of political chaos alternating with government authoritarianism. The legacy of political experiment resulting from the French revolution, and the economic forces at work in much of industrialized Europe, did not leave Spain untouched. They increased the political and social awareness not only of the middle-classes but also of the peasants, the industrial workers and the miners, and led to the development of conflicting political ideologies and extremist organizations.

The first republic proved a productive period for the growth of the most extreme of these views and in particular for those propounded by the followers of Michael Bakunin after the split between them and the supporters of Karl Marx. By 1873 it was estimated that the Bakuninites, who were later to be called anarchists (while Marx's adherents became known first as internationalists and then as communists), totalled over 300,000 in Spain and were active in 270 centres. The Spanish Bakuninites constituted the largest and most powerful group of their kind in Europe, for they possessed strong appeal in a country which had suffered so long from incompetent central government. Their aim was to overthrow by revolution the apparatus of the state and to reorganize society in self-governing units based on municipalities, occupations and industries. To promote their doctrine they sent the most persuasive of their militants from village to village, concentrating particularly on the Andalusian region, spreading their propaganda and selling their news-sheets. The anarchists were well organized at Rio Tinto by at least 1873 when the British company acquired the mines and found them to be the focal point for politics in Huelva province.

When Alphonso the twelfth, the son of the former queen, Isabella, was summoned to the throne in 1874, one result was an official prohibition on anarchist activities. However, the ban was not rigorously applied by the government and the movement continued to thrive underground. Its publications were still allowed to be sold openly and it continued to extend its grip on the working-classes, especially in the south of the country. The anarchists benefited greatly, from 1882 onwards, by the formation of the Confederación Nacional del Trabajo ('national confederation of labour'), a trade union which also supported a policy of government decentralisation. There was co-operation between the two (followed by a formal declaration of unity in 1888), which provided the anarchists with syndicalist ideas and skills, particularly in the use of strike action as a means not only of obtaining improvements in working conditions but also as a way of trying to create a stateless society.

At the same time, the internationalists were growing in strength in the north of Spain and advocating change not through social disintegration but by action at a national level designed to create an even stronger centralised authority. Their industrial arm, established in 1888, was the Unión General de Trabajadores ('general union of workers'). Thus Spain produced two powerful workers' organizations each violently hostile to each other, and in time both were to join in bitter conflict at Rio Tinto. In the 1870s and 1880s, however, the influence of the anarchists was still paramount at the mines and the communists had little success in infiltrating the ranks of the workmen, though their efforts to do so by competitive wage claims and the strikes these gave rise to constituted a considerable nuisance to the company.

The situation was made even more complicated, and more dangerous, since neither side was able to participate in the normal process of parliamentary politics. In 1876 there had been an attempt to reintroduce a parliamentary system but the working-classes in general were excluded from the franchise. And the traditional practice of *caciquismo* meant that they could not put their trust in the eventual triumph of democracy. For that system, whereby Spanish politicians of the establishment accommodated each other for their mutual advantage, made a mockery of the general elections that were periodically held. The procedure was for the king to appoint the leader of one of the main parties, or his nominee, as president of the council of ministers. The newly appointed premier immediately chose a minister of the interior, and together they prepared a list of candidates to be elected to a new *cortes* which would ensure a working majority for their own party, grant the opposition what they thought was suitable representation and provide some seats for a few republicans of the less objectionable variety. As a courtesy the principal politicians of both the main parties were regularly included. The minister of the interior then selected new civil governors for all the provinces to organize the elections.

The actual work in the constituencies was carried out by the *caciques** (the local political bosses) of the two main parties; men who were normally prominent lawyers, bankers, landowners or, as often happened in Andalusia, the stewards of absentee landlords. In most rural areas their influence was such that they were able to ensure that the majority of electors voted to return the candidates chosen in Madrid, though if the voters proved troublesome they could normally arrange for the votes to be miscounted to give the desired result. In

*The word originally meant an Indian chief.

urban areas, controlling the electorate was more difficult and the *caciques* had to acquire political followings through the manipulation of patronage and by appointments to minor government posts.

In the later part of the nineteenth century *caciquismo* was well entrenched in the rural areas around Rio Tinto. But the increasing industrialisation of the area within the mines' boundaries and its rapid population growth both reduced the influence of the principal *cacique*, José Lorenzo Serrano, a Zalamea landowner, and presented him with a challenge. For he wanted the mining town within his orbit of influence, while the Rio Tinto Company was determined to keep local affairs in its own hands. The company had no liking for the anarchists and the communists who threatened to disrupt efficient working, but it had an equal dislike of the *caciquismo* system which it believed would lead to inefficiency and corruption in the town if it were to be allowed a foothold. It was this consideration that forced the company into local politics and when, early in the company's existence, industrial problems arose the British staff found that they had to contend not only with anarchists but with local landowners led by the Lorenzo faction.

Between 1874 and 1888 there were fairly frequent strikes amongst certain sections of the labourers at the mines, but these were half-hearted affairs, badly co-ordinated and soon over, and never achieving their objectives. Some were due to the men's spontaneous anger over certain incidents, some to deliberate decisions by the anarchist organizers to test their strength, and a few to attempts by communists to prove to the workmen that they could be as militant as the Bakuninite leaders. The demands made on those occasions varied considerably in their seriousness but they illustrate well the general discontent with many aspects of life at Rio Tinto in those years.

A particular source of disaffection was the company's policy on dismissals. Though the company was anxious to keep its labour force at full strength, it dealt severely with drunkenness or theft at work. An intoxicated workman could endanger his own safety or that of others and a second offence nearly always meant dismissal. Stealing from work-mates was seen by the company not only as an anti-social act but as a frequent cause of fights and knifings, while company losses by theft in the early years were considerable. It was calculated that in 1874 employees stole a quarter of all copper produced at the mines and a vast amount of movable equipment and tools. Consequently offenders were always discharged when caught even if they were not prosecuted. These offences provided the reason for the vast bulk of all dismissals between 1874 and 1888 though as time went by there were also an increasing number of sackings of men for what the company described as 'repeatedly stirring up political discontent'.

In this policy the British managers at Rio Tinto were following the normal practice of all mining companies in Spain. They did not appreciate that the large amount of drunkenness and theft was related to the rootlessness of the newcomers, and that much of it was symptomatic of the social unrest that inevitably accompanied the creation of a greatly enlarged community. For when time passed and the new arrivals settled down there was a dramatic decrease in the number of dismissals for these offences. However, the men were in no position to appreciate the company's problems and the sacked workers, who had been drawn from all over the Iberian peninsula and had no prospects of alternative employment in the Rio Tinto area, were placed in extremely difficult circumstances, especially if they had wives and families with them. Most dismissed men took themselves off but there were always a few who preferred to stay on, depending on their former workmates for their subsistence and often provoking unrest. In such cases the mines' manager always instructed the *alcalde* to see that these men were sent away from the district and stopped from pestering their fellows, for the company maintained it had a duty to its employees to protect them from scroungers. To the workmen, however, this attitude appeared not as paternalism but as despotism and provided frequent opportunities for downing tools.

Discontent with wages was also often expressed by strikers. In fact in this period the company was forced to pay its labour-force better than most other industries in Spain. An agricultural labourer's wage in the Rio Tinto district, for example, averaged 8 *reales* (1s 8d) a day while a labourer at the mines would not earn less than 15 (3s 2d) and could earn up to 21 *reales* (4s 5d) on average according to his category. Nonetheless, the standard of living of the labouring families at Rio Tinto was still low by comparison with the Spanish and British staff; and though the company replied to strikes for increased pay by pointing out that most men wasted a good proportion of their wages drinking in the taverns, it was not an argument which their employees regarded as valid.

A contributory cause of unrest was overcrowding in the workmen's houses. The company owned virtually all the buildings within the mines' boundaries and was gradually acquiring all those which were not its property. To accommodate the influx of workers it had to build many more and it was virtually impossible for a mines' employee to live anywhere than in a company house, unless he was prepared to live in a village outside the mines' area. Company houses were a great improvement on those which most workers occupied in the days of the crown's ownership of Rio Tinto and were far superior

ABOVE
*At the time of the 1873 sale of the mines many of the workmen lived in houses which were little more than crude huts. A photo of circa 1875.*

BELOW
*The first houses for workmen built by the British company in 1873–4.*

to agricultural labourers' housing in the area around. They were let to heads of families at rents which were comparable with those paid by farm workers in the district. But, though they were cheap to rent and were built to a better standard than was normal, they still deteriorated fairly rapidly with the hard treatment they received from large families. The fact that the company's building programme in those years was never on schedule also meant that many householders took as lodgers young men newly arrived to work at the mines, and this imposed additional stresses on family life since few houses had more than three rooms.

The sulphurous fumes from the calcining grounds were a major cause of discontent. They produced an environment that everyone resented, for the pall of smoke which frequently hung over the area destroyed much of the vegetation and produced constant gloom and dirt. Much depended on the breezes. If there was no wind the sulphur dioxide clouds would lie densely on the surface, so thick that it was impossible to see through them. On one occasion, when despite the fog an attempt was made to keep the railway operating, there was a head-on collision between two trains accompanied by the deaths of the crews and the travellers in the passenger compartments. If the wind started to blow in the direction of one of the villages the people there picked up their potted flowers and plants and carried them indoors, where they sealed the windows and doors and waited for the fumes to move on. It was seldom that their attempts at horticulture were successful, and if a plant survived and flowered it was a cause for considerable jubilation.

This problem lay behind much of the social unrest at Rio Tinto and many of the short strikes. For when the smoke made work impossible the men were paid half-wages. Their demands for full pay were always rejected on the grounds that, even allowing for the periods off work, the company's policy provided the men with average wages in excess of 16 *reales* (3s 4d) a day. Moreover, in the company's view, the men abused the system, deciding that smoke made work impossible when there was in fact no smoke around.

The open-air calcination also caused problems outside the mines' area, for the prevailing wind in winter tended to blow the fumes in the direction of Zalamea la Real where the best agricultural land in the district was to be found. The company paid annual indemnities to farmers whose land was affected but the seeming indifference of both the company and the Spanish government to demands for an end to the burning of the *teleras*, or their removal to another district, caused great anger and led to the formation of the Huelva province Anti-Smoke League. At the time the only practical way to treat the large quantities of ores was by open-air calcination and the company had purchased from the government the right to use this method along with the mines; to give up the system would have made it impossible to operate at Rio Tinto at all. It was also essential, from the point of view of cost, that the roasting be done near the source of the ores. In any case, the Spanish government was not agreeable to fresh calcining grounds being set up elsewhere. Nonetheless, despite the company's attempts both in the *cortes* and in the press to make its viewpoint known it had few supporters and the smoke problem provided a continual cause of indignation and resentment.

The workmen made frequent demands, too, for the abolition of piece-work and the introduction of an eight hour day. The company, indeed, experimented with a system of daily payments in some sections of the mines but found that where these were introduced output immediately dropped. If, therefore, they adopted the practice for all employees they would have had to cut wage rates to allow for resulting decreased production, which the directors knew would have led to further industrial unrest. Moreover, the company maintained that throughout the year the men in fact averaged an eight hour day. Official hours of work were from sunrise to sunset, following the common practice in Spain at that time. But the men were required to work later than four o'clock only if there was essential work to be done. However, the directors were not prepared to give up the company's right to call on its employees to work later than that if it was judged necessary by the departmental heads.

The company's medical service had been introduced on the takeover of the mines, and in 1883 it was enlarged to provide free treatment and medicines not only for the men but for their families and also for the widows and orphans of former employees. In return the company deducted one *peseta* (equivalent to four *reales* or 10d) a week from the men's pay as a contribution to the medical fund which bore the cost of the service. The men frequently demanded the ending of this compulsory *peseta* deduction. Since in 1887 this provided only 101,000 *pesetas*, or £4,200, towards the total cost of almost 400,000 *pesetas*, some £16,000, the directors were sympathetic. It was essential to have healthy workmen and it was obvious that new recruits would in future be drawn largely from the ranks of the employees' families. Thus it was in the company's long-term interest to provide as good a service as possible even though this could be expensive in cash terms. Therefore by 1888 the board had decided to abolish the weekly *peseta* contribution although it decreed 'no change to take place at this moment in consequence of the way in which this matter has been pressed'.

Another frequent claim was for the introduction of a system of indemnification for injuries for men unable to work as the result of accidents, and for the dependants of those killed at work. The company at the time and for many years after persistently rejected this demand. There was a system of *ex gratia* payments and these, the directors believed, were as generous as those to be found at any other Spanish mines. But the company wished to retain the non-obligatory system for political reasons. For the threat to withhold payment, in the event of a pensioner causing industrial unrest by his political or trade union activities, was considered an important device in helping to prevent disputes. But the men resented the fact that such payments were not automatic and the threat itself provided a cause of discontent. Never once in those years did the company withdraw a pension awarded because of death or injury, and it could probably have replaced the *ex gratia* system, as often urged by the British staff at the mines, without either increasing the burden on company funds or appreciably encouraging industrial conflict.

Underlying all the unhappiness and difficulties was another, even more basic, fact. The majority of the company's new workmen had been recruited largely from the agricultural peasantry of the peninsula, and were men unused to the disciplinary demands made on those working in mining and industry. As farm workers they may have been economic serfs, but they possessed a certain personal freedom which was lacking at Rio Tinto. In their former jobs it had not mattered much if some mornings they had started late or on some days they had rested. They had to work hard at times but often they had been able to relax for several days in succession. For the sake of the better wages paid at the mines they were required to forego this more carefree attitude towards hours of work. Industrial efficiency required that they kept strictly to time and worked consistently hard each day. When they had worked in the fields they had been able to take shelter from the intense heat of summer or the cold rains of winter; but in the great open-cast pits where most of them were employed they had to work without protection whatever the weather except when the smoke from the *teleras* lay on the ground. They had, in fact, to learn to accept industrial discipline; no easy task, as Britain's experiences during her own industrial revolution proved. But the British staff, travelling to the mines from their own industrialized country where the workers had already adapted themselves to the demands of the system, failed to allow for the inevitable difficulties of the essential transitional period. Late workmen were subjected to fines that increased in scale with the frequency of their failure to get to work on time. And this was an important and additional aggravation for the men which was hardly appreciated at all by the British management.

In May 1883 there arrived at Rio Tinto the man who was to take advantage of all this dissatisfaction among the work-people and was eventually to organize the demonstrations which led to the tragedy of 1888: Maximiliano Tornet. He had been a revolutionary in the Spanish colony of Cuba. Taken prisoner he had been deported back to Spain where his talents as a propagandist were recognized by his fellow anarchists who sent him to Rio Tinto as their principal organizer there. He immediately applied for a job at the mines and, since the British staff was unaware of his background or the fact that he had to report to the police at monthly intervals, he was given work. He began as a labourer at the refining furnaces but because his education was much better than that of most of the men, for he could read and write well, he was quickly promoted to be a weigher and checker and then a timekeeper, in which capacity he was responsible for calculating the wages to be paid to the labourers employed in loading the waggons. In March 1887 errors were found in his accounts. It was suspected that these miscalculations were deliberate and that he was crediting the men with more than they had actually earned in order to take a percentage for himself. But he was given the benefit of the doubt and transferred as a timekeeper to a new blast-furnace. All this time his political activities were not once suspected by the British staff although, as later emerged, he was actively recruiting men into the anarchist movement, collecting money for his cause and arranging the growing number of strikes.

Then on 26 August 1887, the acting manager of the mines, John Osborne, walked into him just as he was selling copies of revolutionary newspapers to his work-mates and dismissed him on the spot. Osborne sent a report to the provincial authorities in Huelva and five days later the civil governor arrived in person to arrest Tornet and to charge him with selling illegal publications. The anarchist leader was taken away by an escort of civil guards to Valverde del Camino, which was where the court was situated for the area of criminal jurisdiction in which Rio Tinto lay.

A search of Tornet's house brought to light a number of documents showing the advanced extent of the anarchist organization at the mines and a notebook written by his brother-in-law, a celebrated Cuban rebel called Barona. Osborne was surprised and alarmed at the wide scope of the Bakuninites' activities at Rio Tinto, but could do little more than dismiss those who appeared to be leaders of cells in the various works' departments.

At Valverde, Tornet remained in prison for two months, during which time he was permitted, rather surprisingly, to write political articles for the left-wing newspaper *el Productor*. When at last he was brought before the court to receive the verdict, the judge acquitted him and ordered his immediate release on the grounds that the newspapers he had been selling, and which formed the basis of the charge, were printed in Cadiz and Barcelona under government licence and could not therefore be illegal, however revolutionary their contents. Tornet thereupon returned to Rio Tinto, where many of the workmen contributed money each week from their wages to maintain him, and he resumed his political work, this time openly and in consequence that much more effectively.

Almost daily throughout November he was visiting the mines' departments, enrolling recruits to the movement, selling newspapers and holding meetings. Whenever he was seen by the British heads of departments he was escorted to the gates of the workshop or section concerned and told not to return. The warnings did not keep him away, however, and Osborne held meetings with the senior British staff at which Tornet's activities were discussed. It became plain to the acting manager from the reports that were made to him that Tornet was preparing the men for a large scale demonstration or strike, but precisely what it was to be he could not find out from the Spanish workmen themselves, who were either too hostile to the management or too fearful of the consequences to speak about it. On 6 December Osborne summoned the head of the company guards and the *alcade* of Rio Tinto to his office. He told the former that in future Tornet was to be arrested the moment he entered any of the company's working areas and held for the rest of the day. He explained to the *alcade* that since Tornet occupied one of the few remaining houses in Rio Tinto not owned by the company it was not possible to evict him in order to drive him from the area. However, Osborne wanted the official leader of the Spanish community to tell Tornet to confine his political activities to the public streets if he did not wish to be arrested. A letter was sent, too, to the civil governor of Huelva asking him to arrange for Tornet's forcible removal to another district by the civil guard; a request which was refused.

In fact by the end of the first week in December Tornet had finished making his preparations amongst the workmen for what he intended should be the biggest demonstration in Rio Tinto history, and he had no further need to enter the gates of the various departments and thus court arrest. For the remaining three weeks of the old year he was busy in the towns and villages of the mining district of Huelva province. In his determination to produce a clash with the Rio Tinto company that would gain European-wide publicity for his cause he was busy forming alliances with the very men who were most antagonistic to the anarchist ideal: local communist organizers, the officers of the Anti-Smoke League of Huelva province, the *caciques* of the main political parties and the landlords and their stewards.

To have been able to unite such men in bonds of secrecy for a common cause under his own leadership demonstrates Tornet's remarkable organizational abilities. The purpose for which these men agreed to come together was to force the government to prohibit open-air calcination of *teleras* at all the mines of the province. But the investigation that was later held was probably right in its conclusion that each faction in this unlikely alliance had hopes of exploiting it to promote their own ambitions, not only at the expense of the monolithic and apparently unassailable Rio Tinto Company but also of each other. The communists probably felt obliged to assist Tornet because if they had refused to participate in the planned demonstration they would have lost ground to the anarchists in the race for new members. (In fact their attempts to play a prominent part in directing the action on the day was so successful that it was briefly thought by the civil authorities that the riots were an attempt at a communist revolution.)

Most of the *caciques* and the landowners, or their agents, and the members of the Anti-Smoke League had a genuine grievance against the mining companies because of the damage to their properties as a result of the *teleras* for which the law failed to provide a proper redress, and it is understandable that they should agree to order their tenants and the farm workers to turn out to demonstrate on the day. But there also seem to have been ulterior motives on the part of some of them. For the leaders of the Anti-Smoke League not only demanded an end to the *teleras* but also the nationalization of all mines owned by foreigners, by which they meant their sequestration and sale to Spaniards such as themselves. Moreover the *caciques* seem to have had hopes of using the anticipated disturbances to extend their own political influence into the mining towns then dominated by the officials of various foreign companies.

The anarchists intended to use the demonstration both to draw attention to themselves and to reinforce their hold over the workmen, with the additional hope that they could also use it to bring about a breakdown of law and order as a step towards the destruction of the state apparatus which they hated so much.

The alliance began cautiously. It decided to test its strength by calling on the *ayuntamiento* of Calañas to forbid open-air *teleras* in its district. The order was issued and the Tharsis mines were instructed to end their calcination work by 20 February. The *ayuntamiento* of Zalamea la Real immediately issued a similar order. Within the next few days all the district councils within the northern part of Huelva province, with the exceptions of only Nerva and Rio Tinto, had followed suit. These bye-laws were made with varying degrees of enthusiasm. At Zalamea, where the town was by tradition hostile to mining, the local councillors were very willing to pass the necessary resolution. But at the other *ayuntamientos* there was greater reluctance. They were all visited by Maximiliano Tornet, José Lorenzo Serrano the Zalamea landowner and his son-in-law, José Ordoñez, the leader of the Huelva Anti-Smoke League. The pattern was the same in each town or village: first a meeting with the local *cacique*, then discussions with the *alcalde*. When the leaders of the alliance were told, as they invariably were, that the district councils had no powers to make such an ordinance they pointed out that it was only by passing the required resolution that the *ayuntamientos* concerned could ensure that there would be no demonstrations or riots in their districts. Within a week all had capitulated.

Tornet decided that the time was then right to carry the campaign into Rio Tinto itself. On 9 January representatives from other nearby *ayuntamientos*, headed by the *alcalde* of Zalamea, visited the Rio Tinto town hall and demanded that the *alcalde* and councillors there issue the same prohibition. Tornet at least must have known that there was no chance of this happening in view of the company's influence and the obviously disastrous effect that the suspension of open-air calcination would have had on the economy of the town. Indeed he was certainly banking on the town council's refusal for, without waiting for the reply, he started to make plans for a demonstration designed to produce such disorder that it would force the Rio Tinto councillors to resign and the company to suspend operations.

On 17 January the *alcalde* of Rio Tinto issued a statement that he had received legal advice that only the central government could issue a law forbidding the *teleras* and that consequently he could not comply with the demand of the delegation which had visited him. At the same time the *alcalde* wrote to the civil governor of the province pointing out that the entire civil guard for Rio Tinto and Nerva consisted of only ten men, and asking for reinforcements. As a result he was sent another thirty-five guards in case of trouble.

In London the directors of the company had been studying Osborne's letters on the situation. They felt that their acting manager at the mines was exaggerating the dangers of the situation and decided that there was no need to ask the Madrid government to intervene. A new general manager had just been appointed to take charge of the mines and was at that time in England. Summoned to a meeting, he was told of the events at Rio Tinto and instructed that in the view of the board these constituted a minor problem which could be successfully dealt with by firm handling at a local level and in co-operation with the civil governor. The new general manager was William Rich, an administrator whose mining experience had been gained outside Spain. He was unfamiliar with the labyrinth of Spanish industrial politics and knew nothing of the psychology of the Spanish workman. Consequently he took the view that he should be guided by the opinions of the directors, and on his arrival at the mines in the

*William Rich, appointed as general manager in February 1888.*

last week of January he refused to accept Osborne's advice on the perils of the situation. The former acting manager reverted to his assistant's role and having been rebuffed by his new chief decided to offer no further unwanted advice. Thus his own expert knowledge was not made available to the new general manager when Rich made a series of decisions that played straight into Tornet's hands.

On only his second day at Rio Tinto, Rich authorized alterations in the working arrangements of three of the mines' departments which Osborne wisely had felt should be postponed. Charles Davies the head of the calcining ground explained to the manager that new machinery had been set up there which would make it possible for the workmen to load four waggons in the time being taken to load three. Rich authorized the bringing into use of this machinery the very next week, on 1 February, together with a reduction in the amount to be paid to the men for each waggon loaded. The new wage rates were intended to give the labourers average increases of almost a third on their pay (20 *reales* or 4s 2d a day compared with 15 *reales* or 3s 2d) made possible because of the higher productivity. But as there was no system of consultation with the workmen Tornet was able to convince them that the announcement that the rate per waggon was to be cut represented a serious reduction in their wages.

At the same time Rich also consented to a proposal by John James, in charge of the loading of the mainline railway waggons, to take on fifty temporary labourers to help clear a large accumulation of ores for export. It was proposed to pay these workers 17 *reales* (3s 7d) a day for loading 18 tons of ore: the same amount as the permanent employees received for loading 15 tons. Tornet persuaded the latter that they should go on strike if the extra men started work, since the differential pay rates were an attempt to start undercutting wages. James told the men that they had no cause for complaint since at every other mine in the province men doing the same job had to move 21 tons a day to earn 17 *reales*. However, to try and prevent a strike he asked the general manager to approve a system of bonus payments which would give the permanent workmen 22½ *reales* (4s 8d) a day for loading the same amount (18 tons) as the temporary labourers. Rich gave the authorization but the men rejected the offer when Tornet convinced them that if they increased output they would cause their own eventual redundancy. Nonetheless, having reached a decision Rich refused to alter it.

The third case of discontent occurred among the miners in the San Dionisio workings. Constantine Fernan, the departmental manager, had the duty every month of preparing fresh contracts for the teams of workmen, with the aim of providing them with a rate per ton of excavated ores that would ensure them an average wage of 16 *reales* a day. (This represented 3s 4d, some 20 per cent less than was then being paid to similar workers in British coal-mines.) The traditional practice of working by monthly contracts was something that had caused discontent at Rio Tinto a hundred years before when Director General Angulo had tried to change the system. For when the nature of the ground suddenly changed the workmen's earnings suddenly soared up or dropped down according to whether the rock became softer or harder. Fernan knew what he was doing and operating the system fairly. During the whole of 1871 the men had averaged 16½ *reales* (3s 5d) a working day. But in January 1888, when the crisis at Rio Tinto was slowly coming to a head, a change from hard to easy going led to the men's wages rocketing to an average of 35 *reales* (7s 4d) daily for the month.

When preparing the new contracts for February, Fernan calculated the rate to be paid so that it would bring back earnings to the norm of 16 *reales*. To the workmen this represented a marked drop in wages for precisely the same amount of work. There was also the real possibility that another change in the nature of the ground might make the men much worse off. Osborne, when acting manager, had pointed out to Fernan that if the going once again became as hard as it had been in December, and there was a likelihood of this, then the men's pay would drop to about 8 *reales* (1s 8d), and had proposed instead a gradual lowering of the rate spread over two or three months. Rich on taking over from Osborne, however, refused to countenance this and told Fernan to proceed as he had originally planned.

Consequently in these three departments Tornet found it a simple matter to organize a strike starting on Wednesday 1 February when the changes were due to come into force. He also called on all other workers to strike in sympathy on the same day. This would be the first time that an attempt was to be made to bring about a total closure of the mines in this way and he knew that he would have to deal with a number of men who, because they were unused to the idea of trade union solidarity, or unsympathetic to his cause, would wish to continue working. He therefore briefed his assistants to prepare to use force to intimidate such workers. On the Wednesday morning the strike was almost one hundred per cent effective. About thirty men reported for work and these were protected by the *guardia civil* and the company's own armed guards. The lieutenant in charge of the Rio Tinto civil guards informed the civil governor in in Huelva by telegram of the situation. The governor immediately sent instructions that the thirty-five men he had sent as reinforcements were to leave Rio Tinto and return to their own districts.

The guard commander and the general manager were thunderstruck by the order. The first reported that the withdrawal of the guards would probably signal the start of serious disorder. The second accused the civil governor of deliberately encouraging the men for his own political reasons.

Throughout the next day the situation remained tense but was kept in hand. There was still no violence although threats were used to keep the previous day's non-strikers away from work. In the evening the civil governor confirmed his instruction for the withdrawal of the civil guard reinforcements and the thirty-five men left the mines. The guard commander told Rich that since he had a command of only ten men, one of whom was in bed ill, he anticipated that the strikers would take advantage of the situation the next day by attacking company property. Rich agreed that the lieutenant's small force should be kept mobile and that the mines' buildings should be protected by the company guards, leaving the *guardia civil* to deploy wherever trouble looked like brewing. At a meeting with the British staff Rich told them to stay at home the next day and to keep indoors with their families. He hoped that they would not be attacked but in case they were he suggested that they should keep their rifles and revolvers ready. Those who did not own firearms were issued with revolvers and ammunition from the company armoury.

Friday dawned peaceably. It was not until midday that the situation suddenly became ugly. Almost to the very stroke of noon the strikers emerged in their thousands onto the streets of the town. Anarchist cell leaders started haranguing them with inflammatory speeches. The guard commander informed the civil governor, in the first of the fourteen telegrams which passed between the Rio Tinto civil authorities and those of Huelva on that day, that he was trying to prevent the orators from whipping up too much passion by using his nine men to break up the various meetings that were going on simultaneously in the various streets of the *pueblo*, but he could not go on doing this indefinitely.[3]

At half-past twelve the *alcalde* and all the town councillors of Rio Tinto resigned in protest at the civil governor's rejection of their renewed requests for reinforcements for the civil guards. By telegram the civil governor refused to accept their resignations and ordered them to continue in office. He also asked for details about the number of demonstrators and information about the strike leaders. By his own later admission, the civil governor then went to lunch and did not return to his office until six-thirty, so that during the whole of the afternoon he was unaware of what was happening at the mines and of the pile of telegrams building up on his desk.

At three o'clock the judge from Valverde arrived as a result of Rich's insistence that he should make enquiries about the political nature of the strike and order the arrest of the ring-leaders. Faced with the huge mobs on the streets, which by that time had grown to an estimated 8,000 men, a good number of whom were not company employees but townsmen from Zalamea, he seems to have decided that discretion was the better part of valour and confined his questioning to Rich and the harassed civil guard commander. The latter was in an extremely difficult position. With only nine men to control the demonstrators he could not make the arrests which Rich was demanding. His pleas for help to Huelva were bringing no response, and he knew it was only a matter of time before violence erupted. He came to the conclusion that the best he could do was to try to prevent the situation from becoming inflamed by withdrawing his men from the streets where their appearance was causing as much unrest as it cured.

Just before four o'clock lookouts posted by the demonstrators shouted that blacklegs were at work in the south lode open-cast. Unwisely, Rich had agreed to smuggle in a small group of non-strikers and provide them with company guards. A mob of 4,000 angry men charged through the streets of the town and out onto the lower slopes of *cerro* Salomón towards the pit. At the sight of them the guards and the men they were protecting took to their heels. In a fury the crowd smashed down the offices and storehouses.

The civil guard lieutenant decided that if Rich would not make a gesture to bring the situation under control then he would do so. He went into the *plaza* and managed to get the crowd to listen while he made a short speech asking the strikers to appoint representatives to meet him and give him a list of their demands. His hopes that this would take some of the heat out of the situation seemed justified. For an hour the men argued amongst themselves, selecting their delegates and drawing up a statement of their requirements. At the meeting, which was held in the town hall, the guard commander was faced with no fewer than twenty separate delegations making widely different demands but he undertook to pass these on to the company, though he added that if there was any more violence he would order his men to open fire. He went to see Rich in the nearby company office and told him of the men's conditions for ending the strike, pleading with him not to reject them out of hand but to allow twenty-four hours to go by in the hope of some sort of solution emerging.

By that time it was early evening and at last a message came from the governor, promising that thirty civil guards would leave Huelva for Rio Tinto by train at seven o'clock. He had given orders for a troop of cavalry to get ready for the same journey and had also asked the governor of Seville to provide soldiers. The lieutenant cabled back that he had just heard that the next day there was to be a descent upon Rio Tinto by people from neighbouring towns who were coming with firearms to help the strikers overcome the civil guards. In view of this even more men would be necessary than the governor proposed to send. At five minutes past nine the reply came humming along the wires. The thirty guards were on their way. The cavalry troop of fifteen men would arrive the following morning. Twenty soldiers on duty at the Tharsis mines would be transferred to Rio Tinto and would reach there before noon the next day. The captain general of the *guardia civil* in Seville had promised additional men on Sunday. 'Meanwhile', the governor instructed the guard commander, 'tell the strikers that the company manager is consulting London by wire and awaiting the reply. If some subtlety is used I have no doubt that you can keep the men waiting until the soldiers arrive when control can be re-established by any means available.'

The arrival of the additional civil guards late that night helped to restore the lieutenant's confidence. The unhappy Rich, however, having to cope with one of the worst crises in the mines' history when he had been only a week in his job, heard of the demonstration planned for the next day with dismay and cabled to Hugh Matheson in London, telling him of the situation and his fears that there must inevitably be bloodshed. At 8 a.m. the next morning, Saturday 4 February, the *alcalde* of Rio Tinto, the head of the company guards and the civil guard commander all sent telegrams to the civil governor containing the news that had come in overnight about the preparations being made in the nearby *pueblos* for that day's march on the mines. The head of the Rio Tinto civil guards also asked if he was permitted to use force to keep the demonstrators out of the town.

At nine-thirty the governor replied to him that he was himself boarding the train for the journey to the mines and added, without explaining how to do so, 'If [demonstration] riotous and disorderly forbid it.' Another telegram arrived from the captain general in Seville instructing the lieutenant to act firmly and to open fire the first moment that any violence was offered.

Tornet, Lorenzo and Ordoñez had spent the whole of Thursday and Friday riding from village to village organizing the demonstration for the Saturday, and early that morning they arrived at the two places where the contingents from the various *pueblos* were to meet. Tornet took command of those mustering outside Nerva, and Lorenzo and Ordoñez those joining-up at el Campillo. The two columns of marchers, one preceded by Tornet on his horse, the other by Lorenzo and Ordoñez on theirs, and each headed by a brass band, made their way to the rendezvous point. Precisely at noon the Zalamea landowner and the anarchist agitator met face to face on the road by the *plaza de* España, on a high ridge of ground opposite Salomón and overlooking the *pueblo*. As the bands struck up they embraced and then riding abreast wheeled their joint force down the long straight road running down into the town. It must have been an impressive sight and, as the column of about 4,000 men strode along, thousands of the strikers and their families ran out to greet them, waving and cheering them on.

The narrow streets of the old *pueblo* were jammed with people, laughing and shouting as though participating in a *fiesta*. Pushing its way through the climbing streets to the *plaza*, the head of the procession stopped before the town hall. The three leaders dismounted and their horses were taken away down a side street to be tethered to the hitching post at the rear of the company office. Then, accompanied by the *alcalde* of Zalamea la Real and delegates from the different villages, they strode past the civil guards' lieutenant and his men, into the *ayuntamiento* building, and mounted the stairs to the first floor council chamber where the Rio Tinto councillors had assembled.

For almost an hour they argued. Threats were shouted at the Rio Tinto *alcalde* and his colleagues that if they did not pass a decree prohibiting the open-air *teleras* then there would be riots. The *alcalde* replied calmly, time and time again, that he had been told by the *ayuntamiento's* legal adviser that even if he did sign such an order it would have no validity having been extracted by intimidation. Outside, the crowd in the *plaza* and in the streets was growing tired of waiting. Though the bands still played, the mood of the demonstrators was changing. Soon they were jeering at the guards and shaking their fists at them. A few stones were thrown, but the civil guards stood still.

Meanwhile the civil governor had arrived at Rio Tinto station, where he joined the troop of soldiers from Huelva and the soldiers from Tharsis who had met unobserved by the demonstrators. With the lieutenant-colonel commanding the soldiers, he marched at the head of the forty-five men into the town. Forcing their way through the side streets where the majority of the demonstrators were gathered, and pushing their way through the square, they made for the town hall. Lorenzo and Ordoñez saw the soldiers coming from the balcony of the council chamber, and announced that they would withdraw from the meeting to take refreshments in a nearby house and to allow the representatives of the people to continue the discussions unimpeded by their presence. But once outside they rushed to their horses and rode home to Zalamea.

As they left, the civil governor and the soldiers reached the town hall. The governor went inside while the troops compelled the crowd to fall back a few paces so that they could line up in front of the building in two ranks. In the council chamber the civil governor demanded silence while he addressed the gathering but was completely ignored. Trying to shout above the general hubbub he called on Tornet and the *alcade* of Zalamea to leave the town with their followers, but meeting with no response he went onto the balcony to see what was happening outside. The mass of about 2,000 people below him had grown more violent during the few minutes since his arrival; and though the argument inside the chamber was about calcination and anti-smoke decrees, the mob outside was shouting for more pay and less work, and some of them were brandishing rifles. As stones were thrown at him he shouted that the demonstrators must immediately disperse if he was not to use force against them. In the din he could not be heard, and the lieutenant-colonel below repeated the governor's order in his more stentorian tones: 'If you do not leave quietly my men will restore order by force of arms.' Immediately a shout went up from the window of the house nearest to the *ayuntamiento* building: 'We too have arms. Fire!'

What happened in the next two or three seconds is not absolutely clear. Some of the demonstrators later declared that the soldiers opened fire in a panic without instructions. Some, that the commanding officer gave his men the order to shoot. He, however, swore that he did not, but that his men mistook the voice of the demonstrator in the nearby window for his. The colonel's explanation seems more likely. In any event, the front row of soldiers knelt and fired into the mass of people in the *plaza* while the second row stood upright, shooting above their fellows' heads.

The crowd stampeded, trying to run away down the side streets but found the exits blocked by the vast throng which had not been able to get into the square. Several volleys were fired during the confusion before the heaving mass had swept back the human obstacles in its way; and suddenly the square was empty save for forty-eight dead or wounded and the uprooted and broken cast iron benches, some of which had collapsed under the impact of the struggling bodies.

In the council chamber, as the sound of the shots rang out, the angry disputants froze into immobility. The *alcade* of Zalamea, realizing what had happened, collapsed into a chair sobbing quietly. Until the colonel's order to cease firing rang out, there was a complete and utter silence broken only by the stifled groans of the Zalamea representative. Suddenly Tornet came to life and strode onto the balcony. Hurling abuse at the troops he shouted, 'The real enemies of our country are not the English capitalists but the Spaniards themselves.' The commanding officer gave the order to shoot him; but before the rifles could be raised Tornet turned back inside, ran past the civil governor and the shocked occupants of the council chamber, and down the back stairs to the street where he mounted his horse and rode off.

Inside the town hall the civil governor ordered the delegates to return to their *pueblos* and await the news of what was to happen to them. They asked if they could be provided with a military escort in case the people they had brought with them lay in wait and attacked them, but were told they would be given protection only as far as the town boundaries and after that would have to fend for themselves.

Meanwhile the bodies in the square were being picked up and taken to the company hospital. Thirteen were dead on arrival. The youngest was a boy of eighteen, the oldest a man of thirty. Thirty-five received treatment for bullet wounds including three women, a boy aged five and a baby of one year. The injured men were formally arrested and charged by the civil guard commander, and held prisoner

*The town hall and main square of la Mina (Rio Tinto); a photograph taken immediately prior to the riots of 1888.*

in the hospital until they were well enough to be taken to Valverde. But the forty-eight dead or injured represented only a small proportion of the casualties. Many of the severely wounded were dragged away from the square and from the side streets by friends and relatives who were fearful to leave them behind in case they were arrested or in case reprisals were taken against their families. Unable to receive medical treatment, because this would lead to discovery, a great number must have died. How many it is impossible to guess. The company could not find out, and in any case soon decided it was better to play down the seriousness of the whole affair and gave up its attempts to discover the number of the casualties, though Rio Tinto tradition puts the total number of dead at between one and two hundred. Their bodies were furtively concealed in the great piles of waste surrounding the mines, for they could not be buried in the cemeteries by the priests without revealing their identities. From time to time, when it has been necessary to move a waste tip, or when the rain and wind has eroded a heap, skeletons have been found and taken away for a belated and anonymous burial. Doubtless many more still lie there in their forgotten graves.

As the women wept for their menfolk and their children on that bleak Saturday afternoon, the troops were not idle. They searched Rio Tinto house by house, primarily looking for Tornet but arresting any wounded men who had been unwise enough to return to their homes. The anarchist organizer had in fact got to Zalamea where he was seen in the street the following morning, but by the time the civil guards arrived on the scene he had disappeared. During four days Zalamea was in a state of constant search but somehow he stayed undiscovered and unbetrayed. Only when the guards were called away did he emerge from his hiding place and slip across the frontier to Portugal. The *alcalde* of Zalamea, too, decided not to wait to be detained and made his escape to Madrid where he was seen once or twice in the streets by people who recognized him but he successfully avoided arrest. The majority of those apprehended on the day of the tragedy were never put on trial. When the reports were considered and the time came for prosecutions the authorities decided to charge only those whom they declared were the ringleaders, who were largely those whom the civil governor had confronted in the council chamber. There was no question of Lorenzo and Ordoñez joining them in prison, for they were men of substance and social standing, and also of political influence. Later the civil governor and the soldiers were also officially cleared of any responsibility for the shootings.

The events of the Saturday broke the strike and on the following

Monday morning there was a return to work. The men prepared petitions of grievances for submission to William Rich containing all the old familiar demands but they could have had little hope of the company accepting them, for their resumption of work was, in effect, an unconditional surrender. Indeed, Hugh Matheson and his fellow directors entirely rejected the petitions except for a request for the abolition of the weekly *peseta* contribution to the medical fund. Since it had been their intention to abolish this levy on the men's wages they were happy to concede the point, subject to an interval of a few weeks so that no one would think the board had given way in the face of threats. Matheson made the journey to Rio Tinto to conduct his own investigation into the strike and the shootings in addition to those being made by the Valverde judge and the military authorities. He gave his support to all that Rich had done but decided that in future there must be closer liaison between the mines and London. From then until Matheson's death the general manager had little power of decision making. His job almost exclusively became one of carrying out the orders emanating from Matheson's office. The chairman also made a thorough appraisal of the company's relationship with its workmen. Since his view was that the Spanish workmen were simple and unsophisticated, and needed to be treated in many respects like children, he decided that the main lesson to be learnt from the troubles was that there was a need for even greater company paternalism: speeding up the housing programme, improving the medical service, providing workmen's clubs (run by stewards who would ensure that the men did not drink alcohol to excess) and controlling even more closely the work of the *ayuntamiento*.

His diagnosis certainly convinced his colleagues on the board who agreed with him that the riots proved that, unless the company exercised close control over all aspects of life at Rio Tinto, work would break down in confusion. Meanwhile it would have been foolish to take unnecessary risks, so in Madrid Matheson arranged with the president of the council of ministers for the government to station a body of cavalry-men continuously at the mines to be available to deal with any future demonstrations, and the company constructed a small barracks for them on the Mesa de los Pinos, a piece of high ground overlooking the south lode open-cast.*[4]

The riots of 1888 had another effect: they precipitated the company into the middle of the Spanish political arena in a blaze of publicity. Henceforward the company had to accept, whether it liked it or not, that its existence in Spain would give it a prominence on the Spanish political scene which would bring it many problems. Thus, for both the British and the Spaniards at Rio Tinto, the tragedy of 4 February

*It was used by the soldiers garrisoned at the mines for thirty years until, at the end of the first world war, it became a workmen's club.

became endowed with a significance beyond the twisted bodies and the blood-soaked cobbles. Things could never be the same again. The day became a reference point for both communities. For the British, the memory of it loomed large whenever plans were being made for development or change. It represented a crystallization of all the fears and all the uneasiness which lay beneath the surface of life at Rio Tinto. For the Spaniards, too, it marked a break with the past (a past which, in retrospect, seemed happier than it probably was) and became a symbolic representation of the present in which they found themselves. A present in which they were subjected, without any rights of participation, to the decrees of an authority which could in the last resort depend on armed might to enforce its rule.

*Rio Tinto cricket team, circa 1890.*

12

# *Not on Queen Victoria's birthday*

## The British colony: 1873–1902

Elderly Rio Tinto Spaniards who remember the British community which occupied Bella Vista village during the 1910s and 1920s have many happy memories of some of its individual members but, almost without exception, they roundly condemn the nature of the society in which the British lived. They recall how Spanish people, with few exceptions, were rigorously kept outside its gates, and recollect the bitter humiliation felt by senior Spanish employees at not being permitted to share in the social life inside Bella Vista's walls.

But by that time the British themselves were largely the victims of a tradition which originated in the 1880s, when the members of the company's British staff and their families first tried to re-create in miniature the sort of society which they had left behind them in their Victorian homeland. For in the last twenty years of the nineteenth century there was a conscious attempt by the members of the British colony to isolate their homes and their leisure-time activities from the Spanish community on whose labours they so largely depended for their remuneration and their comforts. And although the menfolk were in continuous contact with the Spaniards during working hours, and normally were on good terms with the more senior Spanish staff, once they were outside the workshops and the offices their social conventions and national attitudes led to the erection of impenetrable barriers.

However, it is not often remembered that before the establishment of the *colonia inglesa** at Bella Vista there was another British community which consisted largely of young unmarried men who did not separate themselves from the Spaniards but lived amongst them in the *pueblo* of Rio Tinto (also known as la Mina), mixing freely with them not only at work but also in the evenings and at weekends. Something of the discomforts, and also the pleasures, of life there as experienced by these first British arrivals at Rio Tinto can be obtained from an account written by Alexander Hill of his early years at the mines.[1]

Young Hill was the son of Laurence Hill, a friend of Hugh Matheson's, and in 1875 he was offered the post of technical assistant to

*The 'English colony' The Spaniards failed to make any distinction between the English and the Scots even though the latter predominated in numbers in the early years.

*Alexander Hill in middle age.*

Theodor Blum, the head of the mining and extraction department, with responsibility for setting up a laboratory at the mines. He sailed for Spain from Falmouth aboard the *Fun-Shun*, one of Matheson's boats, together with the chairman himself. Also in the party were Peter Denny, an amiable director of the company, John Dixon the contractor for the Huelva pier and David Forbes the company's mining consultant.

Hill found the journey an embarrassing one since Matheson and his ribald compatriot Forbes were scarcely on civil terms with each other, and the latter frequently addressed remarks to the young man which he must have known would infuriate the chairman. On one occasion, for example, Forbes asked Alexander Hill if he could speak Spanish and, when the youth replied in the negative, advised him to take a Spanish mistress if he wished to learn the language quickly, adding that he personally had always found the method very satisfactory. He was silenced by an angry reproof from Matheson and the atmosphere of the voyage became even more strained. (This exchange had an amusing sequel, for when Matheson made another visit to Spain six months later he took Hill on one side and asked him if he had yet learnt Spanish. When Hill replied 'yes', Matheson enquired tartly if this was the result of following Forbes's advice! Hill was able to reassure him that this was not so.)

On arrival at Huelva the party stayed overnight with Mark Carr and his wife at the general manager's house and the next day travelled by train along the new railway to the mines. At Rio Tinto, Hill was allocated lodgings in a boarding house where he was to stay for six months until the completion of a company house for bachelor staff. His temporary accommodation consisted of an *alcova*, or a recess, without any kind of window or light and separated from the public dining room only by a dirty sheet hung over a rope.

Hill worked long hours and like the rest of the British staff could normally expect to have only Sundays completely free. Like them, too, he received periodic invitations to spend this day at Huelva with the Carrs who provided hospitality every Sunday for a group of British assistants from the mines. (Arising from this friendship with the family, Alexander Hill eventually fell in love with the Carr's daughter and married her. A son of this marriage, Laurence Hill, DSO, MC, also worked for the Rio Tinto Company at the mines and in his turn married Nan Ross, the daughter of the celebrated head of the company's medical service Doctor Russell Ross, and later played a distinguished role at Rio Tinto during the civil war troubles of 1936.)

Often there were hunting parties on Sundays in the *sierra* to shoot wild boars. This was a favourite Spanish pastime, too, and frequently

*Boar hunting: British staff and Spanish workmen at the end of a day's shooting.*

Hill's companion was his friend the Spanish doctor, Jesús Alonso. The doctor's father was a wealthy farmer at Alaja, a village some four hours' ride on horseback over the hills from Rio Tinto, and Hill became a frequent visitor there. He enjoyed the old-fashioned courtesy that he found in the household, where the womenfolk never sat at meals with the men but hovered around the table offering the dishes and the delicacies which they had made; and he admired the vigorous and forthright manner of the old farmer who had made most of his money not from farming but by smuggling goods into Spain from Gibraltar.

*Doctor Jesús Alonso.*

Hill found that the house at Alaja had many secret hiding places. One of those which impressed him most was entered through a well in the farmyard. Fifteen feet down inside the shaft was a door leading into a large chamber cut out of the rock. Whenever a large quantity of contraband had to be hidden, perhaps because of the possibility of a raid by government officers, it was concealed in this room, the door being plastered over with clay and then faced with stones matching the lining of the shaft. Finally the well was topped up with water. This was Hill's first introduction to the smuggling activities which were almost a way of life for the people of the region, but it is hard to imagine that there were many Englishmen whom the local people would have trusted with such secrets.

These young unmarried men who constituted the first British staff at the mines were happy to spend their evenings in the better sort of taverns and in the company of the Spaniards who worked as their assistants. As a result they came to know some of their Spanish friends' families, and invited both the young Spaniards and their sisters to the parties they held. They soon found that their chairman objected to these social contacts with Spanish women and they were careful not to arrange any parties for times when he was likely to be at the mines. One weekend, when Matheson was visiting Rio Tinto, the bachelors were delighted to discover that he was planning to leave for Huelva much earlier than they had anticipated. Immediately they started to arrange a dance for the Sunday evening in the new house in the *pueblo* into which four of them had just moved. While the party was in full swing, Hill was summoned by an orderly to the directors' house where he found Hugh Matheson and Mark Carr who had returned to the mines because of a mishap on the railway. They asked Hill to take them on a tour of the newly built bachelors' quarters which the chairman had not as yet seen.

Unwilling, because of his fear of Matheson, to tell him to his face that a party was being held, Hill conducted his two superiors through the dark streets towards the brilliantly lit house. As the sounds of merry-making grew louder so Hill's apprehensions increased. Matheson, a strict sabbatarian who disapproved of music on Sunday as much as he objected to the British staff mixing with Spanish women, gradually became more and more tight-lipped. When the trio entered everyone fell silent at the sight of Matheson's face. 'Then [Malcolm] Graham, like the cheeky beggar he was, went up to Mr Matheson and asked him if he could introduce him to a partner! This was declined frigidly.' Carr insisted that the party must be kept going for the sake of the guests and took Matheson off, but the dance languished and soon everyone departed in gloom.

*Neil Kennedy.*

The next morning Alexander Hill reported to the general manager and asked him if there was going to be any trouble because of the previous evening's performance. Pulling slowly on his long cigar Mark Carr remained silent for what, to Hill, seemed an eternity and then told him that he did not think anything further on the subject would be heard from the chairman. For as he and Matheson had walked away from the festivity, Matheson had revealed that his distress at the sabbath-breaking had been mitigated by one of its more agreeable details. 'These Spanish girls', he had said to Carr, 'certainly have very pretty ankles!'

Like young unmarried men anywhere, the British bachelors found opportunities to indulge their high spirits. Alexander Hill shared the house with three other assistants, Neil Kennedy, Malcolm Graham and Charles Dale, in which each had two rooms and shared a communal dining room and drawing room. Since Graham and Dale claimed that they were musically inclined they gave musical evenings for their colleagues, though the standard of their proficiency was still sufficiently low for them to need to take regular lessons each week. To give them tuition a short and very stout music-teacher travelled from Huelva. One of his visits ended uproariously. It was the evening before the feast of Santa Barbara, the patron saint of miners, and the four young men had purchased a large store of fireworks for the occasion. Dale and Graham decided to signal the arrival of their music-teacher by firing a rocket out of the window at the top of the stairs. However, it hit the window frame, rebounded into Hill's sitting room where the fireworks were being kept, and set off a chain of explosions in which rockets and jumping-jacks flew in all directions.

Hill and Kennedy dashed up the stairs to rescue their friends, who were burnt and only half-conscious, and got them to Doctor Jesús Alonso. On their return to the house all four of them were amused to see the feet of the music-teacher protruding from underneath the sofa where he had hidden in the belief that the anarchists had risen in revolt and were attacking the house. The man was so rotund that he lifted the sofa right off the floor and made it wobble with each breath he took. Pretending not to see him, and despite the discomfort which two of them were experiencing because of their burns, they all sat down simultaneously onto the sofa, producing piercing shrieks from the teacher underneath.

There was another, grimmer incident which impressed itself vividly on the mind of the young technical assistant. Early one dark morning in December 1875, as Hill was dressing for breakfast, the whole town was rocked by an explosion from the direction of the south lode open-cast. He raced from the house towards the pit with Neil Kennedy following close behind. On the hillside, at a spot where the workmen and the women and children (who were employed to help carry away the overburden) habitually gathered to warm themselves at fires before starting work, the two young men saw by the light of the flickering flames a terrible sight.

Around a hole in the ground were strewn limbs, heads and decapitated bodies. Lying wounded or crouching close by, dazed and shocked, were the speechless survivors. Theodor Blum had already arrived but he was too shaken by the horror of what he saw to take command, and he stood there perplexed and stupefied. Hill himself was so overcome that he could at first do nothing. Kennedy alone was sufficiently in control of himself to take charge and to arrange for the relief of the injured and the collection of the bodies. Hill discovered that the explosion had occurred when a borer, whose job it was to drill holes for blasting, had approached one of the fires with a basket of dynamite slung over his shoulder. Ordered by a foreman to leave for safety reasons the man had refused and, when pushed away, had started to fight, his dynamite basket falling on the fire in the tussle. Though there were twenty dead and forty badly maimed, Hill took comfort from the fact that there had been some amazing escapes. A small boy who was stooping down warming his hands next to the dynamite man had not even been scratched. And though the culprit himself had been blown into unidentifiable pieces, the man grappling with him was only slightly hurt and was able to return to work within two weeks.

The grief in Rio Tinto at this incident was intense. In their distraction many of the people blamed the British staff for letting the accident happen. In the afternoon, when the remains of the dead were taken for burial in the Rio Tinto cemetery just outside the mines' *pueblo*, there was a vast and hostile crowd waiting outside the walls of the burial ground. The procession of coffins, family mourners, priests, municipal officials and British staff passed inside and then the *alcalde* ordered the gates to be shut. The throng outside became incensed and started to climb over them shouting '*Que mueran los ingleses!*' ('Death to the English!')

In a panic, the *alcalde* ordered the civil guards escorting the procession to open fire on the men invading the cemetery. Mark Carr, who had travelled from Huelva for the funeral, countermanded the order and had the gates opened to admit those locked out, and one of the priests, at Carr's request, addressed them asking them to behave reverently, and the service then proceeded. The committal over, the crowd again became threatening and stood across the roadway barring the exit of the British staff. As the tension mounted Mark Carr, with great deliberation, lit one of his large cigars and then slowly walked forward. As he advanced there were a few moments of indecision amongst the people in the front ranks of the multitude and then gradually they started to fall back leaving a narrow lane for him to walk through. He beckoned to the members of his staff and, chatting unconcernedly to those nearest to him, led them through the sea of shaking fists and repeated cries of '*Que mueran los ingleses!*'

In those early years of the company's ownership of the mines, before the directors were forced to the conclusion that the company would have to participate, however reluctantly, in Spanish political life to try and ensure the smooth working of their operations, government officials at times harassed the British staff as a way of inconveniencing the company. The sort of thing that could happen is also described in Alexander Hill's account of his first few years in Spain, where he gave a description of his brush with the local judicial authority following the accidental death of a workman.

*The general manager's house (the casa grande) from the garden.*

*The entrance hall of the casa grande.*

*The study in the casa grande.*

One afternoon, when Hill was travelling on the footplate of a locomotive on his way back to the town from the cementation department, he saw a group of men who had finished work walking along the line ahead. At the engine-driver's blasts on the whistle, they hopped off the track, but then one of them, little more than a boy, took his coat from off his shoulder and stepped back directly in front of the oncoming train, holding out his coat like a bullfighter flourishing his cape at a charging bull. The driver slammed on the brakes but Hill, in horror, saw the lad looking not at the locomotive but over his shoulder, laughingly acknowledging the cheers of his friends. He left it too late to jump clear and was killed on the spot.

The engine-driver was charged with manslaughter and Hill was asked to present himself at the judge's office at Valverde del Camino the next morning to assist in the inquiry. When he presented himself at ten-fifteen he was shown into a room containing the judge and his clerk, and then completely ignored for three hours. At last he was questioned and at the end of the interrogation was told to sign the clerk's notes of the judge's questions and his own answers. Hill asked if he could read the deposition before he added his name. The judge grew angry but since Hill persisted he was eventually given permission to read his testimony. He found that he was credited with having said that he was in charge of the locomotive and that the engine-driver had obeyed his orders in running down the young workman. A heated argument took place and eventually, line by line, Hill succeeded in getting all his replies to the judge's questions written down correctly. Only then did he sign the deposition. By this time it was evening. Hungry and weary, Hill stood up to go and bade the two law-officers good night. At this the judge called out and two civil guards came into the room. Ordering them to arrest Hill, the magistrate told the Englishman that he was committed to Valverde gaol. He gave no reason though Hill assumed, since he had called the judge a swindler, that it was for contempt of court.

At the prison he asked the gaoler to send a telegram on his behalf to José María Parejo, a Spanish lawyer who looked after Rio Tinto Company interests in Huelva. When this had been despatched the gaoler offered Hill dinner and a clean bed for the night in his house across the street. Next morning, while still in bed, Hill was wakened by the gaoler who asked him to return to his cell as quickly as possible as the judge was about to visit him there. He had just got back when the judge, gracious and apologetic, appeared with Parejo and told Hill that he had not realized who he was and that the whole business had been an unfortunate mistake. A few minutes later over breakfast Parejo explained to Hill that the judge had wanted to demonstrate his power by making difficulties for the first Englishman to appear in front of him, thus giving the company notice that, if it wanted no further problems with the Valverde court, it should adopt the customary practice of giving the judge his expected gratuity. The company decided to bow to tradition and neither Hill nor the engine-driver heard any more of the matter.

In the years for which Alexander Hill recorded his impressions of life at Rio Tinto (that is, 1873–9) there were great differences compared with the later part of the century. One of these was in the proportion of Spaniards employed as senior staff members. The staff list for 1877, for example, shows a total of fifty 'principal employés' in Spain, of whom twenty-nine were British, three were German and eighteen (approximately one-third) were Spanish.*[2]

The most senior positions of all, the headships of the departments, were with one exception held by the British. But under the departmental heads the Spaniards and the non-Spaniards worked alongside each other, the technical jobs being done in general by the British and the Germans, and the administrative posts being filled by the Spaniards. There was, however, one Spanish engineer employed as head of the maintenance department. Interestingly, there does not appear to have been, in 1877, any significant difference in pay between similar jobs, whatever the nationality of the men doing them. Thus in the mines' accounts' department there were eight clerks, of whom four were British and four Spanish. The British received respectively as annual salaries £176, £150, £125 and £100, while one Spaniard was paid £150 per annum and the other three £125 each, and all eight occupied company accommodation. The situation was similar in other departments, though for those not living in company houses a rent allowance was paid of between 100 and 166 *reales* a month, with a no noticeable dissimilarity between what was allowed to Spanish and British staff. (The British bachelors would undoubtedly have been better off than any of the married Spaniards on the staff receiving similar pay because they had no families to maintain. But that was incidental.)

Twenty-five years later, however, the picture had completely changed. In June 1902 the company's principal employees in Spain consisted of sixty-four British staff and only six Spaniards, and though the latter were highly paid they were all specialists, such as lawyers and doctors. There was not at that time one senior Spanish clerk in any of the mines' departments.[3] While this process of replacing Spaniards by British staff was in progress, marked changes were taking place in the British community which effectively isolated its members from much that was happening around them, and these were responsible

*The Huelva branch of the company (comprising the general manager's office, the traffic department and the pier) employed fourteen of the fifty members of the staff: eight British and six Spaniards. The rest were at the mines.

for a great deal of the resentment which the Spaniards at Rio Tinto came to feel.

The reasons for the changes were many and are difficult to analyse with certainty, but there can be no doubt that during the 1880s and 1890s the British colony at Rio Tinto became completely transformed. Much of the metamorphosis was undoubtedly connected with the arrival of a large number of married men, accompanied by their wives, and the building of Bella Vista village to accommodate them. Of course, in the nature of things it was impossible for any of the British ever to feel that they belonged in Spain. There were always people coming from and returning to England, children being sent home to school, eagerly-awaited letters arriving from relatives and, above all, dreams of retirement to the peace of the green English countryside, to remind them of where their hearts really lay. That the large gabled houses built for them in Bella Vista should be virtually replicas of the Victorian villas then being built in places such as Aldershot and

*An unidentified member of the British staff at Rio Tinto, circa 1885.*

Camberley, save that Spanish stone replaced English brick and that lemon trees were more common in their gardens than laburnums, was perhaps not so odd as it may seem to us separated from them by this distance of time, for the architecture of Bella Vista may well have helped to provide an antidote to homesickness.

Nonetheless, many of the British staff spent most of their working lives at Rio Tinto. Often their children were born there and some of these intermarried and stayed on, the young men joining the company as their fathers had done, the young wives setting up home and caring for their families much as their mothers had done. For these Rio Tinto became the centre of their universe even if Spain did not evoke in them the same sentiments as it did in its own people. And so Bella Vista came, in the closing years of the nineteenth century and at the beginning of the twentieth, to have an ethos of its own, completely unlike that of any Spanish *pueblo* and equally dissimilar to that of any English village of comparable size. If anything, its ethos was more akin to that to be found in the British communities that were growing up at the same time in the colonies of the empire.

*The wife of a member of the British staff in the grounds of the casa Colón, Huelva, 1894.*

Certainly, the faded photographs that survive from that period of Rio Tinto history are more evocative of the British raj of imperial India than of middle-class England: the men in their white sun-helmets and jodhpurs, mounted on well-groomed horses, or striding out from the *club inglés* in white flannels onto the scorched-earth cricket pitch, with the hills of the *sierra* as a dramatic back-drop; the women strolling under the palm-trees of Huelva or Punta Umbría, where they went in the hot season with their children, dressed in the same voluminous silk dresses which they would have worn in England, making no concession to the Spanish sunshine apart from their parasols, and taking tea on lawns never quite as smooth and as close-cropped as those of home.

As in the Indian cantonments, much of the life of the new community centred on the club, with its well-equipped billiard-room, its tea-tables, its sports' cups and trophies, its large engraved portrait of Queen Victoria and the back-numbers of the *Times*, the *Morning Post* and the *Illustrated London News*. The *club inglés* stood (and stands still) close to the *casa grande* where the general manager resided with his family in a style appropriate to a colonial governor. Normally he remained aloof from much of the social life of Bella Vista and on those occasions when he attended communal functions he was treated with more reverence than familiarity. Responsible for the entire administration of the mines and of the welfare of all the people inside the Rio Tinto boundaries, his word had almost the force of law. His displeasure was a cause for dismay; his approbation was essential for advancement. Though surrounded by his aides in the persons of the departmental managers, he bore the burden of knowing that in the last resort it was his ability, or inability, to make decisions which controlled in great measure the fortunes of both the British colony and of the thousands of Spaniards who served the mines and their economy.

That life at Bella Vista came to have a quality similar to that of the stations of British India, though without the uniforms and the military splendour, is not surprising. The men who went to Rio Tinto in the 1880s and 1890s, and the wives who accompanied them, were of similar stock and outlook to most of those who were going out to administer the empire. Herein lies one reason for the transformation of the British colony at Rio Tinto in those years. For these men were drawn from the newly prosperous middle-classes of Britain for whom empire and overseas enterprise provided novel and ever increasing work opportunities. They were educated specifically for these by the British public-school system which had sprung into existence in the mid-nineteenth century to cater for the educational needs of the more increasingly affluent middle sections of British society.[4]

This system, modelled so painstakingly on that established at Rugby by the great Thomas Arnold, was intended to produce Christian gentlemen, self-disciplined and self-reliant, capable of ruling over, and of setting the finest examples for, 'the lesser breeds without the law' of whom the empire was so largely composed. The tradition which the schools created was in many respects a fine one, but there were accompanying defects one of which was that many of the pupils that emerged from the public schools in those early years of empire saw in themselves and their fellows an élite, trained (or even born) to rule, set apart from the generality of mankind, both by their experiences and by their acquired outlook on life.

And, in the same way that the flood of former public-schoolboys into the Indian administration corresponded with a widening gulf between British rulers and Indian subjects in the 1880s and 1890s, so the growing numbers of their compeers at Rio Tinto coincided with the growth of an apparently impassable void between the *colonia inglesa* and the Spaniards. It is indicative of the lack of sympathy between the two communities that by 1900 the word 'native' was often applied at Rio Tinto, not in its simple sense of one who belongs naturally to a place, but with the new colloquial connotation, which first came into use in the colonies, of an inferior coloured person.[5]

The growing separation between the two communities, which was almost entirely the responsibility of the British staff, was re-inforced by some of the customs which they took with them to Spain.

*Queen Victoria's birthday celebrations in Bella Vista: the egg and spoon race. (Photo circa 1895.)*

The union jack was flown from the company office on the king of Spain's birthday, but public celebrations were reserved for the anniversary of the birth of Queen Victoria on 24 May. It must have been incomprehensible to the Spanish workmen that on Britain's empire day they were given a holiday and provided with feasting and sports, while the town was bedecked with British flags; and that the *alcalde* and his colleagues from the town council called at the company office to convey their congratulations to the general manager and to drink with him the British loyal toast. Hardly less puzzled must have been any Spanish traveller who, wishing to travel by the Rio Tinto railway, consulted its time-table and saw the footnote: 'Trains do not run on Sundays or on Queen Victoria's birthday.'

But the photo-albums and the programmes decorated with union jacks, which were printed for the various festivities that punctuated life at Bella Vista, show only one aspect of British society at the mines during the high-Victorian period. For life was not simply golden sunsets and croquet and balls at the *casa grande*. The village had to be brought into existence quite literally from nothing. Jobs had to be done and life made bearable in strange surroundings. The members of the British community had always to remember that they constituted a highly privileged minority in a foreign land where their actions were liable to be regarded with suspicion. And in the process of creating the British colony there were strains and social problems, often of great complexity, to be solved.

The turning point in time between the two types of British colony came in the year 1879 when Mark Carr resigned as general manager and was succeeded by Charles Prebble. The board of directors decided that Prebble should set up house at the mines in view of the expanding mining projects, and control at Huelva was transferred to a commercial manager. Prebble decided not to live in the *pueblo* but to have his house built almost two miles away on top of a hill of ancient slag with a fine view (*una bella vista*) of the *sierra*. The board then decided that the hilltop would make an excellent site for building a village for the British families that were beginning to arrive at the mines as the pace of development increased.[6] Gradually the *colonia inglesa*, including the bachelors, was rehoused, and the Spanish hilltop converted into a quasi-English village. Bella Vista, despite its fine views, was not built on particularly fertile ground but the British applied themselves to gardening with their usual remarkable devotion and gradually it became tree-shaded and bright with flowers. (However, so poor was the soil that it was not until 1933 that anyone had success in cultivating a decent lawn that survived the fierce Andalusian summer. This prodigious feat was achieved in the garden of number 12.[7])

Thus began the physical separation of the British and Spanish communities, although reasons for the directors' decision to isolate the homes of the British staff are nowhere stated. Presumably the intention was to provide the colony, for the sake of its womenfolk in particular, with superior and more spacious housing than could be made available in the mines' town itself. But, as Bella Vista increased in size with the arrival of new staff members and their families, so the social inhibitions and restrictions of middle-class Victorian Britain became more firmly established, and even magnified. Many of the newcomers were Scottish, with emphatic views as to what constituted acceptable social behaviour and a strong sense of hierarchy. With a marked tendency, instilled by kirk and school, to receive uncritically the decisions of those set in authority over them, they normally accepted uncomplainingly the decisions of the general manager and developed an intense loyalty to their company. That this was so was probably fortunate, for it helped to give them a self-discipline with which to face the problems that came with living in a small and insular community.

Where self-discipline was lacking it was a cause for disapproving gossip and expressions of hope that the general manager would enforce it on the racalcitrant. Those members of the staff who became too addicted to the bottle, apparently a particularly easy fault to acquire in such a society as Bella Vista, once their vice was made known to the general manager were immediately sent packing. Insubordination had a similar result.[8] The least suspicion of dishonesty led to a speedy interview with the general manager and the humiliation of instantaneous dismissal. A storekeeper whose keys were apparently used for theft was dismissed on the spot. Book-keepers who indulged in fraudulent activities were on the train for Huelva with their families and en route for England within minutes of being on the carpet.[9]

At times some unmarried members of the British staff 'went

native', to use the disapproving Bella Vista phrase, and fell in love with, and frequently married, Spanish girls. But this, however natural it may seem now (for the girls of this region have long had a reputation not only for great beauty but also for good housewifery), was then a source of criticism and even alarm. To the British, and perhaps more especially the women, it implied a rejection of both the conventions which they regarded as essential for their group and of them themselves, for it was accepted that no member of the British staff who married a Spanish girl could live in Bella Vista but had to set up house in one of the Spanish *pueblos*; and even though the husband might continue to receive invitations to social functions his wife would not be asked to accompany him. Inevitably such married couples, if they stayed on at the mines, became absorbed into the Spanish community and their children were raised as Spaniards.

The general manager normally saw it as his duty to try and dissuade any member of his staff who was contemplating such a marriage, not only because of the social problems it would cause, but because of the probable difficulties it would lead to at work. For such men often found their new Spanish relatives, who were usually employees of the company, asking for favours or for special treatment. The Spanish tradition which then existed, of looking after one's relatives and favouring them at work, could place a member of the British managerial staff who married into a Spanish family under considerable strain, and perhaps make it impossible for him to maintain discipline. If this happened he was told to resign.[10]

There were occasional scandals, too, when British employees took Spanish girls as paramours. One such case in 1886 caused considerable agitation and resentment in Bella Vista. The wife of a senior manager left him because of his flagrant misconduct with one of their Spanish housemaids, and he openly installed the girl in his wife's place. Ignoring the snubs which he and the young woman received from the British womenfolk, and rejecting the requests of his colleagues to turn the girl out, he was summoned to appear before the general manager who appealed to him to move house into a Spanish village. He refused. The directors considered the matter at a board meeting in London but hesitated to dismiss him because of his value to the company. They told the general manager to call a meeting of the British staff to vote on a resolution, which they were first to have the chance to talk over with their wives, calling on the man to leave Bella Vista. The motion was carried unanimously but still he refused to behave as expected of an English gentleman. The problem was resolved only when he obtained a post with another company outside Spain and deserted the girl, leaving her to return to her own village to give birth to the baby she was expecting.[11]

In 1881 Heinrich Doetsch, the company director who lived in Huelva, drew Hugh Matheson's attention to the existence of an excellent beach near Huelva and close to the small and poor fishing village of Punta Umbría. Doetsch's partner, Sundheim, had already erected a bungalow there for himself and the former suggested that the company might like to turn the beach into a convalescent resort for those members of the non-Spanish staff and their families who were recovering from illness. He thought it would be a particularly good place for those suffering from malaria and from chest complaints due to smoke from the open-air *teleras*. Matheson suspended one of his frequent arguments with Doetsch, over the commission which the German director was making on his monopoly on the sale of foodstuffs and wines to the mines, for long enough to give him a contract to build and furnish a number of bungalows which were brought into use in June of the next year.[12]

Because Punta Umbría was intended as a sanatorium it came under the supervision of the company's medical department at Huelva, the head of which drew up the rules for its use. By 1892, however, the company's village there had become largely a weekend and holiday resort for the British staff, though visitors were still subject to the rules

*The beach at Punta Umbría*

of the medical department which were printed and deposited in each bungalow.[13] By these, male servants were completely excluded from Punta, and all the Spanish maids accompanying British families had to reside together in one of the bungalows and were not allowed out at night save for attending their mistresses if the latter were ill, in which case certificates had to be obtained in advance from the medical officer for the maids to show to a company watchman if challenged.

Other rules provided for a weekly inspection of the houses by a member of the medical department to ensure that a proper standard of cleanliness was being maintained, and governed the use of the sailing-boat which was available for visitors. Presumably this was intended not only for general recreation but also for duck shooting in the nearby marshlands since one stipulation was that 'No guns must be taken if ladies are on board.' By the medical department's rules heads of families were held responsible for the conduct of their wives and children: 'Persons should not be allowed to promenade the platforms of the houses, hold musical entertainments, nor engage in any occupation in the houses by which noise is caused, after eleven at night.' (Doubtless a reasonable provision as far as the early-to-bed British were concerned, but one that must have been wearisome to any maidservants who liked to follow the Spanish customs of eating late and of not getting to bed until the small hours.)*

Spanish maidservants were admitted to the company resort since they were very necessary to do the shopping in Huelva, for there were no shops in Punta and the only way to get food was to take the company's steam-launch which left for Huelva at 7 a.m. each day and returned from the company pier there at 10 a.m.; but Spanish wives of British employees were not accepted for some years. Occasional attempts were made by the general managers to get the board to alter this rule, but it was not until 1896 that the first Spanish wife was admitted when the board decided: 'Mrs Symes (Spanish), wife of F. K. Symes, clerk in the mines' accounts, to be allowed Punta privileges of English staff.'[15] This concession to Mrs Symes, fourteen years after her marriage (which had taken place in 1882, the same year that the resort was started), was the beginning of a change of policy. From then on it was easier for a Spanish wife to get to Punta with her British husband, although each application had still to be considered by the board and often permission continued to be refused.

This state of isolation, and the restricted social life which was lived by the members of the British community in the last twenty years of the century, seems largely to have been of their own choosing. The board never issued instructions forbidding social intercourse with the Spaniards. The directors, whatever their own inclinations may

*The company purchased the part of the sea-shore at Punta Umbría on which its dozen bungalows stood in 1896. In 1917 it bought more land and built more houses, and also acquired *La Peña* hotel nearby for conversion to apartments to provide extra holiday accommodation for British members of the staff who were unable to take their home leave during the first world war.[14]

have been on the subject, restricted themselves to endorsing the prohibitions which the majority of the members of the *colonia inglesa* or its spokesman, the general manager, requested. Even if the directors sympathized with the majority attitude they could hardly have been blamed for it.

This policy of withdrawal and exclusion, however, not only helped to stiffen British backbones and upper lips; it also carried with it certain factors which at times helped to undermine the happiness of the community which the policy was designed to create and preserve. Inside the walls of the village minor disagreements were often magnified out of all proportion, and an argument between two Bella Vista wives, when it reached the stage when they decided to cut each other, could cause extreme embarrassment in the community. Innocent friends could become embroiled in the petty squabble, and tea parties and conversaziones could be wrecked by temperamental outbursts if the adversaries were invited.

There were also the dangers of boredom and the development of marked eccentricities, although some oddities of behaviour were perhaps accepted more readily at the mines than they would have been in an English village, at least until such time as they outraged the moral sense of the community. An English doctor at Rio Tinto in the 1890s, for example, provided a fruitful source of gossip by his periodic disappearances for three or four days at a time; all the more so since nobody could find out where he went or what he did. But the sudden critical illness and death of a member of the British staff during one of these absences, at a time when the doctor should have been available for duty, produced a wave of anger and social ostracism from which the doctor was freed only by his dismissal by the board a month later. The directors noted when making the decision to discharge him that no members of the staff or their wives were prepared to have dealings with this physician in future.[16]

Another doctor who caused considerable controversy was Ernest Morrison, an Australian who was later to achieve fame as the Peking special correspondent of the London *Times* during the Boxer rebellion. When Morrison was appointed as assistant medical officer at the mines in 1888 Hugh Matheson (according to the doctor) expressed the hope that in addition to his performing his duties he would also 'help bring the word of Grace to the heathen in Rio Tinto.' But on his arrival Morrison quickly acquired a reputation for promiscuity with young Spanish women and almost ostentatiously absented himself from the British prayer-meetings and social gatherings at which his presence was expected. Not unnaturally, his colleagues were hesitant to let him attend to their wives if they were unwell. (Though they needed to have had no fears: the young doctor found the British women uncongenial.) Added to all this he easily lost his temper and refused to show the usual deference of the British staff to their superiors.

In 1889 he had a head-on encounter with the company hierarchy over the provision of drugs for the hospital. Originally these had been purchased in Spain but Doctor J. Sutherland Mackay, the company's chief medical officer, and Doctor Jesús Alonso, the principal assistant medical officer, had discovered that the Spanish supplier (a Seville businessman) was adulterating the supplies to increase his profits. The company therefore had cancelled the contract with this man and given it to another Spaniard on condition that all supplies, despite the extra cost, were bought in England and shipped to Spain. On arrival at Rio Tinto, Morrison became friendly with the former supplier and was convinced by this man that the company was paying unnecessarily high prices for its medical supplies with much of the purchase money finding its way into the pockets of company officials. He saw the general manager, William Rich, and accused a number of employees of defrauding the medical service, proposing a restoration of the contract to the former supplier.

Rich interviewed those involved and instructed the mines' accountants to investigate the hospital's accounts. Morrison, who was not content to await the outcome of the enquiry, sent a rashly-worded letter to Hugh Matheson in London, accusing Rich and a number of the hospital officials of stealing company money. In further letters he accused Rich, Alonso and others of being liars, and told Matheson that he was failing in his duty as chairman to protect the shareholders' money and the health of the employees. Rich's report, when completed, completely exonerated the hospital employees of fraud, and found that the funds had not been abused although there was a sum of slightly less than £100 which could not be accounted for in the accounts of the druggist. (The board of directors decided that henceforth the druggist should no longer keep his own accounts but that his books should be kept by the accounts' department.)

A letter was sent to Morrison informing him that the board accepted the general manager's report in view of the lack of evidence to substantiate Morrison's charges and censured the young doctor for the 'violence of your language and the baseless allegations you have made'. He was instructed to send no more such letters to the chairman. On receipt of the reprimand he immediately resigned. Later he wrote:

'I had discovered an extensive series of frauds in the druggist department involving several thousands of pounds. On sending the company a report on this fraud, I received an astonishing letter . . . expressing the gravest disapproval of the directors at the unbecoming language in which my report was couched. I resigned two hours after getting the letter. It appears that the frauds I discovered were insignificant compared with other frauds that soon after came to light, of which even some of the directors who had admonished me had not been guiltless.'[16]

There seems little doubt that Morrison had his facts wrong. It seems hardly likely that Sutherland Mackay and Jesús Alonso, together with their colleagues who created the highly-praised Rio Tinto medical service, were the thieves that Morrison believed them to be.* (Nor is it possible to discover what Morrison meant by 'other frauds that soon after came to light', unless he was referring to the arguments at board level about the arrangements some of the directors had obtained whereby other companies in which they held interests had been given contracts to act as the Rio Tinto Company's agents for sales and purchases at rates of commission that yielded big profits.) The storm would appear very largely to have been caused by those differences of personality and temperament which so easily divided the British community into warring factions, although in this case Morrison constituted a minority of one.

The strain of such a circumscribed life produced not only cases of alcoholism and eccentricity but sometimes mental breakdowns which led to a man or his wife having to be sent home. On occasions the board was even asked to note than an employee had quite literally run away from Rio Tinto and taken a boat for England without staying to tender a written resignation.[17] At other times the pressure of life was such that there was a sudden rash of resignations sent in by people who felt they could take no more of Rio Tinto, and these were frequently accompanied by strongly-worded accusations against the senior staff of dictatorial attitudes and complaints against the general manager, or the directors themselves, of failing to make proper provision for the staff's welfare and security. These desertions and resignations normally occurred when industrial unrest was intensified and added to the problems of life at the mines. It was no coincidence that following the 1888 riots sixteen of the British staff resigned simultaneously, almost a third of the total.[18]

As the nature of the British colony at Rio Tinto started to change in the early 1880s, so the matter of religious observances began to occupy Hugh Matheson's mind. Between 1873 and 1882 there was no regular religious provision made for the British staff. The archdeacon of Gibraltar made occasional visits to the mines at the company's invitation and administered the holy communion to the Anglican members of the staff. Matheson himself also conducted Sunday services when he was at Rio Tinto, in his capacity as a presbyterian

*The work done by these men is described in chapter 13.

lay-preacher, which all staff were expected to attend. At these his rousing 'hell-fire' sermons would last two hours or more, to the annoyance of at least the younger members of the staff. But the establishment of Bella Vista, and the need for a formal religious structure which would help the British community to maintain its self-discipline, led Matheson to propose the appointment, in 1882, of a full-time chaplain. With the agreement of the board the appointment was left to the chairman's discretion and Matheson appointed David MacDonald, a Scottish presbyterian minister.[19]

Since the majority of the mines' staff and their families were Scottish people and of the presbyterian persuasion this was natural enough, although some of the English members of the community found that the services were uncongenial, and preferred to await the visits of the archdeacon before making their religious observances. There appears to have been a short period, covering the 1880s, when the colony consciously felt itself internally divided into English and Scots. This was emphasised by the religious division but this was in reality not as important as the social cleavage, for as the Scottish peopled arrived in increasing numbers they brought with them stricter conventional attitudes and a more puritanical approach to life than those of the Englishmen already at Rio Tinto. In particular the influence of the presbyterian minister, in his effort to halt the back-slidings of the community, was so all-pervasive that it caused feelings of resentment amongst some of the English who disagreed with him as to what constituted sin and self-control. There was a belief, too, that the watchful eye of this representative of the Scottish kirk led to the display of a certain amount of religious zeal in public which sometimes served to conceal private inadequacies. In fact some of the English were able to accuse the Scots of hypocrisy.[20]

One newly arrived eighteen year old was invited by his departmental superior, a middle-aged Scotsman who was prominent in the religious life of the mines and one of the elders of its kirk, to accompany him on an evening tour of Huelva. Surprised to see that the man, who had a reputation as a strict teetotaller, consumed great quantities of alcohol as the night wore on, the boy was relieved when told that the time had arrived to visit an old friend. They went to a house where the Scotsman was warmly embraced by an enormous fat woman and several girls. After drinks had been taken, the older man announced that he was going upstairs for a while and that the boy was to wait for him and 'enjoy himself'. No sooner had he disappeared from sight than one of the girls climbed onto the boy's knee as he sat in a chair and started to embrace him. Unable to speak Spanish he was reduced merely to pushing her off his lap but immediately the fat woman smilingly brought another girl up to him. It suddenly dawned on the horrified boy that the house he was in must be a brothel, an institution about which he had been given the direst warnings. Wondering if his out-of-sight superior was aware of this, and praying silently for his quick return, the young man felt obliged to wait as he had been instructed. With growing apprehension he shook his head as the lady of the house paraded girl after girl in front of him. The woman's smile had changed to a frown, but suddenly a beam of comprehension illuminated her face. She fired off instructions at a couple of the girls and they ran out into the street.

For a few brief minutes the lad was undisturbed and he was just thinking that his ordeal must be over when the two girls returned, bringing with them half-a-dozen boys whom the woman pushed forward for his inspection. Horrified, he jumped up and began shouting in what he hoped would be understood as insulting English. The boys were sent packing and he was left sitting alone for the rest of the night until, with the dawn, the Scotsman appeared on the stairs, dishevelled and with his arm around a girl. Sadly the young man realized that his boss knew very well the sort of house which they were in.[21]

In 1891 the decision was taken to provide a church at Bella Vista. Previously the presbyterian services had been held in one of the schools, but there was plainly a need by that time for a recognizable spiritual centre for the community. In consequence the board authorized the building of a church, provided that the cost was no more than £1,150 and that 'the work is carried out without any demonstration or ostentation.'[22] When it was erected the building resembled both externally and internally a Scottish presbyterian kirk. With the church built regular church-going, by the Scots at least, became a feature in the life of Bella Vista, though even in this activity the congregation could not forget the company, for seating was to a great extent allocated in accordance with work-status, the general manager occupying the front pew with his family and being the first to leave while the rest of the congregation held back to permit an unimpeded passage.[23]*

At the end of the village a burial ground was laid out for the community to provide the last resting place for those who would never make the journey home. The English cemetery was surprisingly un-Victorian. A small garden, set around with high plant-covered walls, full of flowering shrubs and brilliant blossoms between the gravestones, it was (and is even now, for it is still lovingly tended by an aged Spanish gardener) a place of poignant beauty. Much of the social history of British Rio Tinto is contained within

*In 1926 the presbyterian chaplain was replaced by a church of England priest and services henceforward were taken from the English prayer book. Now that Bella Vista is occupied almost entirely by Spanish families the church has become a Roman catholic chapel of ease.

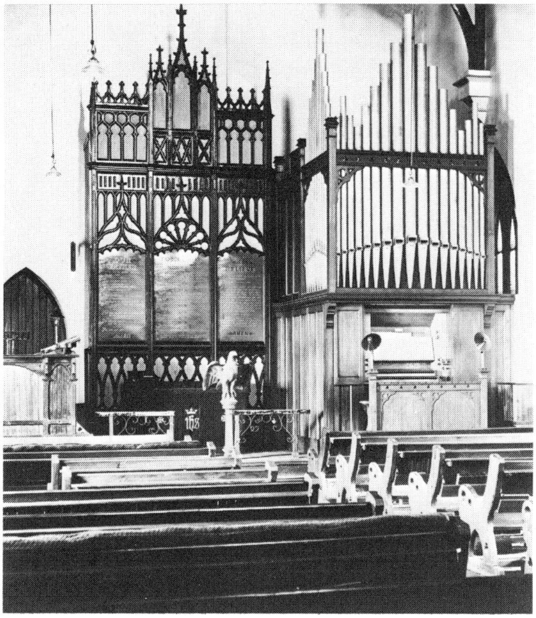

ABOVE:
*The church at Bella Vista: exterior.*

ABOVE RIGHT:
*The church at Bella Vista: interior.*

those walls and recorded in the stone and marble monuments. For here lie the mortal remains of young women who died in childbirth, and young men who achieved heroes' deaths rescuing their workmen when disaster struck underground. Here, too, are the graves of the middle-aged and elderly who were unable to accomplish their dreams of spending their last days in the familiar surroundings of home. And, perhaps, most moving of all, the memorials over the little graves of children killed by the scourges of typhoid and cholera.

Buried there, too, are some of the protestant Spanish. It was an unforeseen product of building the church at Bella Vista that its simple ritual and plain interior should attract some of the Spaniards away from the elaborate ceremonial and gorgeous furnishings of their own churches. Those of the British who regarded their small church simply as a part of their social life, attending perhaps merely so that they should be seen by the general manager or their subordinates, could not understand why some of the Spanish people sought permission to participate in its services. In fact there were a number of Spaniards who were deeply religious people but had an aversion to Roman catholicism springing from the continuous conflict in nineteenth-century Spain between the liberal political parties and the hierarchy of the Roman catholic church. Doubtless there were some who wished merely to enhance their status in the eyes of their own people and at the same time to identify themselves with the British managers, or else to seize the one chance that permitted them to cross the rigid social barriers which otherwise separated them from

*A picnic excursion into the sierra: unidentified British couple from Rio Tinto.*

life at Bella Vista. As far as possible, however, the company's general managers discouraged these applications from Spaniards. They did not wish to provoke the hostility of the local Spanish clergy. Yet Matheson was a true son of the Scottish reformation, and felt considerable sympathy for any Spaniards who were apparently anxious to join a reformed religion. In the event a number of senior Spanish employees found their way into the congregation.*

Health and illness were often matters of over-great concern to the middle-class women of Victorian England, but at Rio Tinto there were good reasons for any apprehensions which the womenfolk felt, though in general the British community was spared most of the epidemics which periodically ravaged the Spanish *pueblos*. When there were exceptionally severe outbreaks of cholera and typhoid at Rio Tinto in the July and August of 1890, which killed many of the Spaniards, the board noted with relief that the British families had remained untouched and recorded 'with pleasure' that the Bella Vista drinking water was abundant and pure. The smallpox epidemics of 1898 and 1899, which took heavy toll in the Spanish villages, claimed only one British child. The 1900 typhoid outbreak was exceptionally severe but only two of the British community were affected (one of them being the daughter of the British doctor, Russell Ross) and both recovered.[24] But the inevitable annual visitation was awaited by the womenfolk at least with great anxiety, for they knew that death could not always be kept outside Bella Vista's walls. It was at times such as these that the British had most cause to be thankful for their isolation, for it improved their chance of emerging unscathed through the ordeal of the summer months.

Their separate village, too, gave them some protection from the violence that was rife in the Spanish community around them, although occasionally the turbulence intruded into Bella Vista, causing alarm and concern and underlining the insecurity felt by the British in a country where political and social controversy was often accompanied by a resort to force. But, despite the fears and anxieties, whenever there was a strike or a riot among the workmen there were never any British casualties. At times of disorder, of course, the civil guards and the company's armed watchmen protected Bella Vista from possible demonstrators; but the trade union leaders of those years took vigorous steps to prevent violence against the British, even though they were given little credit by the company for their efforts.

Despite this concern of the more responsible Spanish workmen to protect the British during periods of strife, there were still sudden and unexpected attacks at other times on individual members of the British staff which plunged the *colonia inglesa* into gloom and anxiety.

*In time those Spaniards who rejected their own church became largely organized into a protestant episcopal church with its bishops consecrated by the bishops of the church of Ireland. That they were particularly strong in numbers in Huelva province was no doubt due to the presence of the non-Roman catholic churches provided by the foreign mining companies. Though officially suppressed at various times by government authority, the Spanish protestants have now been allowed to reorganize themselves under the law of religious freedom, and on 1 January 1969 a protestant bishop of Huelva was installed in his pro-cathedral.

*Rio Tinto polo ground, circa 1885.*

These were normally surprise assaults by individual workmen harbouring grudges against specific managers; and infrequent though these were, for they occurred on average only once in three or four years, their effect was unsettling on the small community. Normally they took the form of stabbings, though on two occasions pistols were used. None of the British employees was ever killed as a result, but there were some bad cases of wounding which necessitated lengthy periods of sick leave of sometimes as much as a year. Occasionally, too, there were unexplained explosions, though these did little damage beyond producing nervous flutters among the British womenfolk.[25]

Thus by the end of the nineteenth century the British colony, and its social organization as expressed by Bella Vista, had taken root. Its way of life, its attitudes and customs, strange though these may have been when contrasted with the normal standards of either Britain or of Spain, had been developed and were to persist as a powerful influence to mould and to shape the lives not only of those arriving from Victorian England, but those who were still to come in after years. When the company later decided to erect additional accommodation for newcomers, as the size of the British staff gradually grew (it reached 124 in 1912), it built the new homes not in Bella Vista but as close as possible to the various works' departments: a grim block of twelve flats contemplating the railway yard and caustically named *Vista Alegre* ('joyful view'); additional houses close to the smelter; and, when the concentrator was built, more houses in the nearby village of Naya. Yet the Bella Vista ethos went with the colonists to these new villages-in-miniature. For with all its faults and drawbacks it provided some sort of security, some source of moral strength on which they could draw when faced with the tribulations of life in an alien land.

*The making of the south lode open-cast pit: watercolour by Malcolm Graham, 1876.*

*A typical family group of circa 1880. The miners and their families in this photograph had migrated to Rio Tinto from Portugal.*

13

# *Los mohinos*

## The Spanish community: 1873–1914

The people of Andalusia have a certain facility for inventing nicknames both individual and generic. The mustachioed Gordon Douglas, who became surveyor of the north lode open-cast in 1881 and rode like the wind through Rio Tinto on a jet-black horse, was dubbed by the miners *Don Diablo* ('Sir Devil') while one of his colleagues became known, less respectfully, as *el chivo sin madre* ('the kid with no mother') on account of his constantly lost and quizzical look. For a century or more the people of Rio Tinto have called their Nerva neighbours *los chorizos* ('the pork sausages'), a name which presumably originally referred both to the physical characteristics of the nearby villagers and to the pig-farming and its associated activities which used to be an important element in the local economy.

It was during the late nineteenth century that the inhabitants of Rio Tinto acquired from the people of the *pueblos* around the appellation of *los mohinos*. The word *mohino* has no simple English equivalent, for it embodies a variety of meanings and a diversity of connotation: a show-off; one who stands alone against all comers; full of animosity; obstinate; begotten of a she-ass and a stallion; peevish; sad. The modern residents at the mines no more merit the nickname than the present inhabitants of Nerva deserve theirs. But the misnomer of today was in some respects strikingly appropriate for the Rio Tinto miners of the late nineteenth century, and it provides an expressive reminder of the way they were regarded by the farming folk of the *sierra*.

Certainly the mineworkers felt themselves to be almost a race apart from their neighbours. The normal Andalusian village consisted of little more than a couple of hundred people, normally poor peasant families, scratching a precarious living from the thin soil of the hills. The mining villages of Rio Tinto, however, held large numbers of men who worked in steady employment for regular wages, existing in a cash economy and enjoying a standard of living far superior to that of the agricultural labourers about them. They had a pride in their work-skills and a certain arrogance derived from their knowledge that they constituted an occupational *élite* among the working-class of southern Spain. It was natural for them to feel scorn for the submissiveness with which the peasantry of the region

OPPOSITE PAGE
*Nerva.*

accepted the abuses of authority and resigned itself to economic and social drudgery. They felt themselves superior to their neighbours and they showed it.

Any self-doubts they may have had about their primacy arose from the problems and strains that were a concomitant of their working lives. For they had a fierce resentment of the industrial discipline imposed by their British employers and an intense anger at the squalor of their environment. The continual frustration of their political aspirations, too, was a cause of constant vituperation against the company and the government. And it was difficult for them not to feel the insecurity which went with the daily possibility of sudden death or accident in the mines. It may have been their subconscious efforts to drown these fears that gave them yet another of their characteristics, an enthusiasm for life which appeared to many of the British to be an extravagance in living.

Thus the Spaniards at Rio Tinto came to form a society apart and distinct from that of their fellow-countrymen in the hills around. They were separated from them by their conventions and prejudices just as the British in Bella Vista were cut off from the mineworkers by theirs. As the *colonia inglesa* was untypical of England, so was the Spanish community untypical of Spain. One reason for this was the population explosion that took place at the mines in the years following the Rio Tinto Company's purchase. At the end of 1873 there were just under 1,000 employees, living mainly in Rio Tinto and at Nerva just outside the mines' boundaries.* By 1882 the number of mineworkers had grown to about 10,000. It stayed around this figure until 1905 when it started to climb yet again. By 1909 it had reached almost 17,000.†[1]

This large scale immigration of workmen swamped the old-established mining villages and overflowed into a number of the farming *pueblos* of the area. By 1879 the mines' employees were to be found in eleven other villages nearby: Zalamea, el Campillo, las Delgadas, Montesorromero, el Ventoso, Traslasierra, Campofrio, Membrillo Alto, Membrillo Bajo, Marigenta and Madroño. Most of the new recruits at first had to live in lodgings outside the company's control, for although by 1879 the house-building programme was well under way, the company then still owned only 841 houses, of which 778 were within the mines' boundaries and 63 were outside.[2]

Only a small proportion of the thousands of employees recruited by the company between 1873 and 1882 could come from the villages around the mines, for these were too small to provide the numbers needed. The great majority of the new men walked to Rio Tinto from other parts of the Iberian peninsula and particularly from the

*At that time Rio Tinto and Nerva both had populations of about 2,000 apiece.

†Thereafter it declined, reaching 10,000 in 1920.

coal mining region of the Asturias where there was a high level of unemployment. Most of the new arrivals were young and unmarried. Often groups of brothers arrived together, such as the five Vasquez boys from Solveira in Orense who tramped the 300 miles to Rio Tinto in 1886.[3] All five spent their entire working lives at the mines but only two were able to find themselves wives, for there was a disproportionately high number of young males within the growing community and a lack of young women. This great influx of young men gave the mines' villages something of the appearance and characteristics of mining camps and was accompanied by a number of social problems, in particular excessive drinking, gambling and violence. Difficulties arose, too, with the arrival of prostitutes in considerable numbers.

This lack of balance between the sexes started to decline as soon as the company was able to provide housing on some scale, but a

*Spanish ladies in the bullring at Zalamea.*

significant adjustment did not take place until the late 1890s and early 1900s. By then any mineworker wanting to marry and have a company house could expect to get one after a wait of a year or two. By that time the new arrivals had put down roots and the early social problems had been largely overcome, but the company was having to face a new difficulty, that of providing for large numbers of old and solitary men without families to care for them in the time of their infirmity.

Right down until the time of the first world war the mines' population bore the marks of an immigrant people, with its appearance, its speech and its customs testifying to the fact that its individual members were as much strangers to Rio Tinto as the British in Bella Vista. The lack of social cohesiveness passed only gradually, and only as new generations were born which were able to regard Rio Tinto as home and to create settled communities out of the confusion of the mining camps.

The violence that was rife in the Spanish community in the last quarter of the nineteenth century owed much to the poverty of the surroundings and the rootlessness of the people. Fights were a frequent occurrence in the 1870s and 1880s and were often associated with heavy gambling and drinking. Much of the alcohol drunk by the workmen was crude potato-spirit and under its influence men were quick to become quarrelsome and to draw the knives which most of them carried in the sashes they wore around their waists. Stabbings were almost a daily event and soared at weekends, especially in the bull-ring area. When the company acquired the mines there was a ground where bulls were fought just outside the Rio Tinto *pueblo* at the place known as el Coso (a word which means an enclosure for bull-fights) close to the present Coso railway station. In 1879 it was decided to extend the railway line through the site towards el Valle and consequently the company had to offer a new location and compensation to its owner who built his new ring just outside the *pueblo* in what is now the *plaza de España*. It was a fine building; a smaller version of the famous Seville *plaza de toros*, and came into use on 12 August 1882 during scenes of great festivity, with two of the greatest bull-fighters of the time, Currito Cuchares and el Morcilla, fighting at the opening.[4]

The new ring, however, and the houses close to it, quickly became the focal point for much crime and violent behaviour. At the bull-fights, riots between different factions and drunken arguments between individual spectators often terminated in bloodshed.

Prostitutes solicited and fought each other for customers. In the evenings the brothels and taverns adjoining the ring were the scenes

*The Rio Tinto bullring in 1882.*

of frequent stabbings. The death rate from malicious wounding was almost unbelievably high given the size of the local population, and in the autumn of 1883 murders averaged one a week.[5] Two years after the bull-ring had been built the company bought it for 125,000 *pesetas* (£5,200) and pulled it down, despite the protests of some of the British staff who were among the most regular supporters of the bull-fights, in order to put an end to some of the vice and crime associated with it. At the same time a large number of prostitutes and thieves who had flourished there were rounded up and escorted to the Rio Tinto boundary and told not to return if they wished to avoid imprisonment.[6]

Political arguments and fights were also common. The majority of the workers were politically naïve and considered fair game by the anarchist and communist organizers who visited the mines. The workmen had little contact with the outside world save through visitors and the newspapers. Their dissatisfaction with the raw mining villages in which they lived, surrounded by great piles of slag and waste, and frequently enshrouded in the sulphurous smoke from the *teleras*, made them receptive of the more sophisticated arguments that came via the travellers and the printed papers. Many were easily persuaded that a system which apparently produced such living conditions should be overthrown, if necessary by the use of force.

To contain any violence against company property or the British colony, reliance was placed on the company guards and the *guardia civil*. The company watchmen, recognizable from their dark-blue overcoats reaching almost to the ground, with a waist-length blue cape worn over the top, and their peaked caps bearing the letters GM (for *guardia de las minas*) watched over the Huelva pier, the railway, the stores and the administration buildings and provided the guard for Bella Vista village. Normally they carried revolvers and javelins, though at times of unrest rifles were issued to them from the armoury. But despite their weapons the British staff had little faith in their courage, and if there were prospects of serious disorders, as in 1888, the local detachments of the civil guard had to be called in. Partly military, partly police, the *guardia civil* was able to ensure that the behaviour of the mine workers was kept within bounds most of the time, though occasionally they had to summon soldiers as reinforcements.

It was the *guardia civil*, too, that had the responsibility for dealing with ordinary crime at the mines and investigating the frequent rumours of political conspiracy. The civil authorities always regarded the miners with some apprehension because of their easy access to dynamite, and thefts of the explosive always caused considerable concern. Such acts were frequent and it was seldom that six months went by without a report of one being sent to the general manager:

'Re Raimundo Delgado Banda. . . On Sunday morning last at about 8.30 a.m. he was seen in his underground clothes stealing away from the Alfredo mine. He had been seen coming up in the cage at 5.30 a.m. so the shift boss asked him where he had been all the time. He said he had forgotten his *libreta* [a small book in which a man's hours of work were entered] and had gone down again for it. . . . On Monday morning when the miners went to work they found that their boxes had been opened and over 30 kilos of dynamite taken besides 5 or 6 boxes of caps. A search was made and the dynamite found on another floor where this chap was to work this week and hidden away in different places – his intention no doubt to carry it away little by little.'[7]

*company watchman, 1910.*

This incident occurred just before a strike and there was little doubt in the minds of the British staff that this theft was in preparation for an outburst of violence.

Normally such men were prosecuted and sent to gaol if a case could be proved against them. Often, however, it was obvious that the stolen explosive had been taken not for political reasons but for resale. In these cases action stopped at dismissing the men:

'The man who was caught carrying the 56 sticks of dynamite from the quarry last Saturday, Modesto Dominguez Pérez, has been discharged from work. The boss of the quarry who is evidently mixed up in the affair is named Francisco Espanda Malquez. I have told him that he is discharged for at least one year and that the shortness of the period is due solely to his long and previous good service with the company. He has a company house. His eldest son – about 17 years old – works in San Dionisio. If you think well of it I propose that we should allow the son to keep the house, as I believe the house is full of children.'[8]

*Rio Tinto civil guard commander, 1910.*

Dismissal was no easy let-off, however, for there was little work in the Rio Tinto district which a man could obtain outside the company's employ. If he also lost his company house he could suffer a great deal. Such a man paid heavily for his foolishness. A workman from the Alfredo underground workings stole a box of ninety-five detonator caps for blasting stones on a plot of ground which he wished to cultivate at Dehesa village. He gave some to a friend for a similar purpose and the man reported him to the company. Homeless and out of work he tried to get a living by selling wood in Nerva, he and his wife sleeping in the open when they could not afford the price of a bed. Twenty-four months elapsed before the general manager agreed to his re-employment and rehousing.[9] Such men together with others classified as trouble-makers, were placed on either black or red lists kept at the mines' office. The black list was for those who were not to be employed for at least a year and without reference to the general manager; red, for those who would never be employed again under any circumstances. The latter list consisted mainly of those who attacked and wounded their workmates or engaged in disruptive political activities.[10]

The gradual development of family life among the mine-workers helped to stabilize the community with the result that crime and violence, after reaching a peak in the late 1880s, slowly declined. The company never stopped building houses in its attempts to accommodate the new arrivals and to rehouse those whose homes had deteriorated to a bad state of repair, but family life was not easy in those years. The British staff were well aware of the loss of efficiency in men who were badly housed and were also concerned at the bad effects, caused by the perpetual overcrowding, on the families of the poorer paid workers.

When the British arrived at the mines they found that much of the workmen's housing which the company had acquired was in a lamentable condition. It consisted of 174 houses, many of them one-roomed hovels, made largely of mud walls and with wooden planks and turf for roofs. Within six years these had been replaced by over 800 new houses, most of them built on the Alto de la Mesa, and during the remainder of the century the company seldom built less than 100 a year. These were far superior to those of the normal Andalusian village and represented a great improvement in living

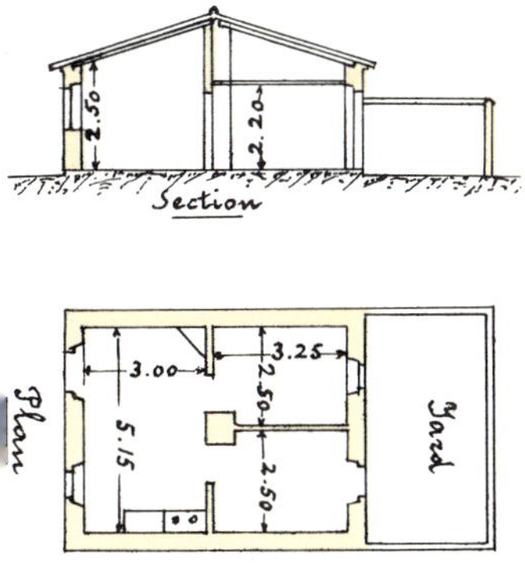


Rio Tinto company housing, 1875, nd ground-plans of two and hree-bedroomed houses.

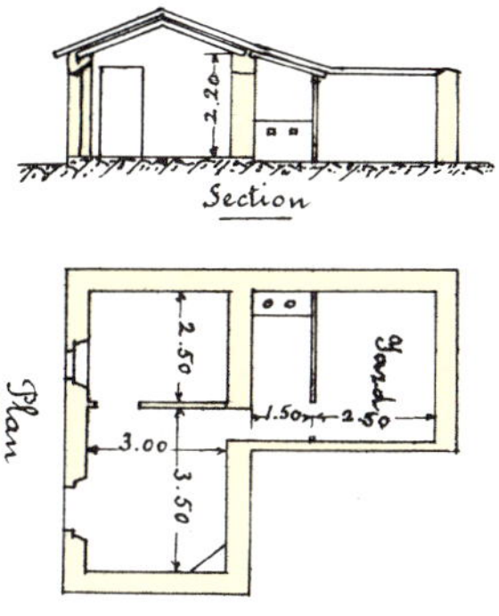


n 1954 there were still many ithout these facilities.

conditions for the majority of the company's workers. The houses had a minimum of two rooms, while those for the better-paid workmen had three or four. The new houses were laid out in rows on a rectilinear plan, each row with its back to the one behind. The smaller houses cost about £25 each to build and were let at 7 *pesetas* (5s 7d) a month, which was the average rent paid in the Andalusian farming villages. But despite their superiority over the peasant houses of the *sierra* they were a long way from being luxurious. It was not until the 1920s that running water and drainage was laid on to any of them.* The privies were in the areas between the backs of the houses while water had to be either purchased from the water-carrier or collected in earthenware jars (which the women were adept at carrying on their heads) from the company's free supply points.[11]

Although the company built houses with a minimum of two rooms to allow husbands and wives a greater degree of privacy than most couples of the Spanish labouring class possessed, many of the workmen realized that there was money to be made by letting sleepingspace to new arrivals. With children, married couples and young unmarried male lodgers sleeping in the same rooms, and often the same beds, there was a considerable amount of sexual promiscuity and frequent domestic strife and fighting. The company's lands' department, which was responsible for housing, was always under considerable pressure, for its staff were the recipients of a never-ending stream of complaints and requests. One head of the department, William Nash, who was himself married to a Spanish woman, spent much time trying to solve disputes between families sharing houses, and declared in an exasperated note to the general manager, 'Whenever two Spaniards share a house there is a row on.' He often had to tell the disputants that he would rehouse them as soon as he could but that until that time they had to live quietly in the same house otherwise both men involved would be dismissed from their jobs.[12]

At times Nash was called upon to dispossess workmen for moral reasons: 'As this man has chucked out his wife and has taken unto himself another woman. . . . I would be glad if you would take the house away from him.'[13] Whenever a new group of houses became available he and the heads of departments were beseiged by applicants for them. He briefed one departmental chief:

*'Sunny Jim' the Bella Vista water-carrier.*

'I have excluded from this list [of registered applicants] those who have already refused houses when offered. . . . Please use all care in making selection as they will all more or less lie as to members of family, numbers in house, etc. Don't consider any who have a bad mark against them. . . . Please put after the names you send me the name of the place where they are to go, and if any refuse let me know and I will note it in the register and pass them over [next time], unless of course they can show good family reasons.'[14]

Although Nash retired in 1917 after thirty-two years in charge of company housing he is still spoken of at Rio Tinto with a certain degree of asperity; though the same would probably have applied to whoever had to do his thankless job during the years when the labour force was expanding so rapidly.

*Women with water-jars, a photo of 1904.*

The food of the Rio Tinto labouring families was simple in the extreme. The midday meal, normally taken to work in a large handkerchief, generally consisted of bread, an onion, and an orange or a slice of melon. The family evening meal was usually a stew of vegetables and pork fat, or a dish of rice and fish, accompanied by cheap wine. On special occasions such as feast days there would be a meal of roast pork. Many families kept sucking-pigs for this purpose and, when the time came to kill them, roasted them in the street over brushwood fires.[15] But living in a cash economy most of the food the Spanish families needed had to be purchased. The company established its own stores at the mines which until 1894 were supplied with foodstuffs and wine by Sundheim and Doetsch the Huelva agents. Flour was also brought from Huelva by rail, for supplying the village bakers, as was fish. Eventually sea-foods became such an important part of Rio Tinto's food supplies that the company in 1891 gave the *ayuntamiento* the land for setting up a fish market under municipal control.[16]

The company adopted the policy of having its own shops in order

*Pig-singeing, 1904.*

to prevent local farmers trying to create monopolies of supply and to ensure that food was available at reasonable prices. The company paid Sundheim and Doetsch the cost-price of the goods, plus their $2\frac{1}{2}$ per cent commission, and transported the supplies to the mines on the company railway free of charge. Thus it was possible to retail them to the workmen's families at only a fraction above wholesale prices; but at times of bad harvests even the charges in the company's stores rose steeply and it was necessary for the board of directors to authorize their subsidization. On one occasion slowness in obtaining this authority led to bread riots in Nerva by some of the women who attacked the company's store there. Thirty-three of them were arrested by the civil guards and held in prison until the company interceded on their behalf and obtained their release.[17]

By this use of food subsidies the company was able to forestall a number of demands for higher wages, a point that was appreciated by the trade union leaders who frequently allied with the tradesmen of the surrounding villages in demanding the closure of the company stores. The directors persistently refused to take this step, just as they turned a deaf ear to demands by Spanish doctors in the area around that they should close down the medical service. The provision of cheap food and free medicines for employees was considered essential by the company. It acknowledged that it gained benefits itself by this policy, especially a reasonably fed and healthy work-force and freedom from too frequent demands for wage increases to meet fluctuating living costs at times of scarcity. But there were real benefits for the workmen, too, especially those with large families. However, many Spaniards saw this company paternalism simply as arrogant behaviour designed to keep down wage rates.[18]

Almost all the labour force at Rio Tinto consisted of people in the labouring category, and in the first few years these included women and children. Hugh Matheson had wished to employ only men but the problem of getting enough workers to help remove the overburden of the south lode open-cast made him give way to the demands of Theodor Blum, the technical manager, to use women and children. They worked alongside their menfolk, often carrying head-loads as large as those of the men, and earning a valuable increment for the family income. The work was not easy. The open-cast pits were exposed to the strong wind and heavy rain of winter and the fierce sun of summer. The men wore normal working clothes with split leather aprons tied beneath each knee, like the chaps worn by American cowboys, and waist-sashes wound tightly around the stomach and lower part of the body as a defence against chills and lumbago. The women wore the normal voluminous skirts of the period and the traditional small narrow apron. Only men were employed underground because of the intense heat and the hardness of the work, and normally they stripped down to leather aprons and boots.

*Rio Tinto working-clothes, 1904.*

In 1888 when the average agricultural wage was the equivalent of 1s 6d a day the unskilled labourer at Rio Tinto was earning 3s 4d on average.[19] The following year this rose to 3s 7d (18 *reales*).[20] Skilled and semi-skilled workers, such as railwaymen and machinists, could earn considerably more. In consequence the standard of living at Rio Tinto was higher than in most of the area around and the company frequently claimed that it paid the best wages in Spain.[21] It was this, the company's officials declared, which allowed the workmen's families to drink wine at meals and enabled most men to own shotguns and dogs and thus indulge in their favourite sport of hunting the wild boar and small game, such as rabbits, in the nearby *sierra*. In 1878 an attempt to recruit Rio Tinto workmen by another company, which wanted men for Venezuela, met with little response even though wages of 30 *reales* a day (6s) were being offered.[22]

Starting in 1885 the company paid its men by cash-tickets. A workman was allowed to go any day before his actual pay-day to the central pay office, which was open every evening for one hour, and collect and draw up to 80 per cent of his earnings in cash by exchanging some of his tickets. He could also use the tickets themselves to purchase foodstuffs and other goods, either at the company's stores or at any private shop in the villages of the area, without any discount or having to pay any commission. Matheson introduced this system when he discovered that many workmen, when being paid their cash in the departments on pay-day, often spent much of it in the taverns on their way home. His new measure was intended to help the men's families 'in getting the greatest value and comfort out of earnings.'[23]

In 1915 the company stopped this system which had been made illegal by the Spanish government because of the abuse of similar systems by certain employers.* The replacing of the tickets by cash payments led to protests by many of the men's wives since it was accompanied by an increase in pay-day drinking. The *alcaldes* of the various villages asked for pay offices to be set up in each *pueblo* to ensure that the women had a chance to receive their house-keeping money before it was spent. Once this new method of payment was established in 1916 it was easy for a wife to accompany her husband to the village pay office and very quickly the men stopped going for their wages and sent their wives instead. It soon became unthinkable for a man to stand in a queue with women, and even the bachelors found either female relatives or friends to draw their wages for them. This unique custom worked remarkably well. In the thirty-eight years between the start of the practice and 1954 there was not a single case of misappropriation nor one claim by any workman that he had not received his money.[24]

In 1884, too, Matheson set up a workmen's club in every village where employees lived. His idea was to give the men a place where wine could be bought cheaply and where they could enjoy recreational facilities without the need to pay for them, while at the same time providing them with somewhere else to go other than the taverns with their inducements to drunkenness and their temptations to gamble. Each club had its own president and committee elected by the members, with a steward as manager.

In 1887 Heinrich Doetsch had induced the stewards and committees to agree to sell only wines supplied by his agency. Very soon the quality of the wines dropped and the workmen started to protest to the company. Matheson had yet another argument with Doetsch but it was not until the latter's death in 1894 that the board of directors felt able to declare that the contracts were void and that the club committees were free to buy their wines wherever they pleased.†

There were very few work opportunities for women once the south lode open-cast was completed. Domestic service in a British household provided the only situations open to them, and though household staffs were large by modern standards this could not have provided work for more than a very small proportion of them at any one time. Consequently domestic posts were readily snapped up. A young woman employed by a British family was in a fortunate position. She would have a little cash of her own and a wardrobe of decent clothes provided by her employer. She would eat well and have good

*Cash-tickets were again introduced by the company during the civil war emergency, in May 1937, to overcome the money shortage arising from the war's effect on the economy, and finally withdrawn in July 1940.

†The workmen's clubs were all closed (with the exception of the Casino de la Unión in the old *pueblo* of la Mina) as a result of the civil war. The new government regarded them as possible centres of discontent and consequently they were converted into temporary food-kitchens and later recreation centres under the control of the local branch of the *Falange*.

accommodation. She also received a sound training in household management that improved her chances of getting a reasonable husband when the time came to marry and to return to the Spanish community.

In 1915 the first wife of general manager Walter Browning, disturbed by the lack of work for single women, founded an embroidery school at Alto de la Mesa to provide work for the daughters and widows of workmen. The embroidery reached a high standard and the goods made at the school sold well in Spain and in Britain, and the money earned by the women, though never a great amount, was a useful addition to their families' other incomes.*

In the middle-years of the nineteenth century the Spanish people were amongst the most illiterate in Europe, 75 per cent of the population being unable to read or write, compared with 33 per cent in England and 40 per cent in France.[26] The only European countries with higher illiteracy rates than Spain were Italy (76 per cent) and Russia (over 90 per cent). By 1900 this high Spanish percentage had been reduced to 56 per cent though the Spaniards still remained close to the bottom of the European literacy table.[27] This improvement in the years 1850 to 1900 was associated, in Spain as elsewhere, with the increasing rate of industrialization.

When the British staff of the Rio Tinto Company were recruiting new employees in the 1870s and 1880s they found almost 100 per cent illiteracy among the labourers, and gave the few that were able to read and write the better-paid posts as timekeepers or as weighers and checkers. Schooling, of course, was not necessary for the vast majority of the labouring jobs which had to be filled. Some of the British staff, however, felt that the education of the Spanish workmen would in itself be a desirable thing and one Englishman felt that it was an essential if they were ever to be able to read the Bible in their own language.

This man, together with his sister-in-law who lived with him and his wife, began evening-classes in his house for some of the children and younger men in 1874. He aimed at providing instruction in the most basic elements of reading and writing and used a Spanish translation of the gospels as a text-book. After two years his ill-health and the strain of teaching night after night led him to suspend the classes. It is doubtful if they would have been soon restarted, for the two weary voluntary teachers had been disappointed by the large numbers of pupils who gave up their lessons without trying to persevere. However, Hugh Matheson on a visit to the mines learned what had been happening and decided that the work should be continued. Perhaps because he felt there was little chance of per-

*Washing day at Rio Tinto, a photo of 1904.*

*The school still survives and now occupies new premises in el Valle, though with increasing job opportunities elsewhere it has only a few embroideresses compared with its early days. Some of the school's embroidery was presented to our present queen on her marriage.

suading his board of directors to spend money on education, he built a day-school at Rio Tinto at his own expense, and also paid for the necessary equipment and the wages of two Spanish teachers whom he recruited. Within three years he had provided another school out of his own pocket and, since most employees were at first reluctant to take advantage of the elementary instruction these schools provided, any Spanish workmen who wished were able to enrol their children.

In 1878 Matheson persuaded the board that the company should finance the schools. He argued that in an industrialized community such as they were creating at Rio Tinto, people had to work in often totally new ways, shouldering new types of responsibilities. Only through education could the company hope to get the responsible work-force it needed with a supply of literate men to fill the supervisory and lower-clerical grades. He also pointed out that the skills of reading and writing were essential if safety precautions were to be effective. Inability to read the rules controlling the use of explosives and machinery could result in death or injury and involve the company in the payment of compensation. Matheson took great interest in the work of the schools. For him their principal justification was that the Bible was taught in them and expounded to the children. The headmaster, headmistress and a couple of the teachers were British and, inevitably, the only Spaniards who could be found to teach in them

*A music class in one of the company's schools, circa 1890.*

were protestants. Apart from such schools as these, education was in the hands of the Spanish church authorities who strictly regulated the religious education provided, and though Matheson's view on this subject would not win the approval of the Spanish clergy they were similar to the views of those who found it difficult to teach under church control. Consequently the company's schools developed a strongly protestant character. One result of this was that many progressive Spanish teachers, who did not like the restrictions which hedged them around in their own schools, sought employment at Rio Tinto, and although most of the pupils spent only three or four years at their studies they were able to obtain an education far superior to that normally given in the Andalusian region. (When Rio Tinto boys were called up for their military service they often found they were able to get good jobs as clerks in their regimental orderly-room.) By 1885 there were 1,200 pupils in attendance.[28]

In 1918, however a member of the British staff, R. S. Low, chief of the labour bureau at the mines and a devout Roman catholic, objected strongly to the religious teaching given in the schools and persuaded Walter Browning, the general manager in Spain, that this part of the curriculum should be the same as in ordinary Spanish schools. One result was that the majority of the Spanish teachers then employed immediately resigned and by 1920 there had been a noticeable deterioration in the standards of instruction. Another was that it was decided that the children of the British staff could no longer attend these schools, as they previously had done, and a special school was set up for them where they would be able to receive the religious education preferred by their parents.[29]

In the first five years of the company's operations the British staff had to contend with a high accident-rate among the workmen. Strenuous efforts to reduce the number of casualties led, however, to the desired result and in consequence the company was able to take some pride in its good safety record. The most frequent causes of death in the mines were associated with mishaps with dynamite and sudden rock falls. Unexploded charges, overlooked as a result of the negligence of the dynamite men, could easily claim victims. In open-cast pits the workmen had to be suspended over the sides of the different levels on ropes to lever away previously blasted rock with crowbars, and it was easy enough for a fall of rock to carry a man down to his death, together with the workmate above holding the rope.[30]

In the first six years of the twentieth century there was an average of 183 accidents a year in the mine workings, of which four were fatal, five were serious and 174 were slight. Serious injuries normally involved the breaking of backs, arms or legs which incapacitated a man for a long period or even for life. Slight injuries were due mainly to the misuse of equipment, producing cuts which kept the men off work for a day or two at the most.[31] Occasionally there was a dramatic increase in the numbers of casualties which led to the company having to take immediate preventive action. In 1908–09 there were eight deaths of workmen in the Alfredo underground mine from falls down the lift shaft. In a report to the general manager the underground manager explained:

'There are stops at each level upon which the cages rest when at a standstill. These are operated by a special guard, whose duty it is to attend these stops, prevent any person going near the shaft, call the cage to that floor, prevent men from throwing anything into the shafts. . . . It is hard to realize how brutal the workmen are at changes of shifts. They fight so hard to get into the first cage going up that 8 or 9 years ago we had to build strong picket gates at the relieving floors with a passage only 17 inches wide to allow only one man at a time to pass into the cage. And yet still three men have been known to pass together. . . . The last two accidents from the cage have been due to the men having been pushed out under the bar.'[32]

The only solution to this problem was the installation of a new cage.

Apart from the casualties within the mines there were others connected with the railway, and in the 1880s and 1890s deaths averaged five a year.[33] Most of these occurred in the marshalling yards during shunting operations or when men riding on waggons (a practice which was forbidden) fell onto the tracks.[34] All casualties were treated by the company's medical service, the creation of which was probably one of the Rio Tinto Company's finest achievements. In the days when the mines had belonged to the Spanish government there had been a small two-roomed hospital in Rio Tinto *pueblo* but at the time of the sale of the mines it was in a poor condition, and had also to serve as a civil guards' post.[35] An extant inventory of 1870 lists its furnishings and equipment: 12 beds, 12 mattresses, 12 woollen bedcovers (well used), 9 pillows, 5 pillow-cases (worn-out), 16 sheets, 3 towels, 6 crutches, 3 stretchers, 5 glass bottles, a box of instruments (incomplete) for amputations, and a number of cooking utensils.[36]

*The Rio Tinto hospital, 1873.*

One of the first of all company appointments in 1873 was that of a Spanish doctor, Jesús Alonso, who was appointed on 23 April. He proved a skilled physician and surgeon, dedicated and hard-working. He began his work at the hospital by ordering the equipment (including chloroform which, though it had been in general use for twenty-five years, did not figure in the 1870 inventory) necessary to deal with the increased number of accidents that were bound to occur with the enlargement of the work-force.[37]

Since the company provided not just a hospital, but a general medical service for its workmen, his early years at Rio Tinto must have been extremely busy. There was a British physician at Huelva, Doctor Law, who was retained on a part-time basis to treat employees there but, he assisted Alonso at the mines only at times of emergency such as the dynamite accident of December 1875.[38] Consequently Alonso was responsible for the health of the British staff as well as the Spaniards.

In 1877 a new hospital under Alonso's direction was brought into use at Rio Tinto on the Mesa de los Pinos. For the period it was an impressive building with excellent facilities but by the beginning of the twentieth century the heavy demand made on the medical service meant that it had become too small and extensions were necessary. It remained in use, however, until 1927 when land subsidence made it necessary for the company to build a new hospital on the hill overlooking el Valle.[39]

The changes at Rio Tinto which began in 1879 with the appointment of Charles Prebble as general manager, had their effects on the medical service. It was decided to expand it rapidly and to cater for the workmen's families as well by setting up surgeries in all the villages

*Men's ward in the second hospital on the Mesa de los Pinos.*

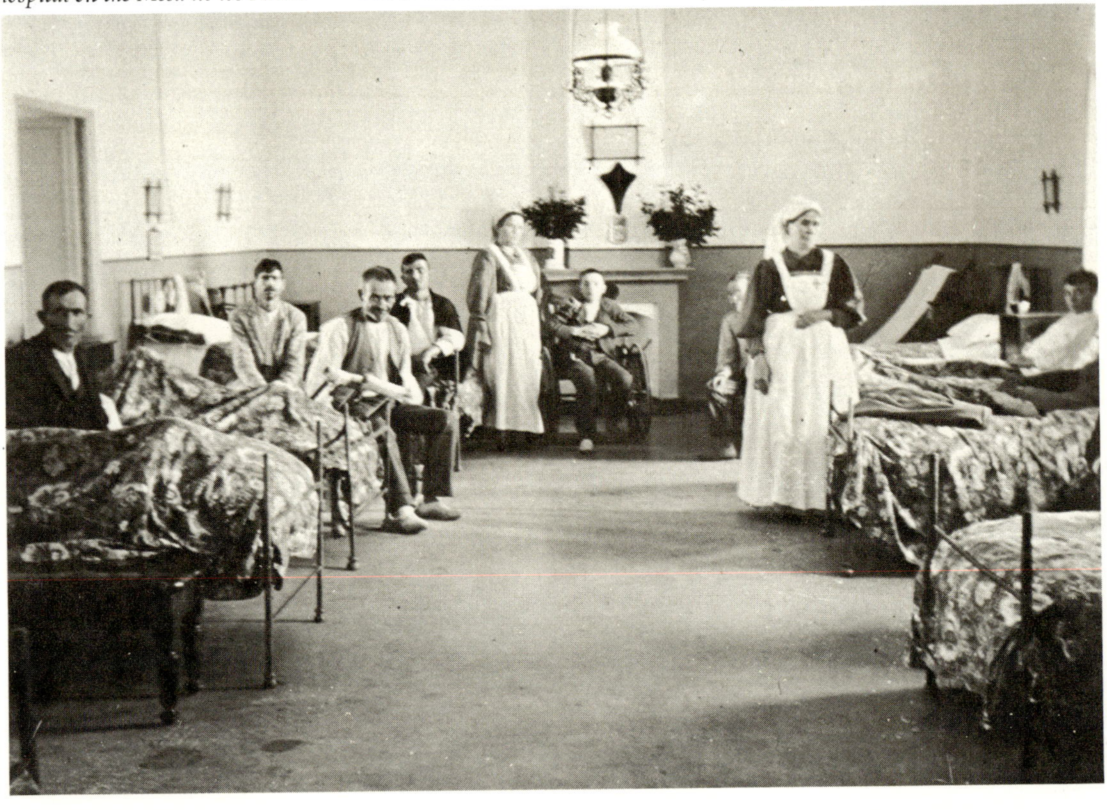

where mines' employees lived. This involved levying a contribution of one *peseta* ($9\frac{1}{2}$d) a week on each workman's wages, a measure which was greatly resented despite the benefits it conferred. This *peseta* contribution was dropped in 1888 when it had come to represent only a small part of the cost of the additional facilities.[40] The expanded service was placed under a new chief medical officer, J. Sutherland Mackay, and Alonso became his principal assistant. Additional doctors, Spanish and British, were taken on to help with all the extra work.[41] Alonso served the company under five British chief medical officers before his death in 1897, which followed an operation in England at the company's expense to remove a cancerous growth.[42]

The first British nursing sister employed at the mines was recruited in 1882 and was a Mrs A. Glass. In 1884 the second, a Miss Blackadder, was appointed as night nurse and stayed at Rio Tinto until her retirement twenty-two years later in 1906. The legendary Miss Ferrier, whose devotion to the Spanish community is still spoken of at the mines, joined the staff in 1891 and retired in 1919 having worked twenty-eight years at the hospital. During that time this extraordinary woman acquired a reputation not only for an intense devotion to the welfare of her patients, but also for a remarkable physical toughness and cool-headedness. (Rio Tinto people still remember how, when male patients in the hospital wards quarrelled and drew their knives – they normally refused to be parted from them on admission – this little woman would stride up to the two contenders and knock at least one of them unconscious with a swift upper-cut or a blow from a convenient bed-pan. No one ever felt any malice as a result, and she was universally loved.)

Another legendary figure remembered by the Spanish work-people was R. Russell Ross who was chief medical officer from 1889 to 1927. In 1900 he persuaded the directors to provide facilities at Rio Tinto for serious surgical operations (apart from those arising from accidents) needed by workmen and their families. Ross afterwards wrote:

> 'Many for want of money could not have been treated out of the mines; and those that might have been able to go to Seville could not have received the treatment (from a surgical, hygenical and alimentary point of view) such as they have in the hospitals of this company, as Spanish hospitals are, with the exception of one or two in Madrid, notoriously behindhand.'[43]

In fact doctor Halliday Sutherland, who assisted his uncle J. Sutherland Mackay, knew of a non-company doctor in Huelva who, as late as 1900, did not use chloroform when operating.[44] The workmen

*Doctor Russell Ross and Nurse Ferrier.*

certainly appreciated the medical service and the treatment it provided and these were among the few aspects of the company's activities which they considered virtually beyond reproach.*

Less appreciated perhaps, because it was not so obvious, was the preventative medicine practised at the mines. When Russell Ross was put in charge of the service in 1889 the greatest single ailment he had to combat was malarial fever, which caused considerable acute illnesses and chronic invalidism. In 1897–8 Sir Ronald Ross discovered that malaria was due to the parasites in mosquitoes. His namesake at the mines read his reports eagerly. Mosquito grubs flourished in stagnant pools, and Russell Ross discovered that the grubs were destroyed if pyrites were dumped in them. On his instructions wire gauze was fitted over windows and doors at Candon station, the most badly affected of all the communities along the railway line to Heulva. In 1900 the people living there remained 100 per cent free from malaria, the first time this had happened since the village had been built. As a result the company fitted similar netting to all its houses.[45]

By 1903 a year had gone by with no new cases of malarial fever amongst the British staff and few amongst the Spaniards. 'The

*In 1921 Russell Ross and Miss Ferrier became the only British people ever to be admitted to the honorary freedom of the Rio Tinto *pueblo*.

diminution of the fever in all the townships of the company exceeded our expectations', wrote Ross in 1904.[46] But the price of freedom from illness was eternal vigilance. Ross's assistant, Doctor S. Forrest (later a chief medical officer) and his wife took up walking as their hobby so that they could keep an eye on all the stagnant pools within the mines' area, and give the necessary orders for them to be drained or treated with oil when it was discovered that the latter gave the same excellent results as placing pyrites in them.[47] As a result of these efforts malaria was completely eliminated.*

Smallpox was another scourge which Ross had to tackle. As early as 1898 he introduced vaccination for workmen and their families, but when cases still occurred he found that the relatives of the victims tried to conceal them. The reason was simply that the existing isolation hospital, a scarcely-habitable iron shed on the Mesa, was loathed by those who had to go there. Consequently he had to persuade the company to build a new isolation hospital for infectious cases on an island in the Marismilla dam. Since the dam contained highly acid water it had the added advantage that there were no mosquitoes in the vicinity. Once this building came into use in 1902 it was found that concealment of infectious cases stopped.[49] Two other killers, especially of children, were diphtheria and diarrhoea. By introducing antitoxins at the mines in 1897, Ross considerably reduced the mortality. He also had a sanatorium for tuberculosis patients built at the mines, though few Spaniards cared to enter it for the prolonged treatment then necessary, and in 1932 this was closed down for lack of use.

By 1929 the medical service personnel had grown to include two British doctors and fifteen Spanish, one infant welfare doctor, three British and ten Spanish nurses, twelve dressers and eight midwives.[50] Their unsparing efforts, and those of their predecessors, had produced what was everywhere acknowledged as the finest medical service in Spain. The inscription of the memorial tablet to Russell Ross on the wall of the English cemetery at Rio Tinto is appropriate for each of them: 'A workman that needeth not to be ashamed.'[51]

The company's provision of welfare facilities could not stop with the medical service. Life could be hard for the aged and the infirm, for widows and orphans. There was no workmen's compensation by law until 1901. The Spanish tradition was that when an individual met with hard times he could expect his community to assist him. And, just as in England in the eighteenth century those falling into poverty were required to return to their parish of settlement, so in nineteenth-century Spain a man was normally expected to return to his place of origin. Consequently in the early years of the company's ownership

*It returned during the Spanish civil war of 1937–39 and persisted until 1943 since it was impossible to take remedial measures due to the shortages of the civil war and the second world war periods. It was eventually brought under control, however, by taking small fishes called *gambuzos* found in the timber floats at Huelva and placing them in all streams, pools and water-tanks where they fed on the mosquito larvae.[48]

of the mines there was a policy of paying those unable to work, and families which had lost their breadwinner, to return to their original homes: 'It appears this man is dying of stomach trouble . . . if he could get some 400 *reales* (£4) he would return home to his family. I beg to advise that he be given the 400 *reales* and sent off.'[52]

But as time went by people were born and grew up in Rio Tinto, and had no possible claim on any other community. Others who grew old after many years in the company's service often had no relatives to whom they could return. Not unnaturally they looked to the company to provide them with the assistance they needed. In the early days the directors were not willing to make payments as of right or to provide a regular system of pensions. Indeed in 1888 they were advised that it probably would be held illegal in the English courts to spend shareholders' money on a pensions' scheme.[53] Consequently there was a system of gratuitous payments. The mines' manager kept the board informed of the needs of persons who were retired, widowed or injured, and it approved weekly payments at his suggestion according to individual needs. But most workmen were reluctant to retire when they grew old. Normally they kept on until they were discharged for incapacity, and long after the time they should have ceased work.

In 1905 R. E. Palmer wrote to the general manager, W. A. Carlyle:

'I enclose a list of twenty-three workmen whom you will see by their service have worked most of their lives for the company. These men must be pensioned off as I cannot do anything with them. To keep them means putting guards over them to prevent their being killed as some cannot see and some can hardly walk.'

Of these men all had at least thirty years' service. Seven were over seventy. Against their names Palmer noted his comments: 'Been fine miner but now can do nothing. Lungs bad'; 'Paralysed in one arm. Been a good man'; 'One son who is no good and has left his father'; 'Too feeble to work. Has one son who is too delicate to work'; 'Been foreman. Has to support his widowed daughter and her five children'.[54]

Though there was no standard rate allowed to such retired men, they could normally expect to be granted 5 or 6 *reales* (1s or 1s 2d) daily, slightly under a third of their pay, although a little more would be allowed if home circumstances merited it.[55] In the case of men incapacitated for life due to accidents, the company normally granted them half pay. After the compensation law of 1901 the company was legally compelled to make the payments which earlier it had made voluntarily. There was no obligation if there had been contributory negligence by the injured workman, but the company continued its policy of paying *ex gratia* pensions or lump sums to such men, or else their families in the event of their deaths. There were also *ex gratia* payments to those who were sick or temporarily unemployed provided they had a record of good behaviour.[56]

The directors felt that the provisions which they made for those workmen or their dependants who were in need were, if not generous, at least fair by the standards then prevailing in Spain. But among the workpeople there was considerable resentment at the non-obligatory nature of the system. They wanted the pensions or compensation as a right in all cases and this was a demand frequently made at times of strike action. It was certainly the opinion of some of the British staff that the company had little to lose financially, and much to gain in goodwill, if it granted this demand. This was yet another occasion where the policy of paternalism was liable to be construed as despotism on the company's part.[57]

In yet another sphere the company found it was likely to be accused of absolutism. This was the field of local government which initially the directors approached with some reluctance. Apart from owning the freehold of all the land and many of the buildings contained within the boundaries of the *ayuntamiento* of las Minas de Ríotinto, and thus having a strong interest in the maintenance of law and order, the company appreciated that it was essential to have competent municipal administration to achieve the stable community which was important for the efficient functioning of the mines. The prospect of having to be responsible for such things as drainage, road-making, lighting and street cleaning was not one that gave the company's directors and staff any pleasure. For it was going to require not only a great deal of money but much time from the manager and his assistants.

Yet there seemed to be only two other alternatives, both unacceptable. If the company left the control of local government to the workmen's representatives it would probably produce confusion and, if the anarchists took charge, conflict between the company and the town council. Yet if the area came under the control of the local right-wing politicians who exercised authority in the agricultural districts around the mines then the *caciquismo* system, and the corruption which was inseparable from it, would also make life difficult for the company. In these circumstances the Rio Tinto Company gradually took charge of local government itself, nominating as *alcaldes* and *concejales* mature and reasonable men who could be relied upon to co-operate with the British staff.

This relationship provided difficulties as well as benefits for the

OPPOSITE:
*La Mina (Rio Tinto) being demolished to allow the extension of the south lode open-cast pit.*

town councillors. The company had acquired many of the rights of the Spanish crown over the area when it purchased the mines, including the power to treat the whole of the land within the boundaries as subject to the exigencies of the mining operations. This meant, for example, that the *ayuntamiento* could not object if the company decided to pull down whole groups of houses or, as happened when the decision was made to extend the south lode open-cast workings at the beginning of the nineteenth century, to destroy almost the entire *pueblo* of la Mina. Even those buildings which were not swallowed up by the quarry had in the main to be evacuated because of the resulting problems of subsidence, their inhabitants being moved to new accommodation in el Valle.[58] Only two streets were left intact.

The interests of the few freeholders of property within the mines' area were declared by a royal decree of 1879 to be 'subordinate to the necessities of the mine-workings and can never contest the rights of the latter nor be put in opposition to them.' Of course, where the company destroyed property which it did not own it had to pay compensation, but this provided grounds for bitter disputes. After this decree the company refused to permit anyone, including the *ayuntamiento*, to build within the mines' area and gradually acquired all the freeholds and leases which were in other hands. Even the town hall was company property.

The local authority derived many advantages, however, from its connexion with the Rio Tinto Company. Under Spanish law the only local tax which the company could be required to pay was an impost of 16 per cent on the value of 3 per cent of the mineral it extracted each year. Instead of paying this the directors authorized annual donations to the town council of sums far in excess of what that tax would have yielded, providing the money for over half the council's expenditure even in the company's bad years. In fact the normal practice was for the company to donate whatever sum the local authority anticipated it would need after allowing for its other sources of revenue. This generosity to Rio Tinto *ayuntamiento* caused some resentment in the other districts around, which contained many of the company's workmen and felt they were entitled to share in the prosperity. In their view the payment made by the company, even though it surpassed the legal obligation, was letting the company off too lightly.

Apart from these grants the company paid directly for street lighting and road maintenance, money which normally the *ayuntamiento* would have had to find. It footed the bills, too, for other items which were normally a municipal responsibility, including the running of the council offices, the educational and medical services,

the town slaughter-house and wash-house, the cemetery and the water supply, all in return for a nominal payment from the *ayuntamiento* of 100 *pesetas* (£4) a year.[59]

Nonetheless, although this participation in the town's administration provided the people of Rio Tinto with some of the best local government services in the country, neither the British nor the Spaniards were completely happy about it. The company's staff found themselves involved with the running of the Spanish community to a far greater degree than they would have liked and in a way that would have been impossible in Britain. And their duties of advising and supervising the town's officials often caused resentment among the Spanish people.

At times the management of the company despaired: they felt it was almost their fate to be perpetually misunderstood. In creating new communities to serve the mines, and in providing housing and amenities on a prodigious scale, the directors and staff felt they had little option to be other than paternalistic. Whatever differences of opinion there were between the British personnel about the way the policy was to be implemented, they were all agreed that it could not be left to the Spanish workmen to provide for their own needs in the fields of education, housing and medicine.

It is difficult to see how the company could have behaved differently. Any other policy would have been detrimental to its own interests and those of its work-people. There is no doubt, however, that the attitude which lay behind the company's actions right up to the 1930s had in some respects an unfortunate result. For many of the British staff and directors displayed at times an arrogance which provoked a hostile reaction. Doing virtually everything that was necessary for the Spaniards at the mines, it was too easy for the British to take a low view of the capacities of the workmen and to regard them as intellectually little more than children. But it must be acknowledged that many Spaniards in their turn often behaved irresponsibly. Some tried to exploit their local government system. Few tried to make it work or to improve it for the future. The majority regarded it as something apart from themselves and treated it with apathy until the tumultuous days of the second republic arrived and the heady wine of revolutionary fervour turned local government into a miniature battle-ground for the political convulsions shaking Spain.

*The new village of el Valle: the procession after the dedication of the church of Santa Barbara.*

*Walter Browning, 'the king of Huelva'.*

14

# *A king in Huelva*

## The rule of Walter Browning: 1908-27

The policy of company paternalism, developed by Hugh Matheson as chairman and William Rich as general manager after the 1888 riots, was pursued in part as a way of keeping the mines free from industrial unrest. In the short term it could be considered to have been successful for there were twelve strike-free years during which the influence which the anarchist and communist factions exercised over the workmen of Rio Tinto was gradually circumscribed.

In striving to achieve this state of affairs Matheson and Rich worked closely together, exchanging views candidly and learning to confide and trust in each other. The chairman himself devoted much time to discovering all that he could about the conditions in which the company's employees lived and worked. These he observed at first hand on his frequent visits to the mines during which he received, to his delight, a warm and friendly welcome on his appearances in the *pueblos*. He sought the opinions of even the most junior members of the British staff on the subject; he set up a system of labour reports to keep him fully informed of all problems as they arose; and by quick action, sometimes by granting concessions to the workmen even before they submitted claims, sometimes by face to face discussions with senior Spanish employees, it proved possible to out-manoeuvre the union leaders and to prevent work-stoppages. So pleased was Matheson with these results that in 1890 he urged as many shareholders as possible to visit Rio Tinto to see for themselves how the effects of this policy were to the benefit both of the company and the workpeople. He told the annual meeting in that year about the social services and recreational facilities being provided by the company; of the increase of 11 per cent in wages which it had been possible to grant in the previous twelve months; and of the friendly relationship which existed between the men and the company.[1]

The death of the patriarchal Hugh Matheson in 1898 was followed by the election of John Keswick as company chairman. He was a man who proved too cautious, too much reliant on precedent, and who was too slow in making decisions to be a satisfactory head of a concern which needed imaginative leadership. Rich soon began to experience a sense of frustration. He wrote stating the conditions he expected to be met if he was to stay on as general manager. Disappointed at their

OPPOSITE PAGE
*Sir Charles Fielding.*

rejection, he retired in April 1900.[2] His replacement was William Carlyle, a Scotsman with a reputation for toughness gained in the United States' mining industry which had become notorious for the frequent recourse to violence by both management and workers. It was no coincidence that his appointment marked the beginning of a change in attitude by both sides at Rio Tinto. The company began to adopt a less flexible approach towards labour problems and the unions began a policy of frequent strike-action to back up their claims.

By the spring of 1904, however, Keswick was a dying man. The Rothschilds, the principal shareholders, nominated their representative on the board, C. W. (later Sir Charles) Fielding, to succeed him.[3] Fielding's accession was unusual in that he had joined the company as a junior clerk in 1880 at a salary of £50 a year. He had been sent in the same lowly capacity, in 1882, to the Huelva office and, in 1883, to the mines' accounts department. Recalled to London a year later he had then gone to the copper smelter which was being leased by the company at Cwm Avon, near Port Talbot, as assistant manager at £150 per annum. In this role he was so successful that within two years his salary was raised to £500 and in 1890 the Rothschilds, noticing his abilities, decided to take him into their employment.[4]

In the ten years between 1890 and 1900 he had made a reputation for himself in the world of finance and then in the latter year returned to the Rio Tinto Company as Rothschilds' nominee on the board. Despite the fact that he had once worked as an employee of the company and thus knew many of the staff in London and in Spain, his former colleagues soon found that he made an uncongenial chairman when he took over that post. The common view was that he was a bully with a needlessly rough edge to his tongue, given to sudden outbursts of temper which were all the more frightening in one of his immense size – he weighed nineteen stone.

He had one habit in particular which caused his subordinates considerable anxiety. If he was attending a meeting, whether in England or in Spain, whether with a company employee or with a government official, and the man he was talking to was called away to deal with another matter, Fielding would allow only ten minutes for his return. At the end of that time he would start to strike the floor violently with his walking-stick, and then the furniture, causing much alarm to the others present, until the absentee returned.[5]

Among his redeeming features, however, Charles Fielding had perhaps the most valuable asset possible for a man in his position: that of being able to judge with a remarkable degree of accuracy the abilities of his colleagues and subordinates. He was accomplished, too, at finding precisely the right man for a specific job.

On becoming head of the company he soon found himself in disagreement with Carlyle, the general manager. Although the two men had similar views on many matters, and agreed that Matheson and Rich had been unnecessarily benevolent and conciliatory towards the work force, both chairman and general manager heartily disliked each other. Disputes arose because Carlyle was unwilling to yield the considerable independence of action which he had acquired during Keswick's last and fatal illness, and Fielding was determined to be master in his own house. As a result Carlyle decided to return to America and in February 1907 sent a cablegram to Fielding giving six months' notice of his intention to terminate his employment. By return of cable Fielding accepted it with immediate effect.[6] Neil Kennedy, who had first gone to the mines as a young man at the time of the purchase and since 1904 had been working in the London office, was once more sent back to Spain, this time as acting manager, while Fielding arranged a successor to Carlyle.[7]

There was no doubt in the chairman's mind as to the man he wanted. On a visit to America he had shared a railway compartment on a long journey with a man whose knowledge of mining and whose extraordinary personality had impressed him deeply. They had exchanged visiting-cards and Fielding noted that his travelling-companion had been Walter Browning, the general manager of the Mazapil Copper Company. Fielding made a mental note to try to obtain Browning's services for the Rio Tinto Company if ever the opportunity presented itself; and thus when the chairman accepted Carlyle's resignation he sent a second telegram. It came to Browning out of the blue and offered him the post of general manager of the Rio Tinto Company in Spain. At the age of forty-one Browning returned to England with his wife, Alicia, for a holiday most of which was spent with the Fielding family. Before Browning sailed for Spain in January of the next year (1908) he and Charles Fielding had become boon companions.[8]

Browning's personal history prior to this new appointment indicates something of the unusual nature of the man. Born the younger son of a wealthy London stockbroker he had gone as a youth to Jesus College, Cambridge. Like many young men in his position he had a tendency to overspend and to run up bills much to the displeasure of his disapproving father. The final straw came when Browning senior was asked to pay one of his son's debts for £10. The boy's allowance was cut off and he was forced to abandon his career at Cambridge after only one year.

He found himself virtually disowned by his father and realized that his best chances of success lay in going abroad to a place where there were opportunities not available in Britain. Shrewdly he decided to take a mining-course at the Liverpool technical college, probably raising the money from his mother unbeknown to Browning senior. From there he sailed to Mexico to become a lone prospector for gold in the *sierra* Madre. The experience transformed him. At first he tried to avoid the trouble which abounded in the Mexican *pueblos* for greenhorns such as himself by carrying in one pocket a currency note and in the other a pistol, offering the former to any belligerent stranger so as to avoid the need to use the latter.

Soon, however, he found that his growing reputation with a gun caused any protagonist in a dispute to back away uneasily. Late in life Browning claimed that his renown as a marksman was due entirely to a fluke. It arose when he loosed off a shot in anger at some vultures, which had gathered to prey on the body of his dead mule and, much to his own surprise, hit one in flight. But even if this account is incorrect there is no doubt that he soon became an expert with both pistol and rifle.[9]

Much of the time living alone in the Mexican hills Walter Browning tested his powers of self-endurance to the utmost, successfully surviving the solitary life with its hardships and dangers. Often hungry for days on end, he had been forced to live on what he could shoot. On occasions he had been attacked by hostile Indians and escaped death only because of his proficiency with a gun. In the process he developed an ability to think and act simultaneously and acquired a remarkable degree of self-reliance: characteristics which were to remain with him for the rest of his life.

The evening that the Brownings arrived at Rio Tinto they were receiving the senior members of the staff and their wives at the *casa grande* when whistles blew and hooters sounded the alarm for an emergency: a great landslide had occurred in one of the north lode open-cast pits. Browning threw himself into the direction of the rescue operations with an energy which was a new experience for the British and the Spaniards alike. The workmen immediately dubbed him *el terremoto* ('the earthquake').[10]

In the following week he proved how appropriate the nickname was. By day he toured the mines and their various departments. By night he sat in his office reading reports and letter-books until the small hours and drafting instructions to his departmental heads. The British staff must have wondered what had hit them. When Robert Palmer, the head of the underground workings, responded to a demand for greater production of pyrites with the comment that it was impossible, Browning told him how it could, and would, be done. The senior staff were told that the monthly reports which they had been submitting to the old general manager were insufficient for the new; Browning required greater detail and fuller explanations. He called for comparative cost statistics from all departments within forty-eight hours and when he got them sent back a stream of sharply-worded memoranda telling heads of departments how to cut out wasteful expenditure. He announced that in future no correspondence would go to London office from the mines without his first seeing and approving it. Suddenly the tone of the letters and reports submitted to the general manager's office became noticeably more formal and more respectful than they had ever been to his predecessor.[11]

Within that first week he established a grip over the mines which, while he remained general manager, he never relaxed. The British staff soon came to the conclusion that he was a complete autocrat who would tolerate no criticism and would defer only to his chairman and friend, Charles Fielding. Not that any deference would be necessary in that direction for the two men trusted each other implicitly, thought as one and ran the company and the mines between them with scarcely a reference to the board of directors. It is significant that, as the correspondence between the general manager and the chairman grew in volume, so the agendas for board meetings became shorter.

But the common assumption by the mines' staff that Browning would never take no for an answer was incorrect. The experience of two Spaniards who came to know Browning well, the late José Sánchez-Mora junior, the company's legal adviser in Huelva, and Francisco Martín-Pernil, the company's agent in Seville, was that Browning disliked yes-men and wanted his subordinates to argue back if they thought him wrong. Unfortunately, most of the British staff had been trained in the tradition of saying 'yes' to the general manager and Browning despised them for it.

Many of the British employees sensed his contempt and resented it, even though they seldom appreciated the reason behind it. This did not prevent them, however, from according him a grudging admiration. Though not tall, he was a strikingly handsome man with a fine figure and bearing. In everything he did he displayed a zest that left other men standing. Curt in speech, trenchant in wit, indomitable in courage, he became a legendary figure in his own lifetime. Everyone that remembers the living man says the same thing: that even in a crowded room his personality was such that he dominated the gathering.[12] Herbert Potts, an assistant in the smelting department noted: 'He was a handsome man with an outstanding personality, and any observant stranger would have had no difficulty in picking him out as the general manager in a room which contained all the staff.'[13]

The stories about Browning and his laconic manner of speech are innumerable. Presumably he had acquired the style when living among those hard, terse frontiersmen and westerners of American folklore whose exploits have come in recent years to fill our film and television screens. Certainly there is a detectable likeness.

Browning (to a head of a mines' department):

'Do you know how to write a resignation?'

'Yes.'

'Then go home and write yours.'

On another occasion (to an Englishman who asked for a house so that he could marry a Spanish servant girl):

'Do you know what time the mail train leaves el Coso?'

'At ten-past two.'

'See you're on it tomorrow. You're fired.'

This story needs to be balanced by another about a young Englishman who told him that he loved a Spanish girl but understood the company disapproved of members of the staff marrying Spaniards. 'If I wanted to marry a *Hottentot*' growled Browning, 'no one would stop me.'[14]

On the death, in April 1911, of the company's lawyer in Huelva, José Sánchez-Mora senior, Browning telephoned his son the same morning saying, I need you to replace your father. Get to his office and start work.' To the young man's protests that he was only twenty-four, newly qualified and without his father's abilities, Browning replied, 'You must have inherited some of his cunning.'[15]

Browning spoke fluent Spanish. He also, in the view of many people, understood the Spanish temperament better than any other Englishman who was at Rio Tinto before or after him. Not that this made the workmen feel any liking for him, for they, too, saw him as an imperious autocrat. But they respected his superb horsemanship and his iron will and described him as *mucho hombre* ('a real man'). It was the ordinary Spaniards, too, who called him by the cognomen by which he is now best remembered. For as his personal power and authority increased over the years they noticed, sometimes in anger but often in awe, how his word acquired the force of law not only at Rio Tinto but throughout the whole of Huelva province, and how high-ranking officials of the civil administration waited upon him deferentially. And thus, by 1914 the people were calling him *el rey de Huelva* ('the king of Huelva province').[16]

His, however, was a tumultuous kingdom, with frequent and violent labour disputes, political turmoil and threats of assassination. There was also the drama and the tragedy of underground disaster. It was on 3 November 1913 that Browning sent to London the details of the 'most appalling catastrophe . . . that has ever happened in the

*'His Rio Tinto Majesty', a cartoon from 'el Socialista', 9 July 1920. Browning ('su Rio Tinto majestad') is shown with the two Datos, prime minister and judge, in either hand and exclaiming: 'Aoh yes! I think that with these 'datos' (the word also means 'facts') I have enough!'*

history of the mines', when, during the bitterly violent strike that was then in progress, some of the malcontents set fire to the Alicia shaft of the San Dionisio underground workings.

The blaze was discovered at four in the morning of Saturday 1 November at the seventh floor level. Three English foremen went down to investigate the damage and reported that the fire was so fierce that the only way to contain it would be to pump in water. But soon the lift cages crashed down to the bottom of the shaft, bursting through the blocks intended to prevent the fire from spreading downwards, and taking the flames with them.

Attempts to send men down the following morning were thwarted by an accumulation of water in the shaft due to a failure of the underground pumps. On the next day two British foremen, John Gilbert and Edwin Wilson, and two Spaniards, Lucas Millan, a pumpman, and Luis Marquez an assistant surveyor, volunteered to reconnoitre by going down the Victoria shaft, which seemed to be free of smoke. At eleven o'clock in the morning they descended to the eleventh floor and started to inspect the tunnels. At this stage they were joined by Robert Sach, the head of the open-cast, and George Wilson, his chief foreman, who had entered the mine by way of some galleries leading underground from the Atalaya open-cast pit.

By the light of their torches the six men gradually worked their way downwards, floor by floor, until they reached the twentieth level where they discovered that the bell wire (used to send signals to the cage operators at the surface) was broken. Gilbert took the cage to the top to report on their survey thus far to Palmer, the chief mining engineer, and then went down again with men to repair the wire. On his return he found that his colleagues had decided to go down still further to investigate the twenty-sixth floor.

This was at twelve forty-five. The investigating team ordered the cage to be stationed at the twenty-sixth floor in case they needed to make an emergency escape, and sent one of the men repairing the wire up to the surface to explain what they were about to do. This man reported to Palmer that George Wilson was dubious about going down to the level where the fire was raging and the poisonous carbon monoxide at its thickest, but that Sach, who appeared to him to be a little light-headed as a result of the gas, was insistent. The last the electrician had seen of the six men had been as they started to walk down the incline-way leading to the floors below. In fact no sooner were they out of his sight than they must have walked into a blanket of carbon monoxide and collapsed, unconscious. Within twenty minutes, or by shortly after one o'clock, they would have been dead.

On the surface above Palmer became more and more worried at hearing nothing from them. Shortly after two o'clock he sent four men down to see if they could find the party. Within five minutes the cage returned with only one of the men who had just gone below. Nearly asphyxiated, he gasped out that on arriving at the twenty-sixth floor level his three companions had collapsed before there had been a chance to do any searching.

Another team of six was sent down with orders that one man was to stay in the lift for a quick return to the surface in the event of an emergency and that the others were to stay together in one group while recovering the three unconscious men. Soon they came up, having been able to find only two of the three, all eight men in the cage being in a bad way and five of them being insensible. Frederick Drewett, an English engineer, volunteered to go with another two workmen down to the same level. Staggering through the noxious smoke they found the missing Spaniard, dragged him to the cage and signalled to be lifted to the surface. On arrival at the top of the Victoria shaft they in their turn were scarcely conscious.

Meanwhile, Palmer, trying to find another way down to locate the first missing party of six men, discovered that the Guillermo shaft was relatively free from smoke. Four British staff (including once again Frederick Drewett) and a Spaniard volunteered to make yet another rescue attempt and descended to Guillermo's twenty-sixth floor. By this time it was almost six in the evening. Only five minutes later the signal came to raise the cage to the surface. It contained a Spanish workman and a British foreman, the latter in a bad condition but able to explain that though the crosscut leading from the shaft was clear the team, as soon as it had turned into the main gallery, had been caught in the poisonous fumes. Immediately a further rescue party of five Englishmen was formed, but soon they called for more assistance and eight more of the British staff joined them. By six-thirty all were back on the surface. But one of them, Frederick Drewett, the hero of the morning's rescue operation in the Victoria shaft, was brought up dead.

Laurence Hill, however, who had been a member of the third team to descend, had managed to get half way through the tunnels to Alicia and had stumbled on the bodies of some of the original six missing men. After quickly gulping down some fresh air on the surface he insisted, despite his weakened condition, on going down yet again with two other British employees. Enfeebled by gases, it was possible for them to get only the body of Edwin Wilson back to the shaft and from there to the surface. They wanted to descend yet again but Browning, who had been directing the operations, forbade it. 'I did not consider we were justified in allowing any more lives to be risked.'

That evening, Browning sat in his office dictating his first report on the tragedy to London.

> 'Breaking the news to the families here (of the dead and missing men) is, as you will realize, very distressing and I have asked you in my telegram to see personally that the families of Messrs Sach and G. Wilson are advised by someone from the London office . . . This terrible accident, coming as it does at a time when the whole country is in a ferment, is most unfortunate; but you can be sure that we are doing everything that is possible to prevent any further loss of life such as has occurred on this lamentable occasion.'

On Tuesday 4 November, the next day, it was still impossible to continue the search for the remaining bodies because of the danger from the gases; but by the following morning the carbon monoxide had to some extent cleared and three members of the British staff accompanied by six Spaniards were able to descend. Using cage-birds they tested all the levels down to the thirty-second floor, finding them relatively free from gas, and then, on the incline from the twenty-third to the twenty-sixth floors, they saw in the light of their torches the bodies of John Gilbert, Lucas Millan and Luis Márquez. Though the search-party had been able to withstand the foul atmosphere they were by that time considerably enfeebled and it was necessary to send for another nine men, making eighteen in all, to get the three corpses to the surface.[17]

The investigation which followed showed, ironically, that the ventilation fans, which surprisingly had continued in operation during the four days, had played a major part in the tragedy. If they had not been working the Alicia shaft would have served as a chimney to conduct the carbon monoxide from the burning mineral and smouldering timbers out into the air above. As it was, the fans drew the smoke and gas away from the shaft and into the workings.[18]

All told, there were thirty-five men who took part in the rescue operations, twenty-three members of the British staff and twelve Spaniards, remembered today as 'the heroes of the Alicia disaster'. The following January, Fielding made the journey to the mines to present them with gold medals commemorating their heroism. Meanwhile, the burning galleries had to be closed off from the rest of the workings by the construction of walls in order to allow the fire to burn itself out without spreading its flames and poison gas to other parts. Almost sixty years later the mineral still smoulders behind the masonry partitions.

The magnitude of this underground disaster reinforced the decision made in 1906 to mine the San Dionisio mass as far as possible by open-cast methods. The introduction of steam shovels at the Dehesa open-cast had shown that the overburden could be stripped at a much lower cost with machines than by hand alone, and a subsidence near the Alonso shaft in the San Dionisio ore body had made it extremely difficult to continue mining at that spot by the underground slicing method. At the end of 1907, therefore, work had begun in making the Atalaya open pit in the San Dionisio mass.[19] Browning's arrival the following January coincided with work getting underway on a vast scale. Under his direction the Atalaya pit became one of the world's largest, being deepened and extended over the years, the steam shovels for the work being purchased second-hand from the makers of the Panama canal.*

In 1911 Browning supervised the making of yet another open-cast quarry, on Alto de la Mesa, a flat piece of table land almost 200 yards wide and almost $1\frac{1}{4}$ miles long. This was to extract iron ore, but since little was sold the operations were halted until 1922, when they were

*Some of these old shovels still survive at Rio Tinto.

*The Atalaya open-cast pit: an aerial photo taken in 1963.*

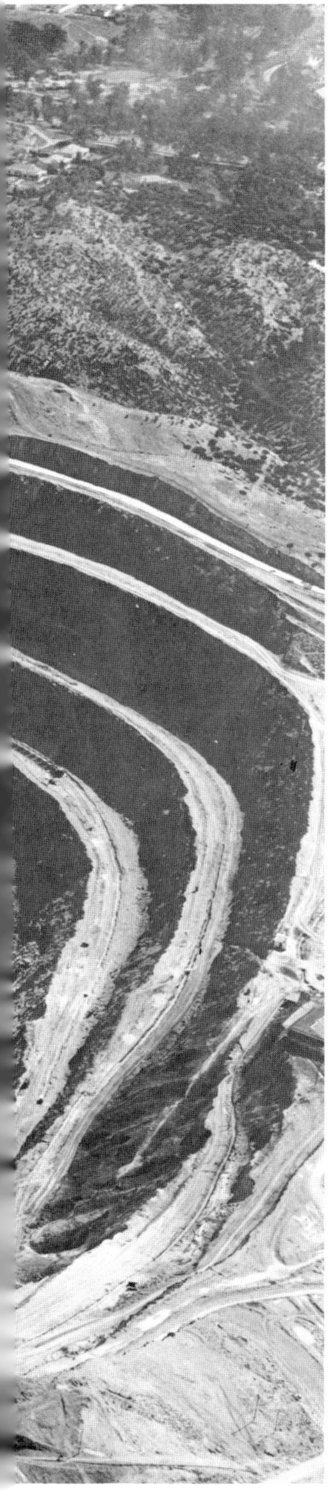

resumed at a time of growing demand. From 1927 there was a five-year boom and an average of 150,000 tons a year were shipped.[20]

Another development of Browning's time, the great increase in mining at lower levels, was beset by a number of problems. One, because of the necessity to move the minerals by rail, was the restriction imposed by the eleventh-floor tunnel which had been constructed in the early days of the company and by 1911 was high above the source of the pyrites. It was decided to build another and in 1912 work began on this at the sixteenth-floor level. Driven from just below the village of Naya through the south lode and the San Dionisio mass the new tunnel was completed in 1918. Its $4\frac{3}{4}$ miles were electrified in 1924, the start of the programme by which electricity replaced steam transport for the conveyance of minerals out of the mines.[21]

The whole period was one of technical change and advance. Originally the company had produced 'lump' pyrites for export, consisting of pieces of up to 8 ins in size which on arrival at their destination were broken down by the purchaser. But, as the roasting furnaces changed and became more efficient towards the end of the century, the buyers required 'furnace' size pyrites of between 1 and $2\frac{1}{2}$ ins. The introduction, as from about 1900, of multi-hearth mechanically operated furnaces required 'fines' of about half an inch in size. To supply both these grades the Rio Tinto Company installed at the mines screening equipment and then, between 1913 and 1917, a crushing and screening plant was built at the spot known as Marin. It contained jaw crushers to break the mineral, and screens to separate the two sizes. Looking to the future, and in the belief (which proved correct) that there would be a demand for even smaller pyrites of less than half an inch, a second unit was also constructed with crushers and rolls able to reduce the mineral to smaller sizes.[22]

The first smelter constructed by the company had been built at Rio Tinto in 1880 to produce a matte (an alloy of copper, iron and sulphur) for sending to Swansea and for sale in the metal market there. In 1901 a second smelter was erected which used a new process to extract the copper from the matte. By this new Bessemer method air was blown through a bath of molten matte, causing the iron to oxidize and form a slag and the sulphur to leave the bath as sulphur dioxide. Each vessel, or converter, could produce only some ten tons of copper before it had to be relined. To enable this to be done it had to be removed from its stand by a 60-ton crane and a recently relined vessel substituted. These converters could also treat the low-grade, high in arsenic, copper precipitate recovered from leaching the pyrites. This change to an easier method of producing copper made

*A Bessemer converter at Rio Tinto in 1901.*

the Cwm Avon works with their sixty furnaces, which the company had leased since 1885, too large for the company's needs.* Consequently, in 1906 the Cwm Avon smelter was closed and a much smaller copper refinery built on the dock side at Port Talbot where the ships from Huelva could unload their cargoes of copper precipitate directly into the works.

The Bessemer plant at the mines required only 9·5 per cent of coke (always a relatively expensive item) on the charge compared with 14.5 per cent needed by the old smelter but a rapid change in copper metallurgy showed a way of cutting down coke consumption still further. 'Pyritic smelting', developed by Lawrence Austin and Robert Sticht at Montana between 1887 and 1896 and introduced at Mount Lyell in Tasmania in the latter year, by reducing still further the amount of coke required provided a means of producing much cheaper copper. A new smelter was therefore constructed at Rio Tinto using this process, and came into service in 1907.† The amount

*In 1887 the company had also acquired the Grange Metal Works at Jarrow but production costs there had proved much higher than at Cwm Avon and the Jarrow smelter had been closed in 1896.

†It continued in use until 1970.

of coke used was, at first, equivalent to between 4 and 5 per cent of the charge of pyrites, but by the end of the first year's operation it was reduced to 3.5 per cent.[23]

Browning, however, considered this was still too high. A. J. Caddick, the assistant metallurgical engineer at the mines, a man who particularly disliked the general manager, was told in 1915 to carry out experiments into ways of cutting the consumption of coke still further. Since he found it difficult to co-operate with Browning he wrote to Fielding giving notice, but his resignation was not accepted and Browning offered to increase his salary from £450 to £750. Caddick thus agreed to stay at Rio Tinto and by 1918 had successfully reduced the quantity of coke needed from 3.5 to 1 per cent. A grateful Browning arranged for Caddick's salary to be increased to £1000 a year and for him to receive a bonus of £1000.[24]

Caddick, however, found that doing the job involved difficulties which he had not anticipated. For this was the time of the first world war, and German agents were busy in Spain engaging in a whole range of activities aimed at injuring British companies there and thus damaging the United Kingdom's war effort. Thus Caddick's experiments to reduce the amount of coke used in the smelting process were represented in German propaganda, disseminated in pamphlets and anti-British newspapers, as being harmful to the Spanish coal-mining industry. Caddick wrote later:

> 'Throughout the experiments, owing to enemy propaganda and outside influences, difficulty was experienced with labour, so much so that alterations in practice were published in the local syndicalist pamphlets, comparing the old amount of fuel used with that which we were using. At times the campaign reached such an intensity that over some months I was frequently threatened on account of what was termed locally *el conflicto de carbon'* [the coal conflict].

It was Browning's time as manager that saw the introduction of electricity on a large scale in the mines. Electric lighting, produced by steam-driven machinery, had been first introduced as early as 1882 in part of the south lode open-cast and underground workings, but in 1895 the machine had been put into the stores and the company had tried, unsuccessfully, to sell it. However, the early disillusionment seems to have passed and in 1902 electricity was used to light the south lode underground pumping station and to work the cages. At the same time, electricity was introduced at the then new Bessemer smelter for lighting and for operating the 60-ton crane used to lift the Bessemer converter shells.[26]

*The central electric plant, Rio Tinto, 1909.*

In 1907 Fielding took the view that 'it is more economical to create power at centres in large units and distribute electric current to the points where it is used, than by the old-fashioned method of numerous little steam-engines working locally'.[27] A power-station was built close to the south dam on the site of the first smelter at Huerta Romana, in a little valley which had once been fertile farm land. It came into operation in 1909 and gradually power was supplied to the various departments and Bella Vista, though the other villages were not lit electrically until 1920. The arrival of the electricity age had other than industrial uses. It made it possible, in 1922, for the company to build a cinema and then, since this proved so popular, to convert the Rio Tinto theatre into another. When, in 1929, the British employees pointed out the difficulty which their wives had in finding servants prepared to work in the intense heat of their Bella Vista kitchens during the summer months, the board decided it might help if electric stoves were to replace coal fires, and agreed to provide electricity for this purpose free of charge.[28]

The growing use of electrical current during Browning's time is shown by the production statistics of the electrical department. In 1910, 3,564,000 kilowatt hours were supplied. By 1912, 8,520,000. This increased to 10,615,000 in 1917 and 12,284,000 in 1922. Between the latter date and 1927 the amount supplied almost doubled: 22,660,000 kilowatt hours.*[29]

*By 1952 the amount had grown to 38,156,000 kilowatt hours. The Rio Tinto central electricity station was shut down in January 1966 and electricity is now purchased from the *Compañía Sevillana de Electricidad.*

The war years of 1914–18 were difficult for all the British staff at the mines but particularly so for Browning. All the younger men volunteered to join the war and went away to the battlefields of Flanders and of France leaving only the older staff to carry on with greatly increased work loads. The general manager became the busiest of all and, with Fielding and the other directors involved in Britain with government committees and other aspects of the war effort, Browning became absolute ruler of the mines. One matter which gave him much trouble during those years was the increasingly strong attack on the company by those Spanish newspapers and politicians who were favourably disposed to Germany, and saw the company and its staff as the principal representatives of British influence in Spain.

The first world war, too, was a time of continual industrial unrest at Rio Tinto. Strike followed strike in rapid succession. It was Browning's responsibility to resolve them, and to limit the violence which accompanied them while protecting mines' property and safeguarding the British staff and their families. Added to these problems were the constant threats against his own life which occasional assassination attempts proved had to be taken seriously. A lesser man, without Browning's never-flagging energy and nerves of steel, might not have coped. But Browning did. To his staff he always seemed to be in ten different places at once, giving instructions, lending a helping hand and himself taking charge of those jobs which carried an element of danger.

Young José Sánchez-Mora, one morning at the beginning of the

BELOW
*Bella Vista memorial to the British first-world-war dead from Rio Tinto. (The column is from the Roman necropolis at Dehesa.)*

BELOW RIGHT
*First world war: roll of honour of British volunteers from the mines in the club inglés. Photo of 1917.*

war, looked out of his window in Huelva and saw a procession of Spanish workmen headed by Browning marching towards the company pier. Eladio Egocheaga, a communist organizer, had called the pier workers out on strike the day before when they were loading a boat with a cargo of copper precipitate for Port Talbot. Browning saw Sánchez-Mora and waved a greeting, shouting, 'Hola! Pepe, how do you like my blacklegs?' When Browning got them to the pier they were attacked by the strikers and had to fight them off. For the rest of the strike civil guards gave protection, but each day Browning travelled from Rio Tinto to Huelva in order to take the strike-breakers into work, for he would not expose them to a risk without sharing it.[30]

Browning had attracted the hatred of individual workmen since his first arrival at the mines, and had always run the risk of sudden assault by an embittered or mentally deranged employee. Between 1908 and 1914 there were at least three attacks made on him by men using knives. He was, however, never injured, for he was always too quick for his assailants. But these attempts were the reason why he was always accompanied by a body-guard in the person of his orderly, Manuel Iglesias, and why he always went heavily armed. As he rode through the mines, his Winchester repeating rifle (the gun which is given the credit for winning the American west) was strapped to his horse's side and always within easy reach. According to Rio Tinto tradition it had always a bullet in the breech and was kept half-cocked, ready for action. On his right hip he carried, slung low on the gun-belt, cowboy fashion, the revolver which he had used in Mexico; and at times of industrial trouble he wore a second. Always, in the inside pocket of his jacket, he carried a Browning automatic which, when he entered his office, or visited someone else in theirs, he placed on the table in front of him since, in the event of a sudden attack, he would not have had time to draw his revolver from its holster.

During one of the wartime strikes a would-be assassin, brandishing a pistol, burst through the outer office, making the clerks dive for cover, and into Browning's room. The general manager, aware of the danger of the situation, sat at his desk reading a report, not looking at the man, but trusting that he would not shoot someone whose eyes were averted, and taking comfort in the fact that the Browning automatic was in front of him if it became necessary to shoot it out. The minutes ticked by in silence. The workman's nerve cracked and he fled without firing a shot.[31]

The year 1919 brought personal tragedy for Browning. His son, a schoolboy at Rugby, died in the Spanish influenza epidemic. Soon afterwards Browning's wife, too, was carried to her grave after an unsuccessful operation for cancer. Browning became a desperately lonely man, living alone with only servants for company in the great house to which he had taken his beloved Alicia as a young bride, and in which they had watched their child growing up through infancy and boyhood. Sometimes in the evenings when the loneliness was too much for him to bear, he would take his guitar and board one of the trains for Huelva. There at the house of José Sánchez-Mora and his mother, he would sing melancholy songs learnt long before in Mexico, and talk of the old days in the *sierra* Madre.

It was a long time before Browning recovered from his double loss, and the young Spanish lawyer was pleased when at last he noticed that his friend was shaking off the intense sadness which had gripped him for so many months and was beginning once again to lead a social life. At this time Browning began to acquire a somewhat different reputation, for although in his fifties he looked much younger and, since his personal magnetism and his importance in the province added to his physical presence, many young Spanish women made clear his irresistible attraction for them.[32]

Thus in the loneliness of this brief but intense interlude of bachelor existence Browning began to spend weekends in Seville, which at that time had a reputation as one of the most easy-going and unconventional cities in Spain, and took to frequenting some of its night-clubs. Earlier he had forbidden his British staff to visit these because of the reputation they had for immorality, but despite the prohibition he

*Walter Browning (2nd row from bottom, 6th figure from the right) and the British staff at the mines after the young men had returned from the first world war.*

sometimes found himself face to face in them with Englishmen from the mines. Since he accepted the social obligations that went with his position, he would invite such men to his table but normally they felt considerable embarrassment at having been discovered and sought an early excuse to leave and return to Rio Tinto.[33] His openness in breaking his own rule and his lack of hypocrisy offended some of the British community but appealed to that element in the Spanish temperament which seems attracted to the absolutism of the patrician ruler and was readily accepted by his Spanish friends.

But it was not long before Walter Browning found that he did not need to go to Seville at weekends to find his recreation. The friendships he started to make with women from the mines' area were distinguished by his way of expressing his regard: most of the company's houses were single-storied, and very often overcrowded, and he ordered the lands' department to add an extra storey to the homes of those young Spanish women who enjoyed his friendship.[34] And so, in this as in everything he did, he lived his life openly and to the full, without fear and oblivious to criticism and censure.

When the war was over, Browning's problems in the field of industrial relations did not cease. The young Englishmen returning to Rio Tinto after demobilization noticed how the atmosphere had altered in their absence. Herbert Potts of the smelting department wrote of his return in May 1919 that:

*Gordon Douglas.*

'The mines were still working at full production . . . and the men were in steady and continuous employment and earning good wages; but I seemed to detect a difference in their general attitude and bearing: they were more restless and less amenable to discipline.'[35]

The next year, 1920, there occurred the longest series of strikes in the history of the mines. From April onwards there was scarcely a week when at least one department was not closed, and the end of the year saw the whole of the mines paralyzed in a bitter conflict that spread over into the early months of 1921. Once again Browning faced a time of great labour and trial. His refusal, as he put it, to be dictated to by the workmen's representatives, his attempts to end the strikes at the conference table and in confrontation with the workmen (when he led the blacklegs whom he had recruited into work) proved yet again that he was fearless and as hard as nails.

One evening during the disturbances as he was riding home to Bella Vista from la Mina, with his assistant managers (Gordon Douglas and George Wynter Gray) in front, and his faithful Manuel Iglesias by his side, a workman stepped out from behind a house at the corner of the street with a raised shot-gun. As the man discharged

*George Wynter Gray.*

both barrels, Browning reared up his horse and pulled it to one side to miss the shots, and then rode the man down. Jumping from his horse, he seized his assailant by the scruff of the neck and the seat of his trousers, raised him into the air and threw him at the feet of the civil guards who were running to the scene. 'This is yours', was his only comment. Then, to the admiration of the gathering crowd, many of whom would have described themselves as his enemies, he calmly mounted his horse and rode away chatting unconcernedly.

A few minutes later, in his home at Huelva, José Sánchez-Mora received a telephone call over the direct line which linked his house with the *casa grande* at the mines. These two telephones had been fitted with early scrambling devices to make conversations between the manager and the lawyer unintelligible to possible 'phone-tappers, and once the scramblers had been switched on Browning said, 'Pepe, there's been an attempt on my life. Rather more serious than anything before. I think we'll have to do something about it. Can you come to the mines straight away?'

Sánchez-Mora took the next train to Rio Tinto where he was met by Gordon Douglas and conducted to the *casa grande*. When they were alone, Browning described what had happened and asked the lawyer if he thought they should prosecute the assailant. Sánchez-Mora insisted that they should, especially since they had two impeccable British witnesses to sustain any charge of attempted murder. Douglas was called into the room and gave substantially the same account of the incident as Browning. Wynter Gray was then sent for. At last he arrived wearing full evening dress and declaring that he was in a hurry to get away for dinner.

Fifty years after the event the lawyer could recall the scene vividly: Browning sprawled in an armchair, jacketless, still in jodhpurs and boots; Gray standing just inside the door of the room elegant in his black dress-suit, casually unbuttoning and pulling off one of his white gloves. The Spanish lawyer knew of the animosity between Browning and his assistant managers (and, indeed, most of the British staff) but he was unprepared for Gray's reply to his question as to whether he could add anything to Douglas's narrative. 'As far as I am concerned,' said Gray, looking directly at Browning, 'I never saw a thing.'

Wearily, the general manager gestured with his hand as a sign of dismissal. Some seconds passed after the door was shut before Browning spoke. 'I knew that he disliked me, but . . .' He did not finish the sentence, but gathered himself together, got to his feet and said briskly, 'Well, Pepe, your job when the prisoner appears in court at Valverde is to tell the judge that I do not intend to proceed with any charges and to ask for the man's release.'[36]

*Sir Rhys Williams, bart., KC, MP.*

As the strikes dragged on, the Rothschild family told Fielding that they wished to send a representative to the mines to try and get a settlement. Their nominee was Sir Rhys Williams, bart., KC, a Liberal member of parliament and a skilled negotiator. Browning told Sánchez-Mora that he resented having the conduct of the matter taken out of his hands, but that it was impossible for Fielding or him to prevent it.

In fact Williams found his task extremely difficult, and Browning's policy of starving the strikers into submission was probably more successful in getting a return to work than any of Williams's efforts. However, the Rothschilds' emissary saw his work as laying the foundations for a future in which a happier atmosphere would be able to permeate industrial relations at Rio Tinto, and consequently his inquiries were wide-ranging. He diagnosed the causes of the unrest, laying much of the blame on outside agitators, but placing a great deal of the responsibility on the company and especially on Walter Browning.

Williams accepted that the general manager was a superb administrator who had unquestioningly and devotedly served the company for many years, but found that Browning was not able to delegate responsibility. 'He is a man of immense willpower who demands from his subordinates the same uncritical loyalty that he gives to the company. His control of every facet of life at Rio Tinto has aroused the dislike not only of the Spaniards but also of many of the British staff.' Williams found that the two assistant managers were so only in name and that Browning completely ignored them. Consequently there was a failure of communication, 'a disconnection and loss of touch in the most important section of the mines' hierarchy.'[37]

Williams agreed with Douglas and Gray that many of the problems at Rio Tinto were due to the wide social gulf between the British staff and the senior Spanish employees. Many of the British believed that it was official company policy to forbid the entry of Spaniards to Bella Vista village for social purposes, and to discourage contact out of working hours. Williams found that some of the British staff would like to see a club of the sort suggested to him by the two assistant managers, where senior employees of both nationalities would be free to mix together.[38] He listened sympathetically when Douglas and Gray compared Rio Tinto with the French-owned Peñarroya mines also in Spain. There were, of course, far fewer differences between French and Spanish modes of living than between those of the British and the Spaniards; but at Peñarroya there were many more Spaniards employed in positions of responsibility than at Rio Tinto. ('For decorative and political reasons', was Browning's marginal comment on this section of the report.) Housing at Peñarroya was allocated by rank alone; there was a common recreational club for all staff; and the same privileges when it came to providing free fuel, lighting and rail-travel. The fact that at Rio Tinto any British employee, however lowly his status, enjoyed many benefits not available to the Spanish staff undoubtedly rankled with the latter.[39]

Despite the criticisms of him, both direct and inherent, contained in the report, Browning was not a man to care overmuch. He was doing his job and doing it superbly according to his lights. Moreover, his friend, Sir Charles Fielding, who presided over the company's affairs in London, would see to it that he was protected. Indeed, Fielding largely suppressed Williams's report to the board of directors, and saw to it that Browning's position was not undermined.[40] From then on Browning had no more trouble with the unions at the mines, for the accession to power of General Miguel Primo de Rivera in 1923 provided rigorous control over organized labour.

Wynter Gray resigned in June 1921 when it became obvious that little would change as a result of Williams's inquiry, though his colleague Gordon Douglas carried on until 1927 when he retired early at the age of fifty-eight.[41] In fact it would have been impossible to impose too many changes while Browning was manager, and the only recommendations in Williams's report which Fielding was prepared to act on were all relatively minor. A joint committee of workmen and staff representatives was set up to discuss workmen's problems. A troop of scouts was formed at the mines as a way of encouraging Spanish boys to lead open-air lives and in the hope of inculcating in them a somewhat better self-discipline than possessed by some of their elders. The board of directors voted to provide each boy joining the troop with a hat and a belt at the company's expense,[42] although another of Sir Rhys Williams's suggestions, that Browning should organize gymkhanas for the Spanish children, was not proceeded with.

The scouts' troop in fact proved highly successful as a social activity amongst the Spanish boys until the civil war of 1936–9 brought scouting to an end. A Rio Tinto contingent of twelve boys sent to a Madrid rally in the 1920s distinguished itself sufficiently for them to be chosen to go to an international jamboree at Liverpool, and the company paid half the boys' expenses. Each scout returned with tales to tell and a medal by which to commemorate the event. That it was a memorable occasion is shown by the example of a man who had been part of the contingent, and distinguished himself years later in the civil war on the side of the victorious nationalists. He was made *alcalde* of one of the mines' villages and on his official appearances, when he wore his *trophées de guerre*, there amongst them was the medal that commemorated his exploits at Liverpool.[43]

*Albert first Viscount Milner, KG.*

The year 1922 saw a boardroom dispute. It seems the Rothschilds were not happy with Fielding as chairman any longer and he had to make way for the distinguished statesman, Lord Milner, KG, though he was to remain a director for another three years. [44] However, Browning experienced little difficulty with the new chairman who was also welcomed by the British staff, with whom he proved extremely popular on account of his interest in staff welfare.

The British servicemen, returned to Rio Tinto from the first world war, many of them newly married, had soon found that they could not live on their pre-war pay. The company had paid them their full salaries during their time in the army but the cost of setting up house at Rio Tinto in the immediate post-war period quickly ate into their savings. Browning arranged salary increases for most of them of 100 per cent spread over two years, but as living costs continued to escalate with inflation during the 1920s they found, as did all the staff in Spain, that life grew rapidly more expensive.[45] The new chairman, Lord Milner, and another director, Sir Arthur Steel-Maitland, bart., spent a considerable amount of time at the mines. On these occasions they were ready to discuss any problem with members of the staff and won much goodwill when they ordered salary increases designed to keep staff pay abreast of the increasing cost of living.

The dance held at the mines in March 1923 to celebrate the company's fiftieth anniversary was therefore a somewhat merrier affair than might have been anticipated a few months earlier, when the staff had been seething with discontent about their salaries and remembering the great profits made by the company during the war. A party for Lord Milner a short while later was happier still since it followed the announcement that the company was establishing a pension fund for the staff, devised by Milner and Steel-Maitland, and that no longer would British employees be dependent in retirement on *ex gratia* pensions. Potts wrote: 'We admired and trusted these [two] men who were gentlemen, and we felt that they did take an interest in the welfare of the staff.' However, in 1925 Milner died and was succeeded by the Right Honourable Sir Auckland Geddes GCMG, KCB, (later Lord Geddes) a former British minister and ambassador to the United States[47] ('that perisher Geddes' as Potts called him). Shortly afterwards Fielding was retired completely from the company's affairs.[48]

Meanwhile (in 1924) Browning had remarried, his second wife being a remarkably beautiful young Englishwoman of whom he was inordinately fond, and whom he met when on holiday in Eastbourne.

For a wedding present, he offered to give her a house, exquisitely furnished and decorated. So he acquired an old monastic building near the village of los Frailes, a beautiful house with splendid views, where she could go when she wanted to get away from Rio Tinto.[49]

To some of the British staff it seemed that Browning began to mellow, though whether this was due to his remarriage, or simply because they were all brought together in bonds of sympathy in their mutual dislike of the austere and chilling Geddes, is not evident. In 1926 Herbert Potts went to see Browning to ask him for an increase in salary. 'What do you want a rise for? I've been here for eighteen years and have never had to spend a penny of my salary all that time', chuckled Browning good-humouredly. 'I wish I had half your complaint', retorted Potts, and in peals of laughter Browning told him he could have his rise.[50]

The conflict between Browning and Geddes began even before the new chairman visited the mines for the first time. The continuous stream of instructions he cabled to Browning, and the latter's resentment of what he considered to be Geddes's interference in matters pertaining solely to the discretion of the general manager, made both men dislike each other before they had even met.

When the time came for Geddes to inspect the mines he made the journey across France and Spain by train. This meant that the chairman had to change trains at Seville and proceed along the public railway as far as las Mallas where there was a junction with the company's private line. When Browning heard that Geddes was on his way he gave orders for the special train with its directors' coach, which was reserved for visiting members of the board, to be made ready to take him to meet the new chairman. But he deliberately delayed the train to ensure that Geddes was kept waiting at las Mallas and when at last the two men came face to face the white-faced and angry chairman verbally lashed the general manager on the platform in front of the Spanish railwaymen who had gathered to see the arrival of this important personage. Unperturbed, Browning shrugged it off.[51]

That night Geddes announced he would be staying for five or six months at Rio Tinto to make the mines' administration more efficient and to impose a series of economy measures. He and Browning, together with senior members of the staff, including José Sánchez-Mora, were in the *casa grande* when the pronouncement was made and Geddes started giving a series of instructions to the assembled heads of departments. Browning stopped him with a protest. 'I am the company's general manager in Spain and in charge of these men.' 'And I,' snapped Geddes, 'am the company's chairman wherever I am, and I give the orders.'[52] The staff members made their

*Sir Auckland Geddes, PC, GCMG, KCB.*

way home, reflecting that there were troubles ahead; for Browning and Geddes, despite their marked dissimilarity of character, were both strong men and both complete autocrats, each used to having his own way.

The following morning the British families heard with some amusement that Geddes, who was staying in the Bella Vista guest-house, had been overtaken during the night by the stomach upset which used to afflict most new arrivals in Spain. He had been compelled to make several rapid visits to the earth-closet in the back yard, which was the only one for the whole house, including the servants. Although the episode put Geddes in a foul humour it did some good, for he immediately gave orders for the installation of water-closets in Bella Vista's houses beginning, of course, with the guest-house.[53]

In the coming months every department at the mines came under the chairman's close scrutiny, with uncomfortable results for most of the British staff, of whom he was severely critical. Their delight was intense when the time came for him to return to London, even though they knew that he would still be keeping a watchful eye on them from there and still visiting the mines once every year with the board delegation.

In 1926 Browning's characteristically authoritarian behaviour caused considerable trouble. The company had entered into an agreement with the Peña company whereby the Rio Tinto railway would transport its mineral. Browning, who disliked the arrangement when told of it, decided he would not tolerate for long the inconveniences it was likely to cause. Soon he had an argument with the manager at Peña and told him that the deal was off. He telephoned José Sánchez-Mora to tell him to handle the legal side. It was the first the lawyer had heard of the agreement, for the chairman had not seen fit to send a copy of it to Spain. When the lawyer asked for a copy he saw, to his surprise, a clause providing that any disagreement should be settled by arbitration in London, for normally disputes over such a contract would have been settled in Spain.

He spoke to Browning on the telephone to tell him of this clause and that his action had therefore been illegal. Browning told him that he did not care and that Sánchez-Mora should prepare a case in the Valverde court to declare the contract void. The lawyer told him it was impossible to work in that way. 'If you won't do it', declared Browning, 'I'll go to Valverde and get another lawyer.' 'Do that', said Sánchez-Mora, hanging up the 'phone. Two months later Browning telephoned again on the subject.

'Pepe, I made a mistake and you were right. We are in a hell of a mess. Because of the arbitration clause, Peña have taken out an English high court writ placing a prohibition on the sale and movement of Rio Tinto minerals within the United Kingdom. And the case I've started at Valverde has back-fired. Peña are claiming vast damages and Geddes is beside himself with rage.'

Sánchez-Mora went to Valverde, withdrew the action there, and negotiated the damages to be paid in consultation with all the lawyers involved in Spain and London.[54]

From then on, everything that Browning did seemed to earn a reproof from the chairman. When Geddes was at the mines in 1927 he discovered that Browning's years of living like a king had been enjoyed entirely at the company's expense, and that it was literally true that, under the arrangements dating back to Fielding's time, all Browning's personal expenditure had been charged to company funds.

In the August of that year Browning was asked to resign. He sent his letter of resignation giving the necessary six months' notice and a month later took his wife and her maids on holiday. No sooner had he arrived in Cornwall to begin his vacation than he was sent a letter, signed not by the chairman but by the company secretary, which informed him that the resignation had been accepted with immediate effect, and that he was not, under any circumstances, to return to the mines. It added that the company would take possession of the house at los Frailes which he had given as a wedding-present to his wife, and that in time the chairman would visit him wherever he might be living in order to present him with some silver plate, and Mrs Browning with some jewellery, to mark the termination of his twenty years in the company's employment (gifts which Browning refused to accept). The board also refused to give Walter Browning his expected pension.[55]

He wrote a bitterly angry letter to Geddes in reply, denouncing the prohibition on his return to Rio Tinto as the act of a mean, contemptible man, all the worse for being calculated. If he were to abide by the order it would present serious difficulties in arranging the packing of his household furniture and effects, and would mean that he would be unable to say farewell to his servants who had been faithful to him during the troubles and tragedies of twenty years. Consequently, it was not an instruction that he would accept.[56] The reproaches this letter contained must have stung Auckland Geddes, for it was not filed but left locked in Geddes's desk for nineteen years.*

Browning also visited the London office to remind the company that he had not drawn a penny of his salary or of the commission which he had earned since the day he had started at the mines. This

*On the resignation of Lord Geddes as chairman in 1947, during a period of bad health, the letter was found in his desk with various miscellaneous documents.

*Los Frailes.*

money had been paid into the company's staff provident fund where it had been earning interest at 4 per cent a year and had built up to over a quarter of a million pounds. In view of the treatment he had received he demanded immediate repayment. In January 1928 the board of directors considered his application and replied that the company could not find such a large sum at a time of financial difficulty and offered him £20,000 on account.[57]

He replied announcing his intention to take legal action and saying that all communications should henceforward be sent to Mayo and Elder, his solicitors. A letter addressed to him at his holiday address by Geddes, asking him to be reasonable, was returned unopened and written across the back of the flap were the words, 'Any correspondence for my attention must be sent through my solicitors.' The company arranged a bank loan and repaid him in full the following November.[58]

When the chairman was next at the mines he gave instructions to P. S. Couldrey, Browning's successor, about the house at los Frailes: it was to be razed down to the ground. Couldrey asked if it could not be kept to provide a week-end rest-house or hunting lodge for the staff, but Geddes was unrelenting. Every wall had to be pulled down, every stone dismantled, every stick destroyed. The British and the Spaniards alike were puzzled at his determination. It was as if Geddes was moved so deeply by animosity for the former manager that he was trying to blot out all remembrance of him.[59]

Browning and his wife returned to Spain and bought an old house in Seville of great charm and beauty, from which he could keep in touch with such friends as he had. He acquired another house in Kent to serve as the family home in England, but he was still only sixty-two and not anxious to retire. He had at one time hoped to continue working at Rio Tinto until he was seventy, for he was still remarkably active and was assumed by those that did not know better to be only about fifty-five. He used his money to start a company to grow rice on the marshland near Seville, but the scheme failed and he was left considerably lighter in pocket.* Although Browning never visited the mines again after moving away, he was not forgotten by the British or the Spaniards and at Rio Tinto they speak of him still with awe and admiration. He lived on to be seventy-seven, dying in Kent at Christmastide in 1943.

His Seville house was bought after his death by an old friend, Jack Adam. It was sold twenty-seven years later when the latter's widow Mrs Erica Adam, and her son John, moved its contents to another house a few miles outside the city. After their arrival they noticed an old trunk which they had not seen before; on its side were the initials

*Although Browning's project was unsuccessful, the nationalist government revived his plans following the civil war and the rice plantations which he had once owned then became a profitable enterprise.

*WJB*. Opening it they found, laid across the top of the clothes, Browning's Winchester repeating rifle which he had brought with him from Mexico: a bullet in the breech and half-cocked ready for action.

*Walter Browning: a photograph taken at the end of his time at Rio Tinto.*

*The Rio Tinto pier at Huelva which played an important part in the expansion of the economy of the province.*

15

# *The labyrinth*

## The company and Spanish politics: 1888-1931

The Rio Tinto Company was not the first large-scale foreign enterprise to be established in Spain. A number of important concerns, including mining companies such as Tharsis, were already operating in 1873. Nonetheless, within a few years of that date the Rio Tinto Company was making a greater impact on Spanish commercial and social life than any other single business. The size of its operations and the amount of money it poured into them soon made it the biggest and most heavily capitalized commercial organization in the country.

In the nineteenth century, Spain was considerably under-developed in certain aspects of its economy compared with the rest of Europe; and the Andalusian region, with its poor soil, absentee landlords and impoverished agricultural peasantry was, despite its developing mining industry, more backward than most of the country. Consequently the rapid expansion of this large and thrusting British business led to considerable changes in the economy of Huelva province and the area around. With the Rio Tinto Company soon becoming the country's largest single private employer of labour and the provider of social services on a massive scale, there were both increasing employment opportunities and the chance of enhanced living standards for many of the labouring population. The needs of the growing number of work-people in the mining villages also gave a general stimulus to regional trade, providing a demand for agricultural produce and consumer goods.

The British exploitation of the mines, therefore, was accompanied by great benefits for the province, and in the early years of the company's existence this was appreciated by many Spaniards of all shades of political opinion save for the most extreme. Successive governments had cause to be grateful for the substantial contribution in taxes which the company was making to the national exchequer, and the award of the grand cross of Isabella the Catholic to Hugh Matheson was in part an expression of this. *El Liberal*, a newspaper which represented an important element of responsible public opinion, took the view that although 'the English foresaw an enormous benefit by the exploration of the mines of Rio Tinto. . . such riches as they have produced have been made by themselves, because in our hands nothing was created.' It also favourably compared the

*'The English vice-royalty: the throne of John Bull at Rio Tinto'; cartoon from 'España Nueva', 20 November 1920.*

treatment which the mineworkers received from the company with that which they had been given in the days of government ownership, and pointed out that 'after the mines were acquired by the Rio Tinto Company the families of some of the deceased miners were paid wages which the state still owed them.'[1]

By 1920, however, there were few Spaniards who felt able to give public support to the company in the controversies in which it found itself enmeshed. The Rio Tinto Company by that date was just as important, if not more so, in the commercial life of the country as it had been in earlier days. But attitudes had changed and the company had lost much of the former goodwill. The two assistant managers of the mines were not wide of the mark when in 1920 they informed their chairman that 'the name of the Rio Tinto Company has come to be detested throughout Spain.'[2]

This shift in public opinion was in itself a matter of little concern to the directors but it was indicative of another change that was of some consequence. For the deterioration of the company's public image went hand in hand with the company's growing involvement in the complicated labyrinth of Spanish politics. The first steps in this process were fatally easy and seemed reasonable enough at the time to those responsible for the conduct of the company's affairs. Since the scope of their activities brought them into contact with Spanish institutions and officials in a way not experienced by most other foreign businessmen in the country, and because they needed to conduct the company's vast operations in a highly flexible manner so as to be able to make the quick decisions often essential in mining, they had to be in frequent contact with the central government and a large number of politicians and civil servants.

Some involvement in politics was probably unavoidable. Right from the start the company's staff had found that they were being frustrated in certain aspects of their work by delays in obtaining various permits vital for building the railway and the pier and for importing mining equipment, and it was found convenient as early as 1874 to recruit a senior Spanish politician to advise the company in its dealings with the Spanish ministers. This was Daniel Carballo y Codesido, *Conde* de Pradére, a deputy in the *cortes* and later, as *Marqués* de Barbançon, a senator in the upper house, who was taken on the company's staff at a salary of £1,200 a year.[3]

The rapidly increasing trade union activity of the period also caused problems which the management could not ignore and necessitated action which, given the Spanish context, inevitably had a political flavour. It was the strike and riots of 1888 which first led the company to participate to a marked degree in political activity. Following the disturbances there were some sustained attacks on the company in the Spanish parliament, most of which related to the smoke problem caused by the open-air *teleras*, though some were motivated by the view that the Rio Tinto mines were the most important of Spain's natural assets and should be compulsorily acquired for exploitation by Spanish commercial interests.[4]

Less than a month after the riots, and as a result of these parliamentary assaults, a royal decree was passed into law to abolish by stages the open-air calcination process. The Rio Tinto directors declared that this was an attack on the company's privileges, for it had purchased the right to use the process together with the mines. Strongly worded protests were lodged with the prime minister and the minister of finance, and petitions were submitted to both houses of the Spanish parliament.[5] An ally was found in the *Marqués* de Oliva*, an Andalusian landowner who was apprehensive about the effect of the decree. For if the company was forced to suspend its operations, or even curtail them, it would have been disastrous for the economy of the area. The company also persuaded the Royal Academy of Medicine to appoint a committee to report on the calcination process. This, surprisingly, declared that the sulphur-dioxide smoke from the *teleras* was in no way dangerous to health and full advantage was taken of the committee's report by the board of the Rio Tinto Company in its representations to the Spanish authorities.[6]

Early in 1889, when the debate was still raging Carballo wrote to the directors resigning his position as their representative since he found he had little sympathy with the case he was expected to argue on the company's behalf. Three months were spent trying to get him to change his mind but he was adamant. However, Oliva was still urging the suspension of the decree in the senate and it was decided to reinforce his efforts by recruiting another company representative in the house of deputies. In March 1889 Enrique Bushell y Laussat, the deputy for Huelva, was taken on the pay-roll at £1,000 a year. Eventually the intensive lobbying of the government by him and Oliva led to the decree being rescinded and Bushell's salary was increased to £1,500 as a mark of appreciation.[7]

Meanwhile Hugh Matheson was having to consider the political situation at the mines. The anarchists were deeply entrenched at Rio Tinto, and their union, the Confederación Nacional del Trabajo ('national confederation of labour' commonly known as the CNT) commanded the loyalty of most of the unionists among the workmen. The communists, however, were spasmodically active at the mines and their industrial arm, the Unión General de Trabajadores ('general union of workers' or UGT), from time to time tried to recruit new

*Oliva was primarily responsible for gaining Nerva's independence from Zalamea la Real in July 1885, and in January 1891 received from the queen regent in the name of her young son, Alphonso the thirteenth, the additional title of *Marqués* de Nerva.

members from among the anarchists. This, of course, was one factor in the demonstrations that led up to the shooting incident in February 1888. Matheson accordingly introduced a policy of increased company paternalism to reduce discontent and thus the appeal of the unions for the workmen. However, although by this means Matheson was able to prevent industrial conflict at the mines he still had to face criticisms from outside, and whereas he was completely indifferent to the attacks launched on him and the company in the British press, he showed a surprising sensitivity to censure within Spain. It seems as if the misrepresentation of his business activities wounded his *amour-propre* and ran counter to that sense of social and religious mission which to a great extent characterized his overseas dealings. It was in an attempt to refute these Spanish criticisms that, as early as 1880, he had tried to persuade ministers to get a member of the Spanish royal family to visit the mines, believing that this would demonstrate the government's friendship for the company and help silence the complaints being made by local objectors to the *teleras*.

In 1882 Alphonso the twelfth himself visited the mines, going into the underground workings as well as making an above-ground inspection. Matheson considered the visit a great success from the propaganda point of view, though it cost £3,300 to arrange, most of the money apparently finding its way into the pockets of government officials.[8]

Later the princesses Isabel and Paz paid a visit which cost the company nothing beyond the provision of refreshments. While Matheson was delighted to receive them, a certain element of sourness crept into the proceedings when the two princesses were being shown some of the archaeological finds from Roman days. They admired a ring of exceptionally fine workmanship, the only one of any value ever found at Rio Tinto, and on their expressing delight at its beauty it was offered to them in the customary Spanish manner, merely as an act of politeness. To everyone's astonishment, and contrary to what was expected, they took it.[9]

Following the king's death in 1885, and his succession by his posthumously-born son Alphonso the thirteenth, Matheson tried to get the widowed queen regent to go to Rio Tinto. After the disturbance of 1888 he redoubled his efforts, obtaining in 1892 a ministerial promise that the visit would take place; but the arrangements were thrown into confusion when the queen was told of them and refused to make the journey.[10]

In 1893 the board of directors considered the Spanish general election that was due to be held. Enrique Bushell was again contesting his seat at Huelva but he had certain apprehensions about whether it would be possible to keep it since the company's opposition to the *caciquismo* system, and the mounting criticism of the foreigners exploiting the Spanish mines, had lost him some support in official quarters and reduced his chances. The board decided to authorize its Huelva director, Heinrich Doetsch, to spend £2,000 of company funds to secure Bushell's return.[11] Presumably this large sum was more than the other candidates could afford and it sufficed to get him re-elected.[12]

However, as the general election of 1896 approached it was decided that Huelva had become too unsafe, and instructions were given that March to company officials to ensure Bushell's election at Valverde del Camino, a town not far from the mines and with a smaller population which was likely to be more tractable than the Huelva electorate. Then, shortly before the election, the *Marqués* de Nerva asked Matheson if he could have the company's candidature and support for his son. By this time the directors had become annoyed with what they considered to be Bushell's lack of persistence in urging the company's views on Spanish ministers. Bushell, when told this, had replied that 'the government did not look upon him as if he had sufficient authority to act as the company's official representative', and had proposed that Hugh Matheson should write telling the Spanish prime minister that 'Mr Bushell is the person in Spain who has the entire confidence and representation of the company.' Matheson told Bushell that he had been 'appointed to a high post, most liberally paid . . . Your remuneration being so liberal would by itself place you in such an influential position as would enable you to protect [the company's] interests sufficiently.'

He had not been prepared to give Bushell the *carte blanche* which the deputy had requested, and the strained relations therefore made Matheson receptive to the *Marqués* de Nerva's request. The chairman summoned Bushell to London and got him to withdraw from the election, assigning him a pension of £400 a year for life.*[13] The *marqués* undertook, in return for the nomination, that both he and his son would serve the company as steadfast allies in the *cortes* and in any discussions with the government.[14]

Nerva and his son Oliva inherited from Bushell an astute adviser in the person of Gabriel Rodriguez, a lawyer with the entrée into the highest echelons of Madrid society. Rodriguez had been appointed a legal adviser to the company in 1873 but his political contacts were as important as his legal expertise. His office was in effect the company's Madrid headquarters and though his retainer was only £300 per annum his value was such that in some years his annual bonus was as much as £2,000.[15] On his death in December 1901 his son, Antonio,

* Bushell drew this until his death on 11 May 1908.

replaced him but after three years it was decided that Antonio should henceforward deal only with legal work, and the political side of his activities was transferred to J. Hervas Valero, a *madrileño* who in 1888 at the age of twenty-two had become a member of the London staff. In 1902 he had accompanied Matheson's successor as chairman, John Keswick, to Madrid in connexion with a Spanish mining commission which was then meeting, and was left behind to become the company's official representative at £1,100 a year.*[16]

In 1898 the company found itself the subject of considerable hostility and press criticism following the outbreak of war between Spain and the United States over Cuba. This was part of a strong wave of resentment that was building up in Spain against foreigners in general, and the board of directors was anxious to show that the company, despite its strong trading links with the United States, was sympathetic to the Spanish cause. Therefore they voted 40,000 *pesetas* to the Spanish national fund which the government had set up to provide relief for Spaniards suffering as a result of the war; and when there were rumours of an expedition of American warships against Spain the company immediately offered its holiday houses at Punta Umbría for billeting soldiers.[17] But because of the strong animosity against the United States the SS *Elwick*, due to take on a load of Rio Tinto pyrites at Huelva for the Pennsylvania Salt Manufacturing Company, was forced to depart empty. Her captain reported that he and his crew had been threatened by the pier-workers and town officials, and finally chased out to sea by naval torpedo boats.[18]

The war also caused problems at the mines, for the United States cut off wheat exports causing a bread shortage in Spain and a sharp rise in prices. The company made arrangements to obtain large quantities of flour from Britain for sending to the mines' villages but before it arrived the women of Nerva rioted for two days against the temporarily increased prices. Thirty-three of them were arrested by the civil guards and though William Rich, the general manager, was ordered by the directors to secure their release and to subsidize the cost of bread in the mining villages from the mines' funds, the company gained little in popularity.[19] In July, Rich was warned by the military governor of the province that the British staff and their families in Huelva town were in danger of attack from the Spanish population and they were withdrawn to the greater safety of Bella Vista.[20]

Another result of the Spanish-American war was a considerable rise in the company's Spanish taxes. The directors resented the first general increases in taxation in 1898, which seemed to them to single out the foreign companies and to make them pay most of the extra

*He became active in politics and remained as the company's representative until 1927, when he was receiving £1,800 a year. In 1935 he became governor of the Banco Exterior.

burden. The board refrained from raising the matter with the Spanish government, however, as it anticipated that the extra taxes would be offset by increased copper prices, but it regretted its silence when in 1900 the company's Spanish tax bill rose from £27,000 to £99,000. The next year there was another rise to £101,000. Additional taxes demanded in 1902 led the directors to challenge unsuccessfully both their legality and the way in which the amounts claimed had been calculated. Between 1900 and 1905 Spain's government experienced financial difficulties as a consequence of the war with the United States, and this not unnaturally led to the imposition of a variety of new taxes on those industries, such as mining, best able to bear them. However, the Rio Tinto Company was concerned at the trend and set about recruiting further allies in the *cortes*, including a number of former ministers. In 1905 Charles Fielding, then recently appointed as chairman, reached a compromise agreement with the finance ministry whereby, in return for a payment of £33,000, the government withdrew its demands and undertook to use a new formula for its future calculations which, for a while at least, led to less heavy claims.[21]

Once involved in Spanish politics, the company's officials found themselves becoming ever more deeply caught up in the system. Closer association with the government led to greater criticism from the opposition, and though a deal could often be struck with critics, at times the company's staff in Spain felt obliged to act against them. In the autumn of 1903, for example, with municipal elections in the offing, Valero wrote to Carlyle the mines' general manager about the annoyance being caused by certain town councillors in the province, and especially members of the *ayuntamientos* of Valverde and Zalamea. The offenders, who had been 'openly opposed to the company' had been appointed through the civil governor of Huelva province at the request of three of the provincial deputies in accordance with the normal arrangements under the *caciquismo* system, whereby an agreed balance was maintained by the political parties by deliberate fraud in the counting of votes.

However, the company, which had rapidly learned how to manipulate the Spanish political system and had no desire to accommodate the *caciques*, decided it would not tolerate any active opponents on the town councils within the province:

'The civil governor [Valero wrote] has instructions from the minister of the interior that there should be no representatives of Sr Burgos or Limon or Sancho Hidalgo . . . a decision adopted by the government to accede to my desires and request . . . The government is not prepared to support [in future elections] any of the deputies above referred to. With regard to the *alcaldes*, please let me know who you wish to remain. I suppose the civil governor has kept the promise he made you in my presence to support the company's candidates against any other candidates, as he had (as you are aware) instructions to do.'

The directors who made up the board delegation which visited Spain in the January and February of 1904 reviewed the political scene in their report:

'For many important reasons, including the controlling of unruly spirits during a strike, the economic administration of municipal affairs, (and consequent reasonable local taxation) and of the mines, and for the efficiency of sanitary matters, it is essential that, if possible, honest *alcaldes* of the towns of Rio Tinto, Nerva and, if possible, Zalamea, should be in office. These officials are at present appointed by a royal order from Madrid, and this royal order is usually granted at the request of the member of parliament, which for our district is that for Valverde . . . The election for the *cortes* is a pure farce. The *alcaldes* and scrutineers who give in the list of votes recorded do so in accordance with instructions and not in accordance with votes cast. Mr Carlyle has had great trouble in getting the appointment of *alcaldes* given to proper persons and is of the opinion that the matter would be much simplified if in future the member for Valverde, while being a satisfactory person to the government in power, be also a man who in local politics would serve the interests of the company i.e. be in favour of honest municipal affairs.'

By this time the Rio Tinto Company was looking for a new deputy for Valverde who would be rather more industrious on the company's behalf than his immediate predecessors. The 'powerful friends of Mr Valero in Madrid' urged the board to give him the nomination, 'and it is undoubted that the company's influence would secure his return. Sr Puigcerver [an influential senator and ex-minister in receipt of a Rio Tinto Company salary as a special political adviser], however, stated his opinion that the company's [official] representative could not with advantage to its interests be a party politician . . . The present minister of finance, Sr Osma, was most emphatic in his opinion that our representative should not be in the *cortes* and this is the unanimous feeling of the deputation.' In consequence Valero was not given the candidature and had to wait a while longer before entering the *cortes*.

Thus by 1905 the Rio Tinto Company was effectively controlling much of the political life of Huelva province, while in Madrid it cultivated ministers and parliamentarians on a grand scale. (One pleasing outcome from the 1904 deputation visit was a promise from Osma, the minister of finance, 'to use his influence to get the coastwise navigation tax remitted'.) As long as such men were content to be on the company's side then Rio Tinto officials in Spain found their task of running the mines, exporting the produce, and administering the villages where their employees lived greatly simplified. But despite the gerrymandering of the then political system in Spain, public opinion was still an important element in the conduct of government; inevitably so in a country with a long history of riots and civil conflict. Bad publicity for the company, on a subject such as labour troubles, if sustained over a long period of time, could easily lead the company's political allies to publicly disassociate themselves from their involvement with it and to reject ostentatiously the company's demands. And this, in the course of the next fifteen years, was precisely what happened.

The years 1900 onwards saw the beginning of a steady deterioration in industrial relations at Rio Tinto. This was due in part to a growing self-awareness on the part of the trade unionists of their potential strength as their numbers increased, but there was also a change in the company's attitude to its workmen which contributed substantially to the worsening relationship between employers and workmen. Carlyle, the new manager, departed from the view, formerly held by Rich and Matheson, that trouble should be deflected before it became serious. He preferred to meet the unions in a head-on collision and to negotiate settlements only when hunger was driving the strikers back to work. His tactics, in the short run, may have lessened the pay increases which the company had to meet, but it also strengthened the appeal of trade unionism for the men and contributed to a rising swell of bitterness.

From 1900 until the accession to power of General Miguel Primo de Rivera in 1923 the company was plagued with frequent strikes. Most were minor stoppages involving men in only one department at a time but they caused considerable trouble to the British staff and involved prolonged discussions with government officials in Madrid and Huelva. Three of the strikes, occurring in 1900, 1913 and 1920, were of major proportions and brought all work to a stop, and were accompanied by displays of violence and damage to company property that caused concern for the safety of the British staff.

The June 1900 strike, the first to take place following the stoppage of February 1888, was called by the anarchist CNT ostensibly in opposition to the introduction of new machinery. In 1897 drilling by compressed air had been introduced and then, in 1900, extended throughout the mines. The union demanded higher wages for the

*Drillers at work.*

drillers which the board refused on the grounds that it was already paying better rates than any other mining company. The communist UGT, seeing a chance to win supporters from its rival union, demanded wage increases in all departments and the CNT, not wishing to be left behind, made similar demands and called for a complete closure of all departments at Rio Tinto and Huelva.[22]

The closure was accompanied by riots, explosions and damage to some of the mines' buildings. The directors instructed the company's representatives in Madrid to demand additional troops to reinforce the cavalrymen permanently stationed at the mines, and 300 infantrymen and 18 officers were despatched. It was noticed by the Spanish press that the troops took their instructions from Carlyle and this caused some criticism, not only of what seemed to be arrogance on the part of the company but also of what appeared to be timidity on the part of the government. Carlyle certainly felt it was his job to control the soldiers, and when the strike eventually collapsed he informed his directors that he had 'allowed half of the troops . . . to leave the mines.' He reported in similar terms when he consented to the withdrawal of the remainder.[23] When the dispirited workmen capitulated and went back to work Carlyle sacked twenty-six of their leaders, and in future such dismissals became common practice after all strikes.[24]

After the strike-free twelve years between 1888 and 1900 the board was surprised at the frequency with which stoppages began to take place. Life at the mines had improved considerably. The open-air *teleras* were disappearing rapidly with the introduction of the natural cementation process, the housing situation was greatly improved, wages were high, and the social and medical services had a well-merited reputation. The directors declared that if the men had genuine grievances they were prepared to remedy them. Doubtless, the lapse of time that had occurred between the 1888 shootings and the start of the twentieth century had produced a new generation of workmen, less timid than those who had witnessed *el tiempo de los tiros*. But there were other factors, too.

One of these was the change taking place in the higher levels of the company's administration. There was a hardening of attitudes at this time in the ranks of the directors and most senior staff and a diminution of the spirit of tolerance and adaptability which had, despite their faults, marked the administration of Matheson and Rich. This was perfectly obvious to some of the British at Rio Tinto who saw clearly the causes of discontent which provided the anarchist and the communists with fertile ground for their activities.

In part, this growing inflexibility in company administration was a reaction to the growing militancy of the labour movement, but it made a significant contribution to the second factor in the new pugnacity: the competing activities of the two unions. Thus the directors were wrong in attributing the troubles solely to political machinations beyond their control. The policy into which the board drifted was one of resistance to union claims with no coherent scheme for trying to prevent the strikes that became ever more frequent.

*icia head shaft.*

Stoppages started to occur every few months. In January 1901 striking anarchists set fire to the Alicia shaft at the ninth floor level. Within a few minutes it was aflame and the few men underground were hauled to safety just in time. The sheet of fire was so great that it reached from the shaft down the tunnels leading off it to a length of over thirty feet.[25] It took twelve months and £1,500 to repair the damage. The following October another underground strike was called in the Contramina department, the men from which ran through the tunnels into San Dionisio throwing lumps of mineral at the workmen there to drive them out and thus simulate the appearance of solidarity.[26]

The departure of Carlyle in 1907 due to his inability to get on with his new chairman, Charles Fielding, and the appointment of the redoubtable Walter Browning as general manager, was followed by a further stiffening of attitudes on the part of both the company and the unions. A strike in San Dionisio in September 1908 led Palmer, the underground manager, to recommend the dismissal of ten strikers, five of whom had attacked fellow workmen who had been reluctant to down tools; but Browning ordered the sacking of twenty to set an example and had three men, who had made threats to kill Palmer, committed to gaol. The discharged men were deprived of their houses and told to quit the mines.[27]

The greater rigidity of the Rio Tinto Company's industrial relations policy under Carlyle and Browning gave the unions new opportunities and provided an incentive to the inter-union rivalry of which the directors complained. In 1909 Manuel Fernandez Arena arrived at the mines with a team of four organizers from the Sindicato Unico and established a branch of the union at Rio Tinto which affiliated to the anarchist CNT in 1911. Early in 1913 a railwaymen's union was organized with branches at Nerva and Huelva, at either end of the company's railway line, and recruited 4,000 members in its first four months of whom 1,000 were railway employees and the rest workmen in other departments. This was a branch of the anarchist Sindicato Ferroviario General de España which sent an adviser named Bacuñana to the mines.[28]

At the same time the UGT sent Eladio Egocheaga, one of its communist officials, to Rio Tinto to repel the threat to that union posed by the activities of the anarchists. He began by infiltrating his own active members into the opposition unions where they were able to exploit the policy differences between the anarchists and to promote a fierce internal struggle for control in which a large number of the local anarchist officials were expelled from office. As a result, the Nerva and Huelva branches of the railway union disaffiliated from their anarchist parent body and joined the UGT. The purged officials then combined with Fernandez Arena's mineworkers' union and began a long and bitter fight to win back members from the rival union which continued without respite from 1913 right through the period of the first world war.[29]

1913 was a year of inter-union strikes. There were two in April, another in June and finally an almost complete closure of the mines for six weeks in October and November. The earlier stoppages were organized and directed by a group of socialists and communists, and the autumn strike by anarchists who made two demands which it was felt impossible for the company to concede: a 25 per cent increase in wages, and the abolition of all control by foreigners and Spanish foremen, i.e. workers' control. The long stoppage was extremely bitter. The Alicia shaft was once again fired by strikers on 1 November and as a result five of the British staff and two Spaniards lost their lives.[30] The thirty-fourth Soria regiment was ordered to the mines by the government at Browning's request and extra civil guards, both on foot and mounted, were moved in. The military railway corps manned the company's trains and brought in the essential supplies for the British community and the non-striking Spaniards. These *esquiroles* ('blacklegs') and their families received effective protection from the troops against threatened retaliation by some of the strikers.

At the end of November some 8,000 out of a total mines' workforce of over 14,000 men were still out. Browning was authorized to make small concessions on the wage rates and the men returned to work. The unions had also demanded a pension-scheme as of right and this the company conceded, since, before the first of the year's strikes was even called, plans had been made for a scheme which gave a pension automatically to any man retiring after twenty-five years' service.[31] But on reflection, many of the workmen felt that they had gained little from the turmoil and seven weeks later, in January 1914, the communists seized the opportunity to call yet another strike. However, on this occasion most of the men were out for only three weeks. They soon became dispirited and when the government offered to arbitrate they accepted the opportunity to return to work. The arbitration award, which made only minor adjustments to pay-scales, was hailed by the directors as fully justifying their attitude of the previous year. In addition, Browning took the chance provided by the men's exhaustion to declare redundant over 900 workers (many of whom had been prominent in the troubles) and to introduce many more pneumatic drills to cut down on manpower.[32]

The outbreak of war in 1914 closed practically all the company's continental sales' outlets and in 1915 the men were put on a short working-week of four, and at times three, days. Despite this the company was able to maintain its profits because of the wartime scarcity of copper and the resulting higher prices.[33] At the start of the war there was a strike over pay as a result of which the company decided to keep all prices in its stores in the mining villages at pre-war levels, to distribute allocations of free food to the poorest people in the mines' area and to give the men a temporary 10 per cent cost of living bonus which it was anticipated would be withdrawn when there was a general fall in prices.[34]

During the next few years discontent among the workmen reached a considerable pitch. The men, though cushioned from the worst effects of inflation, and though full-time working was restored in 1916 (there was, indeed, a small expansion in the number of the workforce), were fearful that the company would withdraw the subsidies and the 10 per cent bonus. They made many demands that these should be made permanent and also made frequent applications for higher wages, backing most of these by short stoppages in various departments.[35] The extra men taken on in 1916 were recruited in the north of Spain and included a number of communists from the coal mines of the north, some of whom were skilled agitators and provided useful reinforcements for Egocheaga.[36]

In 1917 there was a revolutionary strike throughout the whole of Spain which the government had little difficulty in suppressing. At Rio Tinto the civil authorities, with their fear of the miners' propensity for violence, moved in a regiment of Moroccan troops, with a reputation for fierceness, to keep order. The strike lasted only a few days and there was little disorder at Rio Tinto itself. Nonetheless there was considerable trouble in the villages of el Campillo and Nerva just outside the mines' boundaries, where many of the mines' employees lived. In the former *pueblo*, where the disturbances were led by the womenfolk, the troops killed a number of the demonstrators, and in the latter there were also some deaths when the soldiers opened fire to maintain order.[37]

Although Spain stayed neutral in the war, its outbreak in 1914 brought a widespread demonstration of hostility against the United Kingdom throughout the country; and the Rio Tinto Company, as the largest British concern in Spain and a much-quoted example of characteristic British arrogance and high-handedness, was singled out as the recipient of a highly critical press-campaign. In general, the conservative politicians of the country felt a considerable sympathy for Germany in the struggle and in their speeches demanded an end to British political influence in their country, and the nationalization of the Rio Tinto mines and other mining enterprises owned by British companies together with their resale to Spaniards. The Rio Tinto Company was accused in many newspapers and on many platforms of paying low wages, and of subjecting its workmen to inferior standards of living while 'bleeding the country to death' for the benefit of its shareholders.[38]

Associations of Spanish businessmen were formed, such as the Fomento de Trabajo Nacional and the Junta de Producción Nacional, which demanded the promotion of Spanish industry at the expense of foreign companies by subjecting the latter to heavy taxation, although it was not until 1919 that these commercial associations achieved their first great success. In that year the Junta de Aranceles was formed. This was a tariff commission with representatives drawn from the most vociferous of the campaigners, which increased import duties by between 30 and 40 per cent on all the essential articles required by the company to keep the mines going.[39] The following year the prime minister, Eduardo Dato, reluctantly introduced bills to increase direct taxation on foreign companies and to subject them to various trading restrictions.[40]

During the war the German government moved agents and funds into Spain to assist the anti-British agitation. The Russian Bolsheviks sent agents into the country in 1917 with the intention of seizing the imperial Russian embassy and consulates, and promoting revolutionary movements in the larger industrial centres, particularly Catalonia, Bilbao and Rio Tinto. Both the Germans and the Russians were able to achieve very little at Rio Tinto during the war. The *bolcheviques* especially were badly organized and Walter Browning was able to keep a close scrutiny over the activities of their representatives in Huelva province and at the mines. However, the directors watched the situation with some alarm, and became a little too ready to blame some of their industrial relations' problems on the foreign agents, especially following the ineffective revolutionary strike of 1917.

Browning set up his own unofficial counter-espionage service, using his immense influence and wide contacts to obtain normally confidential information about the movement of German and Bolshevik funds through the banks of Barcelona, Seville and Huelva. When Leon Trotsky, the Russian revolutionary, appeared in Barcelona, Browning assumed that one of his future projects might well be to cause trouble at Rio Tinto. Consequently the general manager took steps to ensure that he was kept informed of Trotsky's movements. When the Russian moved to Madrid he was, to Browning's relief, arrested though he was soon 'unfortunately liberated owing to the influence of one of the ministers'. After that, Trotsky was observed to move to Bilbao and then to England, much to Browning's satisfaction.*[41]

The general manager, however, still had plenty to do in trying to counteract the wartime campaign that was mounted against the company in many of the Spanish newspapers. This was in fact a sustained and skilful crusade, conducted with great energy, in which almost every accusation that could possibly be made, whether justified or not, was levelled against the Rio Tinto Company. Of the eighteen principal Spanish newspapers being published during the war, fourteen were considered by Browning to be persistently attacking the company. Foremost amongst these, and constantly appearing with sensational anti-company headlines were *ABC*, *Sol*, *la Libertad*, *el Día* and *el Socialista*.

*ABC* was a conservative publication owned by the *Marqués* de Luca de Tena, a Seville aristocrat and businessman. It was the best-produced

*From England, Trotsky returned to Russia

*The Marqués de Luca de Tena.*

of all the Spanish dailies and had the largest circulation, being markedly pro-German and strongly anti-British. Since Seville was close to the Rio Tinto mines and Luca de Tena was able to observe personally the extent of the company's influence over the economy of the region, and Browning's own wide personal power, its attacks tended to be more on target than most. Browning, however, informed his directors that *ABC*'s criticisms could be disregarded. He claimed that these were inspired by a monthly payment of 60,000 French francs by the wartime German government to Luca de Tena.[42] Even if Browning was right in this assertion (presumably he received this information from his contacts in the banks) it should not have blinded the directors of the Rio Tinto Company to the fact that Luca de Tena was too good a newspaperman to distort facts, and was hardly likely to let such payments influence his editorials. *ABC*'s reporting of events at Rio Tinto was more accurate than any other paper's while its comment was fair in that it represented not just its owner's view but reflected the concern of an important element in Spanish life.

Of the other papers, *Sol* was liberal in politics and when established in 1916 had at first been pro-British. By the end of the war, however, it had changed its attitude, largely as a result of its disapproval of the activities of the Rio Tinto Company in Spain and of the behaviour of the United Kingdom government in Ireland.* *La Libertad* inclined towards socialism. Owned by Alba, a former minister of finance, it devoted a great deal of space to articles hostile to the company, many of them written by Egocheaga the communist organizer who had taken part in the labour disputes at the mines. *El Socialista*, too, made a speciality of articles hostile to the Rio Tinto Company. *El Día* was run by Germans (and Browning understood that it was kept going almost entirely by subsidies from the German government). Consequently it was violently against Britain and seized every opportunity it could to criticize the company and thus, indirectly, the United Kingdom.[43]

Browning could find only four newspapers that were prepared to carry articles at all sympathetic to the company: *Epoca*, 'a dull newspaper belonging to and edited by the *Marqués* de Valdeiglesias', *Diario Universal*, owned by Romanones, a liberal politician, *Correspondencia de España* and *Nuevo Heraldo* (also known as *Hoy*). None of these had a particularly large circulation or any influence.[44]

The ending of the war did not diminish the newspaper criticisms, though one of the papers concerned, *el Día*, ran into financial difficulties and was eventually forced to close down. Browning was amused by an approach that was made to him by its proprietors in 1920. They offered to sell *el Día* to the Rio Tinto Company, pointing out that it would give him and his directors a useful means of replying to the

*In 1919–20 it carried a number of Sinn Fein articles provided by its correspondent in Ireland.

very propaganda which they were disseminating.[45]

At the end of 1918 another union set itself up at Rio Tinto. This was the Sindicato Católico which was growing in power throughout Spain with support from the bishops and the parish clergy. Under the direction of *Padre* Navarrez, it was opposed to the use of violence and generally moderate in the way it presented its demands. The company hoped that it would be successful in its recruitment campaign at the mines. When two branches were formed at Nerva and Zalamea they quickly obtained over 1,000 members. However, the union's moderation was not suited to the temperament of the mineworkers and by the end of 1919 its strength had declined to a couple of hundred men.[46]

At the beginning of 1920 some of the company's Huelva employees formed yet another union, the Sindicato Provincial de Obreros de Río-Tinto. This was set up by some of the older and maturer members of the clerical and supervisory Spanish staff in reaction against the anarchist and communist leaders who seemed to them to put inter-union rivalry before anything else. They confined membership exclusively to company workers and prohibited all external intervention in their affairs. In March and April 1920 the labourers' section of this union conducted a four-week strike. The demands were handed in writing to an assistant manager of the company and when Browning had considered these he met the men's representatives. During the negotiations he found them far more serious and businesslike than any other union leaders he had met in similar circumstances.

Around a table they hammered out an agreement for a return to work. In return for the abolition of the 10 per cent bonus granted in 1915 and the closing of the subsidized company store in Huelva, the workmen accepted a 50 per cent increase in wages for boys earning less than 12 *reales* a day, and an extra 6 *reales* for everyone else. This was one of the rare occasions when all parties appeared satisfied. The union calculated that the store subsidy had been worth just over 2 *reales* a day on average to each man, while the company reckoned its value had been 4 *reales*. Browning, indeed pointed out that the better-paid workmen would be worse off according to his calculations, for a man earning 30 *reales* a day plus 3 *reales* as his bonus and 4 *reales* in food subsidies had been getting the equivalent of 37 a day, whereas as a result of the settlement he would return to work for 36 *reales*. The union, though, was prepared to accept this in return for the better deal which it obtained for the lower-paid men, and was ready to enter into non-strike agreements with the company for the next five years.[47]

In the summer months of that same year of 1920 Browning was faced with a succession of work stoppages at the mines which might have been anticipated following the settlement with the Huelva union. At the beginning of June the stonemasons struck for higher pay, followed two weeks later by the muckers, or waggon-loaders, in the south lode underground workings. In the last week of the month they were joined by the muckers of the Atalaya open-cast pit. On 7 July the members of the foremen's and salaried staff's section of the Huelva Sindicato Provincial de Obreros de Río-Tinto went on strike. Since this included the railway supervisors it made for considerable difficulties in running the railway. Browning retaliated by laying off all the workmen in the marshalling yards and on the pier at Huelva in the belief that this would lead to pressure on the foremen and staff to return to work. Instead, the railway engine crews struck. From then until November only one train a day, carrying passengers and essential supplies, could be run to and from Huelva. At first it was driven and serviced by British staff, and later by government troops.[48]

On 15 August the employees in the mines' central electricity department ceased work and their example was followed the next day by almost all the Spanish clerical staff and foremen at the mines who formed themselves into another section of the Sindicato Provincial. Browning offered the Rio Tinto strikers a flat increase of 9 *reales* a day in return for the abolition of their 10 per cent bonus and the scrapping of the subsidies in the company shops. He also offered the

*Marshalling shed, Huelva, 1920.*

Huelva employees, who had earlier in the year received an additional 6 *reales*, another 3 in order to bring them into line with the offer being made to the people at the mines. The responses to these offers surprised the general manager. The Rio Tinto men rejected it with the comment that the food subsidy was worth on average to each man 8 *reales* a day (and not 4 as the company contended) and a demand for a flat increase of 12 *reales* all round. The Huelva men responded to their offer with a demand for a 50 per cent increase in wages and the dismissal of all non-union labour.[49]

In the course of the strike there was considerable rivalry between the anarchists and the communists to take charge of its direction. Eventually the victory went to neither but to the leaders of the Sindicato Provincial de Obreros de Río-Tinto whose followers compelled Egocheaga and the other outside organizers to leave the mines and allow them to conduct their own strike in their own way.[50]

In trying to resolve the conflict Browning faced considerable difficulties. The Spanish clerical staff and foremen insisted on a settlement of their claims before there was any discussion of the workmen's demands; a course for which they had the support of the railwaymen and the Huelva labourers. The workmen at the mines were more disposed to reach agreement with the company, Browning thought, and he concentrated on trying to get negotiations started with them in the belief that, if he could get them back to work, the other strikers would have to moderate their claims. But he had no success. Feelings among the men were running high. Large numbers of soldiers and civil guards were sent to the mining villages, and an attempt was made to assassinate Browning which led to a hardening of the company's reluctance to negotiate. In the end, hunger started to drive the men back to work. Some 2,000 out of the total work-force of 11,000 were back at their jobs in the last week of November. A representative of the Rothschilds', Sir Rhys Williams, bart, KC, MP, was sent out at their insistence as the majority shareholders in the company. He arrived in November but was unable to achieve a settlement until the following January (1921) when the strike was losing its momentum.[51] For his efforts he was to receive the considerable payment of £100 a day with the company meeting his travelling and living expenses.

By the agreement which Williams negotiated, and of which Browning disapproved as he was convinced that he himself could get better terms since the strike was on its last legs, there was an overall increase of 37 per cent in pay, with the company agreeing to meet the men's income tax, in return for the withdrawal of the 10 per cent bonus and the ending of the system of maintaining pre-war prices in the company stores. An arbitration committee, composed of representatives of the

*'Organised labour overthrowing capitalism.' The newspaper 'el Socialista' greeted the strike at Rio Tinto in August 1920 with this cartoon.*

company and of the workmen, was set up to allow for appeals against dismissals and to advise on future pay claims. Spanish staff and foremen were granted forty days a year for annual holidays and sick-leave. (Browning pointed out, quite correctly, that the effect of this provision would be that everyone would save up their forty days' allowance until December and then take six weeks off, bringing much of the administration to a standstill.) The company conceded, too, that promotion would be by seniority, and that any dismissed men would receive two months' wages as redundancy pay. There was also an undertaking to introduce a better pensions' scheme.[52]

The effects of the strike were felt far outside the mines. There was bitter criticism of the company in the Spanish press and considerable denunciation of the Spanish civil authorities for arresting the chairman of the Sindicato Unico and five other unionists. Though there was coverage of the more sensational incidents which took place, such as bomb attacks on the houses of non-strikers, the general tone of newspaper editorials was to give the strikers the fullest possible backing.[53]

A wave of sympathy for the workmen swept Spain and donations poured in to help keep the strike going. The members of the strike-committee of the Sindicato Provincial published the accounts of the strike-fund in the newspapers, thus winning approval as responsible men, even though there was some disbelief among the strikers themselves at the printed details of expenditure.[54] There were also many offers from throughout Spain to care for the miners' children

*Cartoon from 'la Libertad' of December 1920, when the Rio Tinto strike was beginning to collapse. Browning is shown with a mongrel, which has the face of Eduardo Dato the Spanish prime minister, at his feet. The thin and ill-clad workman comments: 'It's already known that what ruins men is evil company.'*

for the duration of the strike, and over 3,000 children were despatched to different cities to take advantage of this generosity. Their appearance was headline news in most European newspapers, though Browning was amused to notice that they were dressed for the occasion in 'unusually clean and ragged clothes, presumably to arouse national sympathy'.[55]

It certainly did this. The *Sindicato de Actores* (the actors' union) had sent its chairman to Rio Tinto in November 1920 and, acting on his report, its committee approached a number of eminent novelists, musicians, painters, poets and scientists who petitioned the king for government action to bring about a settlement of the strike. The actors' union published a manifesto directed to 'Spaniards, men of heart and conscience', which was published in all the papers:

'These men, who to keep faith with their comrades, are separated from their children to avoid the tragedy of seeing them starve to death, ask only that the company gives them a fair hearing . . . they seek only work and a standard of living appropriate for men and Spaniards. They do not want to feel like unprotected foreigners in their own country . . . They appeal to us, the artists and men of science who, united with them in their sorrow and distress, seek with heart and soul a fitting solution, with no thought of reward other than the satisfaction of having fulfilled a duty.'[56]

The *Sindicato de Actores* also arranged a number of charity performances one of which, for the benefit of Rio Tinto children, was attended by members of the royal family and of the government.[57]

*Concha Espina at the time she wrote 'Metal de los muertos'.*

Another important voice was added to the chorus of condemnation. Concha Espina, one of Spain's leading novelists, had visited Rio Tinto in the summer of 1919 to gather material for a book. She spoke to the people and was shown the scenes of earlier industrial conflict. She went away and wrote her powerful and moving novel *El metal de los muertos* ('the metal of the dead'). In it she caught the tension and the anger of the Spanish workmen, described the lives led by them and their families and reflected their determination for change. Its publication during the troubles of 1920–21 directed still more public attention to the problem of Rio Tinto.

The book was bought avidly by those who wished to know something of what lay behind the newspaper headlines and it won great sympathy for the men in their struggle. Of course, she was not interested in the company point of view and it is doubtful whether many of the workmen would have understood her repugnance for their work, or agreed with her description of the Atalaya open-cast pit as *un cancer inaudito abierto en las entrañas del titán* ('a strange open cancer in the entrails of the titan'); but her pen won the strikers much sympathy among those of the educated middle-classes who might otherwise have condemned them as anarchist or communist revolutionaries, and the picture of life at Rio Tinto which she drew for her readers was one which stayed fixed in their minds and contributed to much of the hostility expressed against the company in later years.

Sir Rhys Williams in his role as negotiator saw it as his duty to try and keep the mines strike-free in future. Some of his proposals to achieve this were somewhat original. While, for example, many of the men were still on strike in January, he had a secret approach made to *Padre* Navarrez of the Sindicato Católico asking him to go the to mines with his officials and to demand an interview with Browning. At this he would make demands on the workers' behalf which had already been agreed beforehand as acceptable, Browning would 'capitulate' and Navarrez, as the new workmen's champion, would order the men back to work. Rhys Williams also proposed that the company should pledge itself to co-operate with this union to help it win members from the others for, since Navarrez was bitterly antagonistic to these because of their anti-clericalism, he could be relied upon to co-operate with the management in the future. The plan was put into effect, but *Padre* Navarrez's 'triumph' did not materialize as the unions still kept a fairly tight control over such men who were still on strike (6,000 in January) and insisted on being a party to any settlement. And although some men joined the Sindicato Católico they soon left when they discovered the moderation with which its affairs were conducted.[58]

Williams appreciated that one of the problems in the field of industrial relations at Rio Tinto was the attitude of Walter Browning and the way in which his views were so completely endorsed by Charles Fielding. In his report to the latter he pointed to the difficulties this had caused, and expressed general agreement with most of the opinions contained in another report which he also submitted, and which had been prepared by the two assistant managers at the mines, Gordon Douglas and G. W. Gray.

The two assistant managers began by quoting the words of Fielding, that 'in the field of labour relations we have always taken such steps as to be entirely beyond the possibility of adverse criticism', and added that very few of the British staff believed this. They attributed the directors' self-deception to the company's administrative structure which placed all power over the Spanish operations in the hands of the general manager and gave him complete control over what information would or would not reach the board.[59]

They accepted that the company had persistently paid the best wages of all similar enterprises in Spain, and had provided the workmen with excellent additional benefits in the way of cheap food, a fine medical service, and subsidized clubs, games and sports. Nonetheless, they and other members of the British staff were critical of the way in which the company had given wage-increases and other benefits. In their view much trouble would have been avoided if Browning's policy had been to discuss demands instead of rejecting them until such time as a strike took place. Because of this the workmen were convinced that the only way to get concessions was to strike first and talk afterwards.[60] Douglas and Gray found, even amongst the more responsible Spaniards, doubts about the company's sincerity and a distrust of its intentions, and they drew attention to the prominent role the Spanish staff had played for the first time in the recent strikes.[61]

The two assistant managers argued that Browning's policy of trying to delay wage increases generally caused more trouble than it was worth. Under any given conditions, they maintained, there was a standard of wages which would give the best results in operating the mines efficiently and economically. Any increases above this standard would unnecessarily increase costs. Equally, the two men argued, if wages were allowed to remain below this standard for too long there would be discontent, inefficient working and eventually a shortage of good workmen.

The two assistant managers and a number of other senior British staff had set up their own committee at the mines in 1918 to study wages, and for two years had collected information about conditions in the labour market and data on wages in other mines and industries. They had studied, too, the cost-of-living statistics, had tried to work out a satisfactory basis for recording fluctuations in prices and had prepared comparative figures on a quarterly basis. They maintained that the chairman was wrong when during the 1920 strikes he had declared that the men had received 'extra pay far beyond the increase in the cost of living', since their calculations showed the opposite. For the majority of Spanish employees the money paid for the unit of work done had not increased to the same degree as living costs and until the settlement in January 1921, the men had been in real terms worse off than they had been prior to the outbreak of war in 1914.[62]

The committee's members did not feel that it was up to them to say whether the company should or could compensate its employees for the rising cost of living, but they were worried that the company's chairman could have misconceptions on this subject, and that the general manager refused to attend their meetings and pass on their recommendations to the board. Rhys Williams supported this viewpoint adding that there was a strong case, too, for calculating wages on a scale linked with a minimum compatible with proper nourishment and adequate physical efficiency. A medical examination of 9,856 workmen in 1918 (admittedly a year of wartime scarcity) showed that only 33 per cent were considered physically fit by the medical staff for the work which they were doing.[63]

Douglas and Gray rejected the view that the causes of all the trouble at Rio Tinto lay with the way in which the workmen could be easily persuaded to embrace any revolutionary opinions by the paid union agitators and they declared that the principal fault lay with the company for not having a well thought-out industrial relations' policy. They proposed that as a way of preventing trouble a joint committee of workmen and staff should be set up to investigate complaints, though this 'would require a spirit of tolerance and adaptability (to which we are unaccustomed at Rio Tinto) on the part of the management.' They recommended, too, the acceptance of the idea that the company should devise a system of consultation and co-operation with the unions as a part of its ordinary working procedures, the appointment of Spaniards to senior positions with similar privileges to the British, including housing in Bella Vista, and greater care in selection of British staff and the removal of those who proved unsuited to controlling Spanish labour.[64]

The chairman, Charles Fielding, received the two reports in March 1921 and in turn reported to his board colleagues. He made no mention of the paper submitted by Gray and Douglas and condensed Rhys Williams's sixty-four pages into a ten-page board paper for the convenience of the other directors, prefacing it with the comment that it contained complete justification of the company's past policies and attitudes. In fact, in the summary which he presented he had removed almost all the criticisms contained in the original. Thus a long and highly critical section in Williams's paper on 'managerial matters', explaining the failure of communication between the British staff on the one hand and the general manager and the directors on the other, was reduced to one paragraph containing a proposal for a weekly meeting of departmental heads at the mines under Browning's chairmanship, and a comment that officials in one department knew little of the activities of those in others.[65]

That the company avoided any other strikes between 1921 and 1930 was due partly to the fact that when the men returned to work after the great struggle the unions were too exhausted by the fight to organize another for some time, and partly to the changes which took place in Spanish politics in 1923. In that year General Miguel Primo de Rivera took over control of the Spanish government. He outlawed all political parties save his own Unión Patriótica and then proceeded to regulate the political turmoil which was doing so much harm to Spanish industry. He had the most prominent trade unionists arrested and then offered an alliance to the communists and socialists whereby the UGT would be given freedom to operate, and allowed various advantages over the anarchists, in return for co-operation with his government. As a result a policy of repression was launched against the anarchist unions, and all industrial concerns were required to recognize the legally approved unions operating within the UGT and to deal with them in all matters involving disputes and working conditions. *Comités paritarios* (parity committees) were established for various industries with representatives of employers and unions sitting under government-appointed chairmen and the joint committee accepted by the Rio Tinto Company as one of the conditions for ending the strike in January 1921 was dissolved.

The company, like other businesses, profited considerably from the new arrangements, as indeed did the workmen. For both sides of industry were able effectively to exchange views and co-operate in formulating industrial relations' policies. The fact that neither side could take direct action against the other gave both a respite. There were occasions when the company was compelled to make concessions, whether it liked them or not, to its employees because of government intervention at times when agreement could not otherwise be reached. Even so it was appreciated that the absence of industrial disputes helped both the workmen (since they were able to depend on regular wages without the interruption of strikes called in the interests of inter-union rivalry) and also the company during a period of growing economic depression.

The Spanish government, too, was faced by considerable economic problems following the first world war. This led to frequent disputes between the company and the Spanish authorities over the increasing taxation which was levied. But the advent of Primo de Rivera meant that the company's contacts in Spanish politics virtually lost all their influence. The dictator's ministers were largely men unconnected with government before 1923 and hence there was no possibility of using allies within ministries to urge the company's point of view on the new administration.

The policy of José Calvo Sotelo, Primo's young minister of finance, was aimed at gaining the support of Spanish industrialists for his leader, and he therefore sponsored projects designed to appeal to their patriotism and their business interests. His management of the Spanish economy was remarkably successful in many respects. Under his guidance the country's industrial production was developed at a speed three times faster than before.[66] But to encourage Spanish industry, and to win its backing for the régime, he felt it necessary not only to make foreign companies provide subsidies for domestic business through increased taxation but also to demonstrate quite plainly that he could not be influenced or intimidated by companies such as Rio Tinto. Nonetheless the company felt it still needed political representation and in October 1927 appointed an Englishman as its new representative in the capital. This was Captain Ulick de Burgh Charles, CBE, a man of great sophistication and political astuteness,

*Captain Ulick de Burgh Charles, CBE (left), with the Marqués de Portazgo.*

who was a prominent member of the highest echelons of Madrid society, and on familiar terms with Primo de Rivera. He had gone to the British embassy in Madrid following his service in the first world war, as third secretary in the commercial department. By 1927 he was a first secretary and highly regarded both in Madrid and London. Both through his embassy contacts and his aristocratic in-laws (Charles had married a Spanish woman who had died soon after the marriage) he had many friends in the government and amongst its supporters. He seemed an ideal man to act as the company's official representative and was persuaded to abandon his diplomatic career and to join the company. Charles was given a generous salary of £3,000 a year plus expenses, but in the event he had to work hard for it.[67] The company provided him with a large and elegant flat at number 4 in the *calle de* General Castaños, in the central district of Madrid, where he could entertain in style, and provide accommodation for visiting directors from London. His guests there included not only politicians but even members of the Spanish royal family. It was at a luncheon there in 1929 that King Alphonso told Lord Denbigh that there were only two things standing in the way of excellent relations between Spain and Britain: Gibraltar and Rio Tinto, and the latter was the more serious problem.* Primo himself was always easily accessible and extremely amiable to Charles, who persuaded him to visit the Rio Tinto mines. This was considered a great success by the directors, especially when Primo gave an undertaking not to nationalize the company. Nonetheless it proved difficult to affect the course of his government's industrial policies.

In 1925 the new chairman of the Rio Tinto Company, Sir Auckland Geddes, had declared that the Spanish government behaved as if there was no limit to the company's capacity to produce cash. As the copper and pyrites trade ran into further difficulties in the world's markets so the government's demands increased.[68] In 1928 the board considered, though it decided not to implement, cuts in the expenditure on the social and medical services provided for the mineworkers.[69] Then in that year the ministry of finance delivered Geddes a shocking blow. It made a demand on the company of over three-quarters of a million pounds for underpayment of taxes during the period 1922–5.

Of this, £366,000 related to money owed as a result of false declarations by the company's officials in Spain relating to the copper content of pyrites exported in those years, and followed disclosures by a former employee. The other £438,000 demanded arose from the government tax jury's decision to reassess the profits' tax paid by the company in the same years, under a new law whereby taxation that had been agreed and paid could be re-examined and retrospectively

*This flat is still retained by the company, with the antique furnishings hardly changed. Much of this book was written in the study there at the desk on which Charles wrote his highly penetrating report on Spanish politics for the company.

*General Primo de Rivera visits Rio Tinto and is greeted by Sir Charles Fielding (with his back to the camera).*

increased. There was also the threat of a fresh scrutiny of all the company's tax payments for many years back.

Geddes found the revelations of the tax dodging profoundly shocking. It was, he declared, 'an unpleasant episode'. However, he applied himself energetically to try to prevent any further prosecution relating to earlier years and appealed against the reassessment of the profits' tax.* Geddes and the board decided that the best form of apology the company could make was to give Primo £300,000 for him to distribute to Spanish charities. It was an expensive business, and all told (by 1931) it had cost the company a total of almost £3 million. It also had a further significance: it made quite plain that the Rio Tinto Company was no longer in a position to manipulate the Spanish political system in its own interests.

When the dictator fell from power in February 1930 as a result of the opposition of the liberals and the professional classes, he left behind him no basis for any successor-régime. The king, Alphonso the thirteenth, tried to continue Primo's system and govern through a

*It was not until 1931, after Primo's fall from power, that the decision on this appeal was given and the verdict went against the company.

directory of ministers led by a general, Damaso Berenguer. But, after seven years of strong government control over all forms of political expression, republican sentiments started to reveal themselves throughout the country. There was a republican rising in the north and the Rio Tinto miners struck in sympathy. The company, on the grounds that the strike was illegal, decided to dismiss the men's leaders and ordered them to vacate their company houses. The men urged their sacked workmates to stay and prepared to resist eviction with violence.[71] Andalusian anarchists proclaimed a republic. Revolution was in the air, and the company once again contemplated a future full of potential danger.

*The directors of the Tharsis Copper and Sulphur Company at the Rio Tinto mines in 1925. The great protagonists, Sir Auckland Geddes the chairman of Rio Tinto and W. P. Rutherford the chairman of Tharsis, are respectively on the far right of the picture and 4th from right. Others in the group are: Lord Glenconner, 2nd from right; P. S. Couldrey (Browning's assistant and then successor as general manager) and his wife are 7th and 9th from the right; and Sir R. Paget 11th from the right.*

16

# *Fight the good fight*

## Economic depression and commercial warfare: 1919–32

Between the two world wars the Rio Tinto Company had to cope not only with political instability within Spain, which culminated in the great civil conflict of 1936–39, but also with the sustained economic depression which afflicted the whole of Europe.

The first world war itself had not produced overmuch despondency within the company. In the forty years in which the business had been operating up to the outbreak of hostilities in 1914, it had established itself as the greatest of all the European copper and pyrites producers, and had distributed over £28 million to its shareholders on a total investment, as represented by its share capital, of £3½ million. The war years, too, despite the loss of the company's German and American customers, the rapid rise in the cost of running the mines and the continuous industrial disputes at Rio Tinto, had produced comparatively good dividends since the restriction in supplies had resulted in increased prices. For 1915 the distributable profits had been £1,112,500 (representing a 55 per cent dividend on the ordinary shares for the year); for 1916, £1,862,500 (95 per cent); for 1917, £1,768,750 (90 per cent); and for 1918, £1,018,750 (50 per cent). The confidence felt by the directors in the company's future at that time was reflected, too, by large-scale spending in Spain. At Huelva the company built a model garden housing estate for its workmen there, and in Rio Tinto instituted a house-building and improvement programme. The medical and hospital facilities were extended and enhanced by large injections of cash. Social amenities were improved by the rebuilding of the workmen's clubs, the opening of two swimming baths and the making of public gardens complete with a bandstand for evening performances.[1]

It was appreciated by the board that copper prices would inevitably fall with the end of the war but it was anticipated that with the peace there would come an ever-expanding demand for sulphur for industrial uses. In preparation for this, the Rio Tinto Company had made a secret bid in 1916 of £2 million for Orkla, the large Norwegian pyrites producer, which had, however, been rejected.[2] Nonetheless, the Rio Tinto directors felt ready to meet the demand when the armistice was signed: everything was ready at the Spanish mines to boost production of pyrites at little increase in working costs.

*Part of the model garden housing estate (el barrio Reina Victoria) built by the Rio Tinto Company for its Spanish workmen in Huelva.*

The demand, however, was slow in coming. Former German customers, cut off by the war, had largely gone over to making sulphuric acid from a new process employing not pyrites but gypsum anhydrite which was available from Germany's own resources. The curtailment of deliveries to the United States, due to a lack of shipping, had forced many American sulphur users to change to the Frasch process for making sulphuric acid, and it looked in 1919 as if some of the US manufacturers were planning to invade the European sulphuric acid market. It therefore became obvious that if the Rio Tinto Company, or any of the other large pyrites' producers in Europe, worked their mines to their full capacity there would be overproduction and a consequent catastrophic fall in prices.

After Rio Tinto, the largest of the companies in the field was Tharsis. It seemed to many to be in the interests of the whole industry that the directors of both should reach a mutual agreement on production and sales. An arrangement aimed specifically at overcoming wartime difficulties had been signed by the two companies in 1916 and was due to expire in 1921. At the beginning of 1920, therefore, negotiations were begun in earnest to produce a new formula for

*Swimming pool at Rio Tinto.*

keeping the European market in good order. It was the view of the smaller producers that, if Rio Tinto and Tharsis could agree on what their respective shares should be of all European sales, then they would be happy to accept what was left them by the two giants.

Charles Fielding, the Rio Tinto chairman, was not prepared to reach an easy compromise with William Rutherford his opposite number at Tharsis. Although sales were declining throughout 1919 and 1920, both men calculated that there would be an increase in world consumption which they estimated would reach 2 million tons by 1922. Fielding proposed a quota which would allow his company to provide half of this amount, leaving the remaining million tons to be divided between Tharsis and the other suppliers. Although what Fielding was demanding was well below the annual tonnage sold by his company in the immediate pre-war years, the figure was completely unacceptable to Rutherford.

In 1919 and 1920 Rio Tinto's combined sales of pyrites and washed sulphur ores, partly due to the crippling strikes which took place at the mines, came to less than half this amount. The result was that dividends on the ordinary shares for those years fell to 40 per cent and nil respectively. Rutherford was apparently confident that Fielding was in no position to persist in his demands and refused to budge from the position he had taken up. But Fielding, too, was immovable. He was certain that Rutherford would not risk a price-war, since the Rio Tinto pyrites had a much higher copper content than those of Tharsis, and though the sale of the sulphur content was an important source of revenue it provided less profit than the copper. So if there had to be a keen sales competition, with price reductions for the sulphuric acid manufacturers, Fielding was certain that his company would be able to take them in its stride while Tharsis would suffer greatly.[3]

Consequently the negotiations reached a deadlock. In an attempt to break it Fielding approached the National Sulphuric Acid Association (a servicing body set up in the war at the British government's urging, and having as its members most of the British sulphuric acid manufacturers) with a proposal to supply all its members at a reduced price. The association contacted Rutherford to get his reactions. He, however, was not prepared to make a counter-offer. In fact he called Fielding's bluff, for the Rio Tinto chairman had hoped to stampede the Tharsis company into a quick agreement with its great rival. Fielding had no intention of trying to supply the whole of the British trade to the exclusion of the rival company through the medium of one all-embracing contract with the association. To do so would have forced the price for the sulphur content of Rio Tinto's pyrites down to a minimum, and put the association into a strong bargaining position to force the price even lower when the contract came up for renewal. Thus Fielding was obliged to break off his talks with the association.

In August 1921, with only four months to go before the current agreement with Tharsis was due to expire, Sir Arthur Steel-Maitland, recently appointed as managing director of the Rio Tinto Company, called on Rutherford. They discussed the demand of the Tharsis board that a new agreement should contain a clause stipulating a minimum price to customers. Steel-Maitland announced that his chairman would not accept such a clause: Fielding refused to tie himself down in this way, preferring to keep his freedom to negotiate prices with customers and to maintain flexibility at a time when no one knew how the markets were likely to behave. Having delivered this message, Steel-Maitland went on to propose that it would be to the two companies' mutual advantage in the circumstances to extend the existing agreement by five years. He must have expected the emphatic refusal Rutherford gave. Then, suggested Steel-Maitland, extend it for twelve months to provide a breathing space; but this proposal, too, was rejected.[5]

Fielding, however, was surprised at Rutherford's intransigence.

The Rio Tinto chairman had no doubt that, with its greater resources and vast stockpiles, his company was in the greater position of strength. But the Tharsis chairman may have been aware of the growing opposition to Fielding's policy among the Rio Tinto directors. The view was beginning to be expressed at board meetings that it would be better to compromise rather than to have a free-for-all, though as yet no one was prepared to challenge Fielding's policy in a head-on confrontation. Moreover, Rutherford was himself working on an alternative plan. In secret, Tharsis was negotiating with Orkla for an alliance of the two companies against Rio Tinto. The Norwegian concern sprung a surprise on Rutherford when it proposed an amalgamation with Tharsis which Rutherford felt forced to reject to preserve his company's separate identity. However, the discussions for a sales' alliance between the two continued.[6]

Meanwhile, the old agreement expired and both Rio Tinto and Tharsis were free to sell wherever they could. The ferocity of the Tharsis attack took its great rival by surprise. In pre-war days Tharsis had been virtually excluded from the continent by Rio Tinto, and in the immediate post-war period the latter company alone had made any impression on the market there and had won back a number of its former customers lost to it during the hostilities. Now, by ordering quite sensational price cuts, Rutherford captured from Rio Tinto most of its German contracts as they came up for renewal, including the important Duisburger Kupferhütte agreement. In France, too, many customers were lost by Rio Tinto to Tharsis, among them the St Gobin company, a major buyer. Rio Tinto fought back and by adopting similar tactics won a number of the rival company's British customers. Fielding threw himself into the conflict with vigour, but the board of directors which he had dominated for so many years was no longer behind him. By three votes to two he was removed from the chair and Lord Milner took over.*[7]

It was in character that Milner should disapprove of Fielding's enthusiasm for a trade war. A highly educated man of considerable culture, who inspired great affection in his subordinates both in the spheres of government and business, Milner was dedicated totally to peace in all its manifestations. In South Africa after the Boer war he won the admiration of the 'Milner kindergarten' of young men who worked under him and who, in the nineteen-thirties, became the apostles of appeasement; and as a member of the first world-war cabinet and as secretary for war he had advocated a compromise peace at Russia's expense with Germany. Thus on becoming chairman of the Rio Tinto Company in 1922 he immediately offered to re-open negotiations with Rutherford.

*Fielding remained, however, as director for another four years, advising on technical matters.

Since Tharsis and Orkla were still discussing their alliance, the talks were widened to include Rio Tinto, and centred around a proposed association of European pyrites' producers, to be controlled by the three negotiating companies, which would regulate markets, and allocate sales' areas and production quotas to the other members. Tharsis and Orkla suggested that the association should be governed by three directors, one nominated by each of the three companies. This was unacceptable to Rio Tinto. Milner anticipated that his company's nominee would be permanently outvoted under any such scheme by the Tharsis-Orkla combination and insisted instead on four directors, two appointed by his company and one each by the others. On this point the talks broke down in a flood of recriminations.[8]

The price-cutting competition was resumed but Milner was convinced it would be suicidal to prolong it a moment longer than necessary. The acrimonious bargaining was again resumed, Milner at last accepted that Rio Tinto would not be able to get its way on the subject of a veto and agreed to the formation of the Pyrites Producers Association in July 1923 more or less on Rutherford's terms. Most of the smaller European producers joined and thus accepted the regulations prescribing terms of competition and allocating production quotas.

The advantage to the member companies was seen almost at once. Despite a drop in copper prices, and the fact that the works of the principal German buyers were closed down for the whole of 1923 and thus unable to take deliveries, the Rio Tinto Company's profits started to rise. That year saw a 20 per cent increase in profits compared with those of the previous year. The gains came entirely from savings on operating costs made possible by knowing the precise extent of the company's share of the expanding market, set at 2 million tons for that year.[9]

In 1924 the price of copper fell to its lowest since pre-war days but the pyrites market picked up and under the quota arrangements production of pyrites at Rio Tinto went up to 2,450,000 tons. Though the mines were still not working at maximum capacity the extra tonnage was raised at little extra cost and, despite greatly increased Spanish taxes, profits again rose.[10] During the year, Sir Arthur Steel-Maitland was, on joining the British government as minister of labour, lost to the company and was replaced by Sir Auckland Geddes.

Just over a year later, in May 1925, Milner died and was succeeded as chairman by Geddes. In August, J. M. Macdonald, who had been a member of the board since 1882 and thus provided a link with Hugh Matheson and the other founding fathers of the company, also died.[11] Consequently there were two vacancies for directors and Geddes arranged for them to be filled by J. M. Buchanan, DSO, MC, and Colonel the Honourable R. M. Preston, DSO, and thus started to consolidate his power-base for the great struggles which even then he knew lay ahead of him.

Auckland Geddes was one of the most remarkable of all the men associated with the Rio Tinto Company. Much of the extraordinary nature of the new chairman's personality was already well known when he joined the company. Born the son of a Scottish presbyterian minister in 1879, he had read medicine at Edinburgh University and gone to Canada to become professor of anatomy at McGill, and later became principal of that university. While there he suffered an accident which would have ended the career of many men: one day when engaged in his favourite sport of sailing, he was struck by a spar and rendered blind.

Told that the damage was probably permanent, he applied himself to training his memory and developing his powers of retention. At the end of a year he was satisfied that he had powers of almost total recall, a skill which was to serve him extraordinarily well in the years ahead. It was indicative of his tenacity. Gradually the sight in one eye started to return, although it always remained relatively weak, and by 1917 he had made a complete return to British public life as a member of parliament. In that year he was appointed director of national service, a post which he held until his preferment to the presidency of the board of trade in 1919. From 1920 until 1924 he served as ambassador to the United States, returning to London in the latter year to enter commerce as a director of the Rio Tinto Company.

A story is told of his time in Washington which illustrates another aspect of his personality: his inflexibility. At a presidential banquet for foreign ambassadors, the diners stood until the orchestra completed the playing of the appropriate national anthems. When they sat, Geddes remained standing stiffly at attention. To an enquiry from the president as to the reason for this, Geddes replied that *God save the king* had not been played and that his sense of propriety would not make it possible for him to take his seat until it had been. A hurried search of the sheet-music revealed that, due to an oversight, the British anthem had been omitted. Geddes commented that the music had better be obtained quickly and an official was despatched to get it. Embarrassment mounted until at last the music arrived and the anthem was performed, after which the president's guests sat down to their overcooked dinner. The American public considered Geddes snooty: the American officials who had to deal with him considered him haughty. Yet he proved a highly efficient and industrious ambassador.

He took this rigidity and lack of warmth, which marked his

personal relationships with those with whom he had to deal in public life, with him to the Rio Tinto Company. It was, perhaps, the result of the stern upbringing he received as a child and possibly some form of self-protection against an innate shyness. However, the company's staff found it impossible to penetrate the barrier of reserve.

Apart from Geddes, Buchanan and Preston, the other directors of the company at the end of 1925 were Sir Charles Fielding, the former chairman, and Lord Denbigh. The latter two alone remained from the board which in 1923 had ousted Fielding from the chair and decided on a policy of co-operation with other pyrite producers, and in that dispute Denbigh had sided with Fielding. Consequently, if it was desired, the time was appropriate to change course once again on the question of co-operation with other producers.

The new chairman, in his dealings with the Pyrites Producers Association, found Rutherford of Tharsis uncongenial and his policies objectionable. Having got the association established on the lines he wanted, Rutherford had successfully influenced it to follow a policy of increasing sulphur prices in order to raise profits. A number of sulphuric acid manufacturers, unhappy about rising costs, began negotiations to enable them to adopt the Frasch process and thus free themselves from dependence on pyrites.

Geddes appreciated that if this happened on any scale the Rio Tinto Company would be a considerable loser, since a drop in pyrites sales would lead to a drop in the quantity of copper and other metals being removed, after roasting, from the cinders. Thus a drop in sales, effected at increased prices, would do little damage in the short term to the majority of pyrites producers; but for the Rio Tinto Company, still depending on metals for the greater part of its profit, it could have been disastrous.

Rio Tinto was able to produce copper more cheaply than any other manufacturer in the world, but needed to be able to fix its prices for ore and to keep these as low as possible in order to boost sales to the sulphuric acid makers and thus keep the residues circulating. Consequently, Geddes and his fellow directors decided in 1926 to withdraw the Rio Tinto Company from the Pyrites Producers Association. Unable to function without the largest of all European producers, the association collapsed and Geddes told his shareholders that the policy followed by Milner had been a mistake.[13]

In fact Geddes had no intention of permitting unrestricted competition but he was determined to organize the markets in the best interests of his own company rather than those of its competitors. He thought out his scheme and executed it with the deftness of a grand master seated at a chess-board. The Rio Tinto Company joined together with Metallgesellschaft of Frankfurt, the German industrial giant, to form the European Pyrites Corporation. This new company, in which Rio Tinto held the controlling interest and for which it provided the management, was to organize the sales of Rio Tinto ores and residues throughout the greater part of the continent.

Geddes then persuaded Orkla to enter into an agreement whereby the Norwegian company's produce would also be marketed by the new sales' organization. The next year (1927), Geddes turned his attention to the Cyprus Mines Corporation, a new and energetic company which under its American owners was making a considerable impact on the European markets. After hard bargaining, during which he made an offer (which was rejected) to purchase its mines, he brought the Cyprus Mines Corporation into his alliance. By guaranteeing these two important producers a substantial share of the market and thus considerably reducing the competition they would have otherwise to face, Geddes was able effectively to isolate the Tharsis company.[14] He followed this in 1928 by successfully negotiating the take-over of the sale and distribution of Mason and Barry's ores (this company was a major Portuguese producer) by Rio Tinto, and also of the produce of the Spanish San Platon mine. Next, he entered into agreements with the French Société des Pyrites and the German Meggen and Oberungarische mines, delineating markets and thus eliminating the trade rivalry between these concerns and Rio Tinto and its allies. All competition was thus to be channelled into an attack on the sales outlets of the Tharsis Copper and Sulphur Company.[15]

Thus from 1926 to 1929 there were three years of intense commercial warfare master-minded by Auckland Geddes. In the bitter exchanges between himself and Rutherford he would not yield an inch of ground. If Tharsis were to enter the alliance it would be on Geddes's terms: no price controls and no selling quotas; for as the Rio Tinto chairman saw it, his task was not only to make his company supreme over all its rivals but to overcome the threats directed at the pyrites trade as a whole by the movement of the sulphuric acid and fertilizer manufacturers away from pyrites to other materials. There was certainly cause for concern. In 1914 pyrites had provided 88 per cent of the sulphur used for making sulphuric acid in the United Kingdom. By 1926 this had dropped to 47 per cent.

In 1929 Rutherford accepted Geddes's conditions and an ingenious agreement was negotiated by the two men which was based on three lists of customers. The 'A list' consisted of customers already buying from the Rio Tinto alliance; the 'B list' of those buying from Tharsis. Both sides undertook not to compete for customers on the other's list, and where a large buyer appeared on both lists a specified sales ratio for future deliveries was agreed. The 'C list' was made up of all other known European purchasers of pyrites, who were obtaining their supplies from producers other than the Rio Tinto alliance or Tharsis, and it was agreed that both sides could compete for this business in any way they wished. While scope for competition still remained, these provisions introduced an element of order and control into the markets.[16]

Agreement was also reached between the two companies on the United States' market. An east–west line passing through Norfolk, Virginia, was drawn on a map. The Rio Tinto Company would supply customers to the north of the line and Tharsis those to the south. There is a Rio Tinto story of how the negotiations on how to divide up the potential American buyers occupied two days of hard and unproductive bargaining. Eventually Geddes, apparently in despair and overcome by exhaustion, drew a line across a map of the United States, seemingly at random, passed it across the table to Rutherford saying he would be content to take north of the line and Rutherford could take south. The Tharsis chairman, tired and dejected, agreed. According to the story, Geddes had carefully planned this move before the talks began and had rehearsed the drawing of the seemingly haphazard line time and time again before the meeting commenced.[17]

The new agreement was planned to run for five years until the end of 1934. In that time both sides gained new customers from the 'C list' in a proportion providing roughly equal tonnages. (So well did these arrangements work that before they expired they were renewed for a further period until December 1938.)[18] The new policy of co-operation made it possible for results to be obtained in other spheres. In April 1936, for example, Geddes and Rutherford travelled together to New York as representatives of the European pyrites' producers and negotiated an agreement with the American companies using the Frasch process, limiting US sales in the European markets.[19]

The satisfactory resolution of the conflicts within the trade itself did not solve all the problems facing Geddes. The great economic depression of the inter-war period and the political upheavals in Spain cut deeply into the company's earnings. It seems almost paradoxical that, having introduced a high level of order into the markets by 1929, the Rio Tinto Company found itself thereafter only just able to keep its head above water. From 1931 until 1947 there were to be no dividends for the company's ordinary shareholders* for by the 1930s the worldwide recession was hitting most major companies hard. It was a period when none had hopes of high profits but had to expend a

*The preference shareholders continued to receive their entitlement.

great deal of energy to avoid serious losses.

Added to this constant struggle were other problems for Geddes. His bitter dispute with Walter Browning, the company's general manager in Spain, absorbed much of his time even after Browning's dismissal in August 1927. The 'Pilkington affair' in 1928 provided another cause for concern.

It began with a fifty-one year old clerk in the shipping department at the *casa* Colón in Huelva called Harry Pilkington. For many years he felt, with some justification, that his departmental head was obstructing his promotion, and preventing him from getting annual salary increases appropriate to his work, because of a personal antipathy between the two of them. Pilkington appealed for redress to Geddes on one of the latter's visits to Spain but was rebuked for doing so. The suppressed resentment of years boiled over. In April 1928 Pilkington resigned his job and went to Madrid where he denounced the company to the ministry of finance for irregularities in connexion with the payment of export taxes, supporting his allegations with files from his former office.

The documents covered the years 1922–5 and Geddes was shocked to learn from his own enquiries within the company that underpayment of export taxes went back even further than this. Spanish taxes had always been a problem for the company because of the arbitrary nature of the preliminary assessments by the ministry of finance officials and the need to devote a considerable amount of time to pursuade these men that the liability was less than they claimed. The export tax was calculated on ore content as determined by analysis of the pyrites shipped from Huelva. At some stage in the past, precisely when it was difficult to determine with certainty, company officials had arranged with local officers of the finance ministry that the copper content of the exported ores should be understated.

*José Calvo Sotelo.*

On 15 May (1928) having heard from Captain Ulick Charles, the company's representative in Madrid, that a prosecution seemed certain, Geddes wrote to Primo de Rivera, the Spanish dictator, assuring him of his horror at the revelation and asking for the matter to be settled informally. Charles held two meetings with the minister of finance, young José Calvo Sotelo, but found him hostile to the company and determined to pursue a prosecution. In August, Geddes visited Primo de Rivera and found him sympathetic though non-committal about the possible outcome.

At the end of August a royal decree appointed a *junta administrativa* to enquire into the charges and the company was invited to nominate one of the four members. It chose an eminent Madrid businessman and despatched R. M. Preston, together with the company secretary and two London lawyers, to attend the meetings at which the company was to be defended by two Spanish advocates.

As the hearing got under way the company charged Pilkington with the theft of the documents which he had taken from the Huelva office and had him conveyed to Valverde (where the criminal court for the Huelva region was situated) where he was remanded in custody to await trial. This seems a vindictive and petty move inspired by Geddes. Under Spanish law it was not a criminal offence for an employee to remove, and hand over to the authorities, papers which showed his employer was underpaying taxes; and if there was to be any hope of convicting Pilkington then the company would have had to rely on its power and prestige in the mines' area to get what would have been a perverse verdict.

Thirty-six sittings of the *junta* were held during September and the first week of October. Charles was hard at work behind the scenes hammering out a solution with his politician friends. He was able to inform his directors that the Spanish council of ministers had agreed to a settlement whereby the company would pay the arrears of taxes claimed, amounting to almost £326,000, plus interest. No penalties, such as fines, would be imposed if the company also gave the government a further £300,000 for distribution to charity. Geddes and his fellow directors agreed; the chairman with alacrity, the others with reluctance. The finance ministry had also demanded a further £438,000 from the company as a result of a further hurried reassessment of the profits tax. Geddes hoped to avoid paying this by reaching an agreement with Primo, but was annoyed to be told that the matter would be decided within the ministry of finance.

Meanwhile Primo suggested to Charles that the company should drop its criminal charges against Pilkington, and Geddes agreed to have the case withdrawn at the end of November. While in prison Pilkington had requested that the government pay him 10 per cent of the amount recovered as the reward normally due to an informer. He found the ministry of finance unsympathetic. He was offered only £4,350 and the good offices of the government to secure his release. Whatever his disappointment may have been at the smallness of the sum, it was still the equivalent of his salary for seven years and he accepted it. (For some reason he and his wife later always maintained that he had never received any payment, and thus had reaped no benefit from his 'public-spirited' act.)

On 17 January the following year (1929) Geddes wrote another letter to Primo which was delivered by Charles, proposing that the further liabilities be settled quietly.

'I feel that it is only fair to the company and to myself that your excellency should give instructions that a *private* settlement of these outstanding matters should be arrived at in order to avoid in our mutual interest the lengthy and innumerable difficulties which a similar examination to that made in connexion with the export tax would involve.'

He suggested to Primo that, bearing in mind the company's generous gift to charity, the government should accept £100,000 in discharge of the outstanding claim for £438,000. A meeting was held between Primo, Calvo Sotelo and Charles to discuss this letter. Charles, however, had to report back to his board that the finance minister could not see his way to reduce the amount claimed. Nonetheless, he was instructed once again 'to approach the authorities to see if some consideration with regard to payment could not be extended.'[20] But despite all the attempts at persuasion no headway had been made by the time the government fell from power in February 1930. Finally, in 1931 the company lost its appeal. Moreover, the republican government which took office in the latter year pursued further claims for arrears of taxes and by the time they were resolved the company had been compelled to hand over a total of £3 million.[21]

Faced with the constant struggle to maintain administrative and technical efficiency Geddes strengthened the board in 1929 by the addition of J. G. MacLeod, for thirty-one years secretary of the company, and G. W. Gray, the former assistant manager in Spain under Walter Browning.[22] These two appointments led to a number of changes at Rio Tinto for the workmen and their families. Both of the new directors were familiar in detail with the mines and the problems of life there. Gray in particular, though a technical man, had once expressed considerable criticism of the way the company handled its labour relations. Now he was presented with the opportunity of putting his ideas into practice. To some extent the changes were being forced on the company by the political upheavals in Spain, especially after April 1931 and the creation of the second republic. But in 1930 and the early part of 1931 the board showed themselves willing to back Gray's proposals at a time when Spanish radicalism was still under restraint.

One of Gray's first suggestions to be adopted was for the expansion of the company's medical service both at the mines and in Huelva. With more funds available it was enlarged to sixteen doctors (including a specialist in paediatrics), ten qualified nurses, twelve dressers and eight midwives. A child welfare clinic was established which handled 14,000 cases alone in its first year of existence.

Pensions to former workmen were reviewed and increased. A new

*The new Rio Tinto Company hospital in Huelva, 1934.*

house-improvement programme was got under way. The company's eight schools, educating over 2,000 children, were provided with extra accommodation. Company shops were reorganized and branches opened in all the villages surrounding the mines. Even sport was not neglected: new football pitches were laid out for the company's fourteen football teams.

A determined effort was made to reduce accidents and the first full-time safety officer was appointed. In March 1930 a system was introduced for a continuous review of workmen's earnings and a group set up to study cost-of-living statistics, while the next month the pension scheme for the Spanish workers was converted into a non-contributory one. At the same time arrangements were put in hand to find suitable employment for men who had been physically incapacitated as a result of their occupation and were therefore in danger of being made redundant.

Instructions were given that when British staff were being recruited for Spain only those should be selected who seemed capable of co-operating with Spaniards and were prepared to learn their language. It was also provided that, when decisions had to be made which were likely to affect living and working conditions, the Rio Tinto town councillors or the workmen's representatives (or both) should be consulted. Thus a conscious start was made gradually to erode the hostility which so many of the Spanish employees had felt for so long towards the company.[23] It was a minor tragedy that these reforms, small though they were, which tokened a new determination to solve some of the long-standing problems of life at Rio Tinto, were quickly made irrelevant by the course of Spanish politics from 1932 onwards. They were not enough to prevent the disorders which characterized Rio Tinto during the second republic, nor sufficient to satisfy the requirements of the nationalist government which took office as a result of the civil war.

At the same time as these changes were being put in hand, a considerable amount of money was being put into new equipment in readiness for the day when the general economic climate would improve and production could be increased. New machinery was installed on the pier at Huelva to make it possible to load four boats in a day. The railway loading system was improved so as to be able to cope with 7½ million tons of pyrites a year. At a cost of £250,000 a new mill, flotation plant and sintering plant were built at Rio Tinto capable of treating over 5 million tons of ore annually.[24]

The year 1931, however, saw not a lifting of the depression but a worsening. The Rio Tinto Company, like almost all businesses, found itself acutely short of money. Moreover, the company had recently embarked on a policy of diversification, involving substantial expenditure in Africa, America and other parts of Europe. (See appendix two.) Economies had to be made, and Auckland Geddes showed that he was just as capable of swinging an axe as his brother Eric, after whom the notorious 'Geddes axe' was named. In the autumn of that year a dozen members of the British staff at the mines were dismissed and staff salaries throughout the company reduced by between 10 and 33 per cent. The news was broken to the staff at a meeting in the *club inglés* and caused a considerable shock. Bitterness was expressed both by members of the staff and by their womenfolk. The wife of one junior employee, whose salary was to be cut from just over £300 to a little over £200, made some caustic observations when she was told that the chairman had not excepted himself and would be taking a salary cut from £10,000 to £9,000 a year.* Even the normally reticent Herbert Potts complained that 'the directors behaved with great harshness; no apologies, no regrets were expressed ... the action of the directors caused a tremendous loss of goodwill in the staff which was never regained.'[25]

Geddes certainly behaved with a lack of consideration towards the

* In fact this announcement did not tell the whole story of Geddes's pay, for he was also in receipt of another £10,000 a year from the Pyrites Corporation (a Rio Tinto subsidiary) which remained uncut. Income tax on both salaries was paid by the two companies.

British employees, which was all the more peculiar at a time when efforts were being made to obtain a better relationship with the Spanish labour force. His cuts might have been better received if he had explained the difficulties facing the company. The consumption of fertilizers in Europe, which provided the major sales' outlet for the company's ores, fell in 1931 to 70 per cent of the previous year's total. After the extraction of the sulphur, the iron in the cinders was normally purchased by continental steel manufacturers, but by the end of 1930 most of them had enormous stocks and little business, and were unable to take further deliveries for at least two years. Despite a vigorous fight for a worthwhile share of the market, Rio Tinto deliveries were reduced in 1931 by 22 per cent and the tonnage mined cut back by 23½ per cent. During 1932 deliveries again fell, representing a drop of 48 per cent on 1930. For the first time in its history the company had to sell its copper at a loss. This was the background to the cuts in administrative costs, including staff salaries, for the outlay on these had to be substantially reduced; and it was: from £248,000 in 1930 to £201,000 in 1932.

Despite the fact that there were 3,000 workmen surplus to require-

*The Naya crushing and screening plant. The old washing plant (lavadora) is in the centre-background; the smelter chimneys are in the distance on the left and the leaching heaps are on the right.*

ments at the mines there was no question of redundancy or wage cuts for them; the Spanish government would not permit such measures. Yet in economic terms a cut in the work force was one economy that appeared to make good sense. By 1932 the ore masses of the north and south lodes were virtually exhausted. In preparation for this eventuality the company had started to open up the western end of the San Dionisio mass, the Atalaya opencast. In the course of the preceding twenty-five years, 20 million cubic metres of overburden had been removed, exposing for easy excavation an enormous body of pyrites with a high sulphur content and of excellent burning quality. With the Atalaya pit in production, no longer was there a need for the large labour force previously employed on stripping the overburden, while the concentration of the previously widespread operations meant that far fewer miners would be required than at any time during the previous forty years.

This was the potentially dangerous economic situation facing the directors and management when the second Spanish republic was born and became a threat to the very existence of the Rio Tinto Company.

*King Alphonso the thirteenth goes into exile. Paris, 16 April 1931.*

# 17 *Tumult and civil strife*

## The second republic and civil war: 1931–6

The Spanish monarchy collapsed on 14 April 1931. The election held two days before had shown that the large towns throughout Spain had turned republican and, though the urban electorate represented only a third of the people, the king's ministers were thunderstruck and irresolute. The sections of Spanish society on which the monarchy depended – the army, the church, the landed classes and the political establishment – considered the preservation of their own particular interests of greater importance than the form of government. Few had given the king wholehearted support and most were prepared to give a republic a try. Thus, as the election results became known and the excited Madrid crowds surged through the streets, King Alphonso the thirteenth went into exile and the same night the second Spanish republic was declared.

The new government was at first headed by intellectual liberals and radicals drawn from the professional classes. Its declared aims were moderate in tone and though it conceded freedom of religion (to the annoyance of many conservatives) it pledged itself to maintain property rights. Most foreign businessmen operating in Spain, however, viewed the situation with anxiety. The senior staff of the Rio Tinto Company, although scrupulously adhering to the directors' instruction to display no political bias, privately expressed alarm. These men feared what did in fact later come to pass: the transfer of power within the left-wing alliance from the more moderate politicians to the more extreme, together with the collapse of law and order and the emergence of conditions in which commerce and industry were bound to decline.

Thus the British at Rio Tinto, after the April elections, faced the future with a gloom that was to prove well justified. They did not have long to wait. At the beginning of May anarchist and communist-sponsored riots started throughout Spain. The homes of prominent right-wingers were attacked and churches burnt. On 12 May alone eighty-seven religious buildings in southern Spain were sacked.

Control in Huelva province, traditionally monarchist, had passed with the elections to socialists and anarchists, and the new head of the provincial civil government proudly described himself as 'the workmen's governor'. In Huelva town the first mob-attack was directed against the house of José Sánchez-Mora, the Rio Tinto Company's legal adviser. Unable to smash through the barricaded doors, on the other side of which Sánchez-Mora and his wife waited apprehensively with their newly-born child, the attackers sent some of their number to raid a company store to steal dynamite. Two members of the British staff, who had come to see if the family was safe, saw what was happening and, when the waiting rioters relaxed their guard, crept undetected through the gardens at the rear of the house, attracted Sánchez-Mora's attention and hustled him and his family outside. On all fours the party crawled through a gully to safety.[1]

As governmental control in the provinces gradually broke down, many of the company's British employees in Spain began to feel concerned for the safety of themselves and their families, and had to exercise more than ordinary caution and to avoid incidents that might cause trouble. Robert Wilton Brown, later the head of the mines' labour office, was one day suddenly surrounded by a threatening mob in the street of a town he was visiting. Rescued by civil guards, he discovered that the British regimental tie he was wearing was the cause of the assault, for its colours were similar to those used by a Spanish right-wing church party.[2]

The initial jubilation of the politically active workmen at the Rio Tinto mines following the elections, and the creation of the second republic, soon subsided. These men quickly realized that they had been too optimistic in what they had expected of the new government. For them, and many of their comrades throughout Spain, the left-wing alliance which was governing the country was too moderate. Once again they started to resort to industrial action: the socialists and communists (the dividing line between them was obscure) in the hope that they could propel the government into aggressive socialism; the anarchists in the belief that the disruption of work and the promotion of disorder would help achieve their aim of a collapse of central authority.

The workpeople of Spain had been compelled by the new government to join trade unions whether they wanted to or not. At the mines the men had a choice of only two: the Unión General de Trabajadores (known as the UGT, and socialist in orientation) and the Confederación Nacional del Trabajo (or CNT, the anarchist syndicate). In every department at the mines the union delegates became highly vociferous. Since many of the workmen were apathetic, or even silently preferred the right-wing authoritarian parties, those who emerged as union delegates tended (at least in the view of the British staff) to be extremists, uninterested in compromise and obtaining election more by their oratory than their logic.

The union delegates insisted on discussing at length with the senior British staff all matters, however trivial. Thus much of the management's time came to be devoted to industrial relations. Frequently the talks were followed by short work-stoppages, and soon the frequent strikes produced serious delays in production which were a cause of frustration and exasperation to the British staff. Even when work was in progress there were continuous disputes between many of the younger workmen and their Spanish supervisors about the way the work had to be done, the arguments often arising from refusals to observe the company's safety regulations. Fortunately some of the maturer men had little sympathy with the more trivial disputes, and although they felt obliged to strike even when this action might have resulted in serious damage to plant or equipment in a department such as the smelter, a hint seems always to have been dropped about what was to happen. Consequently the section of the plant to be involved was always shut down ahead of the appointed hour, even on the occasions when there were 'lightning' stoppages.[3]

Further complications from the company's point of view were caused by inter-union rivalry. The old supremacy of the anarchist CNT at the mines had been largely eroded during the years of Primo de Rivera's régime when the UGT, as a result of its compact with the dictator, had built up a sizeable membership. The re-emergence of the CNT led to the two unions competing for members. On occasions the members of one would strike while those of the other remained at work, but often a stoppage called by one union resulted in the other issuing strike orders since it was afraid of losing its reputation for aggressiveness to its rival.

As discipline suffered so did the morale of the British staff. Even the manager of the Tharsis mines, who had to cope with the same problem, was shocked during a visit he made to Rio Tinto when he heard a group of workmen's delegates address a senior official with great insolence. In embarrassment the Rio Tinto man turned to his visitor and muttered, 'We cannot be sensitive or touchy nowadays, can we?' A few of the British staff, however, were successful in maintaining discipline within their own departments. Potts, the head of the smelter, was a stern disciplinarian and would not see the men's representatives unless they approached him in a civil manner. Their respect for him was such that they conformed to his wishes.[4]

Added to the problem of industrial anarchy were others which caused concern in the British community. The year 1931 had started badly at the mines. Early in the year five members of the staff died. Every Monday for five successive weeks the British colony followed a coffin to the English cemetery.[5] New local-government officials, appointed after the elections which had toppled the monarchy, compelled the company to admit Spaniards freely into Bella Vista. At the hour of the *paseo*, when the Spanish took their traditional evening walk, the Spaniards flocked there, thronging the narrow roads, peering into the windows of the houses and causing some alarm to many of the British women.[6] Then in the autumn the effects of the great slump were felt, and a dozen members of the staff were dismissed while the remainder all received salary cuts. These events, together with the political changes, had a depressing effect on the British community.

The world-wide recession and the accompanying loss of markets hit all the mining companies in Spain. The Rio Tinto Company introduced a five-day working week for its workmen, paying the married men 5 *pesetas* for the day not worked, but there were at least 3,000 men estimated as surplus to what were needed. However, the government, sensitive as it was bound to be to the demands of the unions, legislated to prevent companies from dismissing employees. It also compelled them to re-employ men returning from compulsory military service while keeping in employment, too, anyone who had been recruited to fill the place of a conscript. Thus the workforce at Rio Tinto was actually increasing at a time when production had to be reduced.[7]

The government-sponsored parity committees (and later the mixed juries) set up to regulate the relationship between employers and labour were cumbersome in operation and exasperated the company officials. Not unnaturally the miners' representatives knew little of either the difficulties connected with the sale and shipping of pyrites, or of the technical problems of mining and metal production, all of which were important factors in discussions concerned with matters such as employment, wages and operating procedures.

The company's frequent representations to the republican government in connexion with such things as the state's employment requirements, high taxation and political agitation at the mines, met with no success. Apart from the fact that the régime was less disposed to listen to foreign businessmen than any previous government, its own many problems absorbed all its energies. The new constitution promulgated in October 1931 alienated much Spanish middle-class sympathy by its adoption of anti-clerical measures. The protest-resignations of two ministers led to the formation of a second provisional government headed by Manuel Azaña which was more markedly left-wing. It devoted itself to promoting legislation on a vast range of social topics such as divorce, civil marriage, minimum wages, women's rights and labour contracts, and to consolidating its position by suspending right-wing newspapers, closing church schools, suppressing local councils with monarchist majorities and purging the civil service.

As a result, the left-wing alliance began to disintegrate in a flood of recrimination and acrimony. The sweeping changes demanded by the socialists were unacceptable to the liberals. Proposals for land redistribution and agrarian reform aroused intense controversy. Anarchists and communists engaged in outbursts of violence which the government found it increasingly difficult to control without undermining the basis of much of its own support, although the UGT was disposed to co-operate far more with Azaña's government than was the anarchist CNT, and at Rio Tinto it adhered for a time fairly closely to the legal procedures for submitting claims to the mixed juries. The CNT, however, reacted by adopting a policy of even greater industrial disruption, organizing violent demonstrations and calling more strikes.[8]

In the summer of 1933 Azaña held a general election to allow the country to pronounce on the the work of his government. The result was a considerable defeat for the administration. The popular front of left-wing parties won only 99 seats in the new *cortes*. The centre front took 167 and the parties of the right 207. There began two years of right-wing government, disparagingly described by the popular front as the *bienio negro* ('the two black years') in which many of the decrees of the former government were rescinded.

The Rio Tinto Company was given permission to discharge 1,000 men surplus to requirements. Those the company made redundant had all been recruited during the previous four years and all had come from outside the Rio Tinto area. They and their families were given financial help by the company to enable them to move to areas of their own choice, and at the same time the company made grants to local authorities in the mines' area to enable them to carry out public works and thus provide jobs for some of the other local unemployed. This combination of generosity and dismissals was expected by the board of directors to produce a more co-operative (or subdued) labour force. However, the first three months of 1934 were marked by a series of particularly violent strikes involving two attempts to murder members of the British staff.

In addition a Spanish foreman in the Huelva workshops was killed by a striker. The murderer was imprisoned to a great outcry from many of his fellow workmen. His wife was pregnant and the other strikers protested that his wife and '$4\frac{1}{2}$ children' could not live without their bread-winner. (The man was released, like many other prisoners who also held left-wing views, by the popular front government which came to power in 1936. A few months afterwards, when Huelva was entered and occupied by the nationalists at the start of the civil war, he was arrested and shot.)

*F. W. Cooper.*

Then, in April 1934, to the company's amazement, the civil guards were withdrawn from Rio Tinto leaving the mines unprotected. The general manager, F. W. Cooper, was greatly alarmed. The women of Nerva in particular were a special source of worry to him. Highly aggressive, addicted to gathering together in large bands to threaten and attack those they considered their political opponents, they started to make threats against the British families. Some of them declared their intention of seizing the British wives, stripping them and marching them through the streets. Some proposed in addition that after this the British women and their children should be put down mines' shafts, covered with petrol and set alight. The members of the British staff who heard the rumours demanded that Cooper take action. Those of their wives who found out about the threats treated the matter with smiling disbelief. But clearly there was cause for alarm.

Cooper instructed the men to tell their wives to prepare to evacuate themselves and their children but not to let their Spanish maids know of the preparations in case the word got around in Nerva. In fact it was impossible to hide the preliminaries for their departure from the domestics, many of whom wept when they realized what it meant. Though forbidden to discuss the matter with their maids, the British wives went out into their gardens to collect armfuls of the large white daisies which were then in flower, and filled every room in their houses with them as a silent gesture to indicate that even if they had to leave that night they hoped to return in a day or two. Cooper also contacted the governor of Gibraltar who promised him every assistance should the evacuation be necessary, including a secret assurance to send armed British troops if the Spanish government refused to protect the staff and their families. Then Cooper sped to Madrid to protest against the withdrawal of the civil guards.

On arrival in the capital he went immediately with the British ambassador to the prime minister's office. While they were waiting in an ante-room, the Spanish ambassador to Britain, Ramón Pérez de Ayala, arrived also wanting to see his prime minister. When he heard from Cooper of the situation at Rio Tinto he was appalled and marched straight into the premier's office, protesting his anger that Spain should have abandoned her elementary duty of protecting foreigners resident within her boundaries. He demanded that sufficient troops be sent to the mines before ten o'clock that night in the hope that the evacuation, which would be detrimental to the honour and prestige of the country, could be averted. That evening 300 civil guards and 100 assault guards arrived at Rio Tinto, dispersed the demonstrators and mounted guard over Bella Vista village and the company's installations, and permitted the cancellation of the arrangements for the exodus.[9]

The next day Cooper returned to the mines and addressed a meeting of British staff and their wives in the club. Tired and exhausted from his recent exertions and the strains he had been bearing for the past two years, he chose his words badly and criticized some of the British women as troublemakers. Frayed nerves snapped. Many of those present turned on him in anger and accused him of failing to take proper steps to ensure their safety. Unable to disclose publicly the confidential discussions he had engaged in with the authorities in Gibraltar and in Madrid, and the strenuous activity in which he had been involved during the past forty-eight hours, he felt unable to defend himself. It was apparently with a mixture of relief and distress that he heard a few days later that the directors in London had decided, following complaints from some of the people in Bella Vista, that he should be replaced.[10] (His temporary replacement was G. W. Gray who had once been an assistant manager under Browning and had been appointed as technical director of the company by Geddes in 1929.)

In the following months the decline into chaos and violence continued throughout Spain. None of the country's political parties seemed capable of displaying the energy, the compromise and the understanding required to halt the drift into complete anarchy. In October 1934, when three members of the largest of the political parties of the right (the Confederación Española de Derechas Autónomas) were taken into the cabinet without giving the normal pledge of support for the republic, the UGT saw it as the first step in the overthrow of the republican form of government, and its supporters in many cities rose in armed revolt. At Rio Tinto the miners struck on the first day of the rising on 5 October but (though supplies of arms and explosives had been distributed by union officials) the civil and assault guards, reinforced by the 400 men sent in earlier in the year, were sufficiently numerous to keep control within the mines' boundaries. However, in the two nearby villages of Nerva and el Campillo (the latter at that time renamed Salvoechea by the republicans in memory of a nineteenth-century anarchist), the civil guards came under attack from small-arms and home-made bombs. They were able to resist long enough, however, for reinforcements to reach them from the detachments at the mines and after four days the district was quiet once more, though in some parts of Spain the revolt lasted a little longer. The Rio Tinto men returned to work save for 600 held in prison for their part in the uprising (and whom the company was authorized to dismiss by the government) and another 200 whom the company was permitted to declare redundant.[11]

After the failure of the October revolt there was a general hardening of attitudes within Spain: the Spanish middle class in general came to accept that a military dictatorship would be preferable to chaos and economic disintegration; while many formerly moderate radicals moved into more extreme positions, towards communism and revolution. The result was manifested in the elections of February 1936 when the centre parties virtually disappeared and the popular front of left-wing parties and the national front of the right emerged almost equally balanced with a slight edge in favour of the left, an edge sufficient to give it the reins of government once more.

The workmen of Rio Tinto, heartened by the popular front victory in the polls, again began to adopt more defiant attitudes in their dealings with the management. Agreements reached through the mixed juries were ignored and grievances, both real and imaginary, produced a fresh round of strikes. Many of the younger men started an unofficial form of military training after working hours. The directors viewed the situation with the greatest apprehension and once again issued instructions to the British staff in Spain to remain politically neutral and to avoid confrontations whenever possible.[12]

The Rio Tinto Company's chairman, Sir Auckland Geddes, described the situation at the mines at this time as one of 'ceaseless political unrest based on the expectation on the part of the local members of the parties of the left of a coming socialist revolution. The history of the period can be briefly summarized in five words: political turmoil, strikes, increased expenses.'[13] The new government ordered the company to re-employ the 200 men made redundant in the October of 1934. Captain Charles, the company's representative in Madrid, saw the minister of labour and protested strongly, at the same time making an offer on behalf of the company to provide sufficient cash to enable the men to resettle in another region where they stood a better chance of finding work. He also told the minister that in the nine months ahead the company would need to discharge at least a further 700 men and was preparing schemes to help these workmen find fresh jobs in other parts of Spain. The minister, however, was adamant. The discharged men had to be taken back and no more redundancies would be permitted.

Shortly afterwards Manuel Azaña, once again prime minister, signed a decree granting an amnesty to political prisoners who had opposed the previous government. In consequence the company found itself ordered to give back their jobs to the 600 Rio Tinto detainees of the rising in October 1934, even though there was no work for them. As though to rub salt into the wound, the unions called a strike to force the company to pay these men arrears of wages for the period they had been in prison. At the end of a month the company gave way and conceded the point in order to get a return to work. No sooner had £57,000 been paid out to the released men as back-pay than another strike was called by the unions to compel the company to reinstate working agreements cancelled by the management during the revolt and the accompanying work-stoppage. Since the revolutionaries had been officially rehabilitated and declared to have behaved legally, the men claimed the Rio Tinto Company's abrogation of the agreements must in consequence be illegal. The government backed the unions in the dispute, and the directors had no alternative in the topsy-turvy world in which they found themselves but to give way once more.

In the tumultuous weeks which followed it seemed to the British staff that the Rio Tinto communists and anarchists were preparing to play their part in a movement to overthrow the left-wing government which they had earlier helped to elect. Captain Charles in Madrid received assurances from the ministers he saw that they would do all they could to maintain law and order at the mines but they admitted that their powers were limited as the government was progressively losing control over that part of the country.[14]

During the month of May (1936) the mines' management was reduced to the utmost despair. There were frequent fights between members of the rival unions. Workmen not only flouted orders and insulted members of the British staff if they tried to exercise their authority, but often ignored even their own union officials. For the first time since the company had acquired the mines in 1873, essential maintenance work necessary for underground safety was incapable of being carried out. At the beginning of June, in an attempt to regain control over their own members, the two unions made a series of fresh demands on the company.

The new permanent general manager, Alexander Hall, and his assistants met the men's delegates at a series of meetings under the chairmanship of the governor of Huelva province and in the presence of an adviser from the ministry of labour. After nine days an agreement was reached which the directors in London approved by cable: there was to be an all-round pay increase, a special allowance for all married men backdated to October 1934 and an undertaking that the management would fully co-operate with the UGT and the CNT. In return the unions accepted that they would not interfere with the technical administration of the mines and their safety services, and would abide by the existing agreements governing hours of work and conditions of employment. At heated meetings, however, the men rejected the settlement negotiated by their leaders and on 11 June struck once again.

The stoppage took the form of a stay-in. The men occupied the workshops, factories, underground galleries and open-cast pits where they normally worked, sleeping at nights next to their silent machines, beside the cold furnaces and on the rock floors at the work-face. The women carried food to their menfolk and exhorted them to be firm and see the strike through. To begin with it was all very good natured and the men exchanged jokes with the British staff. When the latter used to say that camping out in the open at night was all right in June but would be uncomfortable in November, the workmen would throw up their hands in mock horror declaring: 'Don't say it can last as long as that!'

After a short while many of the strikers would have been happy to have returned home but save in special cases, such as illness, their wives would not permit this. Bands of women, described by the men themselves as 'tigresses', guarded the entrances to the villages and stoned any man who tried to get to his house without their permission. But after fifteen nights even the most stout-hearted strikers had suffered enough and they returned in a body to their *pueblos*, though the stoppage lasted another four days before the workmen returned to their jobs.[15]

Even then production did not return to normal. Though the UGT officials tried at first to get their members to respect the agreement, the anarchists of the CNT decided to reverse policy yet again, repudiated the newly negotiated settlement and encouraged their own men to continue in opposition to the management by refusing to obey orders and adopting obstructionist tactics. The only explanation to account for this *volte face*, in the view of the directors, was that the anarchists wished to keep the working force in a state of turmoil in preparation for a rising against the government.[16] The union officials readily admitted as much to members of the staff though they were quick to disclaim any ill-feelings against the British at the mines, even though they declared that the British staff would have to be replaced by German communists (refugees from Hitler's *reich*) following the forthcoming revolution and the seizure of mines such as Rio Tinto in the name of the people.[17]

Throughout this period, from the election of February 1936 until the outbreak of civil war in the following July, the country's economic situation deteriorated continuously and rapidly. The *peseta* fell in value and wealth started to flow out of the country in some volume as foreigners and Spaniards alike commenced to take precautions to safeguard themselves against the outbreak of civil conflict. The government declared that foreign commercial enterprises which withdrew their nationals from Spain risked the forfeiture of their property within the country, and both Spanish and foreign business lost completely all confidence in the popular front régime.

Everyone appeared to be waiting for the civil war which seemed to be inevitable. The government, in anticipation, began to order the arrest of right-wing politicians and to purge those sections of the armed forces whose loyalty to the régime was suspect. The army was divided in its attitude but the majority of its senior officers were unhappy with the government's policies, a group of them coming to the conclusion that they would once again be called upon to perform the Spanish army's historic rôle of saving the state from the ills inflicted on it by its politicians.

The murder on 13 July of the monarchist politician José Calvo Sotelo (it seemed, at the time, with government connivance) precipitated the crisis that made the generals decide to act; for in the aftermath it was plain that there was a complete and total breakdown of law and order. Mobs equipped with rifles and graced with the title of government militia took over the Madrid streets and began an orgy of murder, theft and looting. Unable to control them the government could only condone their activities. At the same time the band of generals moved into action. On 17 July Colonel Juan Yagüe seized control of the administration in the north African colony of Morocco where the foreign legion and Moorish troops could be relied upon to support the generals. In the next three days the movement spread to Spain. By 21 July it was obvious that the country was facing a prolonged civil conflict. For though in the first few days the military movement achieved considerable success, gaining control over half the country, lack of time for preparation before the rising meant that once the initial impetus had been exhausted, there was no hope for a quick takeover of the whole of Spain.

The first news of the challenge to the Madrid government was slow in getting through. The rising in Morocco had taken place on Friday 17 July, the same day that many of the British families at the mines had gone to Punta Umbría to spend the weekend on the coast. The following afternoon one of the staff, L. U. Salkield, was in Huelva when a Spanish acquaintance told him 'communications with Seville have been cut and I think a revolution led by the army has started. It would be as well if you returned to Punta.' The rumours reached the mines at much the same time. Trevor Gough, the deputy manager in charge of administration, was told by the captain of the civil guards in Rio Tinto that risings had taken place in different parts of Spain and that it might be advisable to concentrate the British staff and their families into Bella Vista village to make it easier to protect them.[19]

That afternoon the mines' workmen declared a strike and organized themselves into militia detachments in readiness to fight the army, for which purpose the civil authorities issued them with arms and ammunition. Laurence Hill, the deputy manager responsible for technical matters, appealed to the leaders of the two unions for help in maintaining the essential pumping and ventilation services in the mines. The UGT and CNT officials, now forced by the crisis to co-operate together, appreciated the importance of this whatever the outcome of the conflict might be, and immediately detailed men to carry out this work under the superintendence of the British staff.

*General Gonzalo Queipo de Llano.*

News started to come in of the mobilization of miners throughout the province and of the apparent acquiescence by the civil guard detachments in the areas concerned, for though the local guards had a reputation for being opposed in general to the popular front government they were so heavily outnumbered that they had little option but to adopt a passive rôle. In Huelva too, the company's employees formed themselves into armed militia and arrested many of the wealthier inhabitants and known right-wingers and imprisoned them in coal-hulks in the harbour. José Sánchez-Mora, the company's legal adviser, was one of those incarcerated in this way.

On the same day General Gonzalo Queipo de Llano, the commander of the *carabineros*, carried out a memorable coup in Seville, the important socialist-controlled city which from its strategic position dominated the region. On the night of Friday 17 July he had stayed in Huelva on his way to Ayamonte, on the frontier with Portugal, where he was due to inspect the customs post. The following morning he called on the civil governor to pay his respects and afterwards was just driving away when an officer stopped him with a message that he was urgently needed in Seville. He started to drive there at once, escaping by minutes a telephone call from the minister of the interior to the civil governor ordering his arrest. In Seville, by a remarkable display of courage and with the help of only four officers, he imprisoned the pro-republican garrison commander and his staff, took over the troops and compelled the civil governor of Seville to hand over the administration of the city to him.

It immediately became plain to the republicans at Rio Tinto that, with Seville in the control of the nationalists, Huelva province was in danger of being cut off from the rest of republican Spain. Immediately they started to organize an attack on the city. The *alcalde* of Rio Tinto sent an order signed by the civil governor of Huelva to Trevor Gough (who was in charge of the mines during the absence on leave of the general manager, Alex Hall) requisitioning four cars, fourteen lorries and 250 kilos of dynamite. The Rio Tinto civil guards were despatched by the *alcalde* ahead of a large column of men to lead the attack on the city. At the same time civil guards were also sent to Seville from Huelva by the republican authorities at the port.

Between 3 and 4 a.m. on the Sunday morning the main body moved out of Rio Tinto led by Cordero Bell, the socialist parliamentary deputy for Huelva. In the procession was a new Packard commandeered by the workmen. This large and expensive car had arrived at the mines only a few days before and was intended for the exclusive use of the chairman and members of the board when visiting Rio Tinto. At Panoleta, just over two miles from Seville, where the old road from Madrid via Merida, joined the road from Huelva to Seville* the column was ambushed by the civil guards from Rio Tinto who had preceded it and who had decided, with their colleagues from Huelva, to support the nationalist cause. The opening shots hit the Packard, detonating the dynamite with which it was loaded, blowing it to pieces.† Twenty-five of the miners were killed in the ensuing fight and the majority taken prisoner.[20]

On Sunday afternoon the British families spending the weekend at Punta Umbría were ordered over the telephone by Trevor Gough to return to the mines using cars since no trains were running as a result of the strike. As the procession of vehicles neared San Juan del Puerto, a lorry bearing a large red flag and full of armed men drove out towards it. The oncoming lorry back-fired, but so excited and drunk were its occupants that they thought the British in the approaching cars had discharged a shot at them and consequently opened fire. Their aim was not good and the only damage done was one shattered windscreen, though in the argument which followed Leonard Salkield nearly lost a toe when another rifle was accidentally fired. Eventually they parted on friendly terms, and were drunkenly wished a *buen viaje* ('happy journey'). When the British reached Valverde del Camino they were again halted and taken to the town hall. It looked as if they were about to be locked up for travelling without a permit when a Rio Tinto workman amongst the militia recognized them and secured their release. The man offered to accompany them back to the mines and standing on the running board of the leading car ensured that they were able to pass safely through the militia posts along the road.[21]

Monday was a peaceful day at the mines. Laurence Hill made a

*Improvements have now completely changed the configuration of the roads at this spot.

†The remains of the car stayed, scattered along the edge of the road, for some time. Members of the British staff, who had resented its purchase at a time when their salaries had been cut as an economy measure, used to point out the débris to visitors with a certain relish!

tour and found about 200 men at the workshops who had seized all the steel plate they could find and were busy converting vehicles into armoured cars. He was amused to see a lorry on which work had been completed; this had steel plates down the length of each side, in front of the driver's seat and over the bonnet, and a steel roof added to deflect any bombs that might fall on it. His own reaction was that with so many tons added to its weight its speed would be so reduced as to make it highly vulnerable and he would hate to travel in it.

That afternoon was very hot but this did not prevent the British from playing their usual tennis. In the evening, however, they were concerned to hear of the day's events in the locality which indicated that the hot-heads were actively at work. News percolated through that a convent and church in Huelva had been burnt down, and that at Zalamea la Real a mob of Rio Tinto workmen led by a vociferous woman whom the British always referred to as 'Miss America' (from the fact that she worked as a servant to an American employed at the mines) had attacked the church and set it on fire. Fortunately the whole of that attractive building was not destroyed by the flames, though a collection of ancient documents belonging to the church and the town, and some old and beautiful vestments, perished.[22]

The daily accounts of the fighting that were coming through over the radio were highly confused, both sides claiming successes and reporting on the progress of the war in a mutually contradictory manner. General Queipo de Llano in Seville well appreciated the importance of the radio as a factor in boosting morale among his own supporters and in demoralizing the enemy. 'Old Kippe' as the British at the mines called him, made daily broadcasts which were compulsive listening for the British in Bella Vista, for the 'radio general' had a certain panache which possessed its own appeal.

The high spirits which flourished in Bella Vista at this time amongst the British community, owed much to the womenfolk. In particular Mrs Gough, the wife of Trevor Gough the acting manager, moved from house to house, giving reassurance and encouragement where it was necessary, and organizing social activities. Thus morale was maintained at a high level and this in turn helped to keep life as normal as possible given the circumstances.

On Wednesday 22 July Laurence Hill wrote in the diary which he was keeping for the information of his wife Nan, who was then in England with their boys:

'Listened to my old pal Q. de Ll. at 10 a.m. The announcer said that in *pocos minutos* his *excelentisimo* the general of the second division *Señor Don* Gonzalo Queipo de Llano was going to speak. Then the band began to play *The British grenadiers*, *Rule Britannia*, *Hearts of oak* and *A life on the ocean wave*. After that the announcer again said that *dentro de un ratito* H.E. the general etc. etc., and then obviously put on the other side of the record as we had *The red, white and blue*, Faust's *Marching song* etc. When eventually H.E. *Don* G. Q. de Ll. came on his voice was all over the place. He squeaked and growled, cleared his throat, coughed, tried again and at last got going. He evidently thought it was a bit early in the morning! He told us that three ships' crews had mutinied [i.e. declared for the republic] and . . . shore batteries and planes sank all three . . . Carmona had been severely punished because the people had attacked the civil guards' post there. Also Alcalá de Guadaira. General Mola from Pamplona, he said, would take a long time to get to Madrid as he would have to make sure of his lines of communications. Everything points to a longish civil war, I'm afraid, and all hopes of leave are indefinitely postponed.'

The British at the mines were also delighted by Seville radio's idiosyncratic attempts to translate the general's speeches into English and in particular of the often repeated phrase *el ejercito salvador de España* which was rendered into English as 'the salvation army of Spain'.

Shortly after Wednesday's broadcast Hill heard that the militia in Huelva had switched their attention from churches to private houses. That of José Sánchez-Mora and another Spanish employee in the town had been looted and burnt, and a doctor's home had been blown up. He was glad his wife and family were on holiday in England, although everything in Bella Vista was still peaceful and morale was high:

'Everyone is very cheery and there is a lot of bridge, billiards and booze! As the bank has suspended payment we are issuing the 6,000 *pesetas* in the club-till to people, on the condition that at least two-thirds must find its way back again! This will make it last quite a long time![23]

On Friday 24 July a British destroyer, HMS *Boreas*, arrived at Huelva with orders to evacuate the women and children from Rio Tinto to Gibraltar. The next morning the workmen's militia at the mines gave

HMS *Boreas*.

permission for the evacuees to travel down the railway to the coast in a special train with four men to escort them. 'Bella Vista', noted Laurence Hill, 'is now a completely Eve-less Eden.' Another destroyer, HMS *Wild Swan* stood by at Huelva in case of the need to effect a quick rescue of other British subjects.

Food now began to run short for everybody. The three local *alcaldes* visited Trevor Gough and Laurence Hill and asked them to pay to the workmen any outstanding money owed. They agreed to settle accounts and to pay a week's holiday pay to the men entitled to it. That same day (Tuesday 28 July) Huelva fell to the nationalist army. Most of the company's workmen there returned immediately to work as the remnants of the republican militia took to the hills and the army began trying and executing such republicans as they could find who had been associated with local atrocities and outrages. Right up to the time of the seizure of the town the local militia had been executing the right-wing prisoners held on the coal-hulks in Huelva harbour. José Sánchez-Mora was on his way to face a firing-squad when he was rescued by civil guards acting in support of the army.[24]

After that it seemed to the British that an attack on Rio Tinto was likely to be launched in the next day or two, but in the event another month was to elapse before the actual assault on the mines took place. The day that Huelva fell, Trevor Gough received a cable from the London office ordering all the British staff to leave the mines and to board the destroyer at Huelva or else to travel by road across the frontier into Portugal. That night Laurence Hill noted in his diary:

'Really, it almost passes belief. They must have got the wind up very badly . . . Actually, of course, we are much better off here than if we tried either to go to Huelva, and so get mixed up with a potential battle, or "made our way" (how I wonder?) to Portugal. Oh well, we shall have to try and concoct something to send off tomorrow but it really makes one feel a bit mad having orders like that hurled at one. What about the Spanish staff? and our prestige?'

*Laurence Hill, DSO, MC.*

In the next two days the troops from Huelva moved on to occupy the towns of Ayamonte and Tharsis. The feeling in Bella Vista was that Rio Tinto was next on the list. Many of the Spaniards seemed to think so too. 'All the mayors and syndicates have bolted,' wrote Hill, 'and the CNT have formed a committee in Nerva whose chief job apparently is to find and kill the bunch that fled!' On Saturday 1 August the army reached the outskirts of Valverde del Camino and started to consolidate their position and form a base from which an attack on the mines could be launched. A week later the CNT committee decided to attack the army's position and 500 Rio Tinto miners were sent into battle, cheered on by 'Miss America' dressed in her best red blouse and carrying a red parasol. Forty of the attackers were killed and many more wounded before the assault was abandoned. On the army's side the only casualties were twelve wounded.[25] Plainly, whatever the workmen possessed in the way of bravery was more than offset by their lack of trained leaders and military skill.

Shortly before the battle at Valverde the two deputy managers had taken stock of the situation. Food supplies for the British community were virtually exhausted; there were no meat, oil, sugar, rice or potatoes left. Bombing and artillery attacks seemed imminent. They decided that they had better try to evacuate most of the British staff to the bungalows at Punta Umbría behind the nationalist lines, staying on themselves with five others to attempt to protect the company's installations and to afford any protection they could to the middle-class Spaniards, such as the assistant engineers and supervisors, who were at risk from their compatriots in the republican militia. Despite numerous requests to the defence committee, however, permission for a train to allow the British men to leave was not granted. It seemed that the republicans were unwilling to see them go, both because they felt it would be a blow to their prestige if foreigners were to seek refuge with the nationalists and also because they thought that the existence of a large number of foreign nationals at the mines might mitigate the force of the army's assault when it came.[26]

On Friday 14 August the London *Evening Standard* announced:

'The Rio Tinto Company in London received news today from Huelva that 38 members of their staff are being held as hostages at Rio Tinto mines by communists who have taken control . . . The Rio Tinto Company have sent a representative to the Foreign Office in London to see what can be done to persuade the Spanish government to take action.'

The following morning the *Daily Mail* demanded in its editorial on the subject 'IMMEDIATE AND ENERGETIC ACTION OF THE BRITISH GOVERNMENT'. The same issue carried a statement made the previous evening by Colonel R. M. Preston, a director of the company, that 'Our representative saw the Spanish ambassador today in London, and after telephoning to the government in Madrid he promised to reply by tonight but so far nothing has been received.'

All the London daily papers expressed grave concern about the fate of the British at Rio Tinto of whom nothing had been heard by the outside world for two days. The worry was understandable in view of the horrifying stories that were coming out of southern Spain. The *Daily Mail* reported that same Saturday on the advance of General Franco's army through Badajoz province adjoining that of

Huelva. In the village of Almendralejo, the newspaper declared, the troops had found the bodies of many right-wing inhabitants crucified, some head downwards, on the prison walls. They had discovered, too, the smouldering bodies of another eighty who had been burnt alive, and the corpses of many more executed by firing-squads. The soldiers had retaliated, executing hundreds of local people in revenge, including a large number of women who had taken a leading part in perpetrating the atrocities. The families in England of the men still detained at Rio Tinto not unnaturally read such news from Spain with alarm and despondency.

But that very morning preparations were at last going ahead for the evacuation of most of the staff from the mines. The Madrid government had broadcast an appeal to the local defence committee and at the same time as the London papers were appearing on the streets thirteen cars were setting off from the mines for the coast. The first set out at 6 a.m. and the others left after that at half-hourly intervals. Each car needed a permit to get it and its occupants through the pickets stationed at intervals along the road within republican territory as far as Zalamea. Then it passed into nationalist-held ground to a warm welcome from the soldiers.

Three members of the staff, Charles Palmer, Charles Julian and Herbert Potts, had to go from Huelva to Seville on official business. When there they received a message that General Queipo de Llano, having heard of their arrival, wished to see them to enquire as to the situation at the mines. They were conducted straight to the general's office and found him chatting on the telephone about the whereabouts of a company of Moorish troops and trying to locate the spot, with some difficulty, on a Michelin road-map which was tacked onto the wall and was plainly the only map he had available to him. His telephone conversation finished, he talked with the three men for ten minutes offering them the opportunity, which they declined, of broadcasting to Britain. On leaving him Potts noticed that Julian appeared depressed and asked him what was the matter. It seemed that the sight of the general's only map had made him very pessimistic about the nationalists' chances of success, but Potts countered by telling him to cheer up and not to worry as he thought it unlikely that the republicans had any map at all![27]

Laurence Hill had travelled with the convoy of cars to Huelva with the intention of getting supplies and returning to Rio Tinto (he had acquired a pass from the defence committee permitting him to do so) as soon as possible in order to rejoin Trevor Gough and the others who had opted to stay at the mines to safeguard company property and look after the Spanish supervisory staff; but in Huelva the nationalist military governor, *Comandante* de Haro, told him that he would not be allowed to go back to Rio Tinto on any account. Disobeying the instruction he set off with his parcels and two enormous union jacks (which he intended to display over the *casa grande*) driving as far as Valverde where he stopped for tea at the house of two British friends, Mr and Mrs Maidment. There he found sentries posted to prevent an Englishman (name unknown) from going on to the mines.

Consequently he decided to drive to Seville instead and see the 'chief of staff, Arambino, rather a nice humorous johnny, who took me in to see the general [i.e. Queipo de Llano]. I explained the situation to him, viz., that we had Spanish engineers etc. in B.V. [Bella Vista] who were O.K. as long as some of us were there but who might be mopped up if we all skidadalled; and he gave me a pass right away.'

Time went by and Rio Tinto remained in the hands of the republicans, though in the week after Laurence Hill's visit to Huelva and Seville the army gradually took possession of Huelva province village by village until at last only the Rio Tinto mines themselves and the *pueblos* around were left. Then four of the remaining staff left for Huelva leaving only the two deputy managers at Rio Tinto. In the middle of the week, on Wednesday 19 August, Hill noted the arrival of 'another weird telegram last night [from London office] to say Gough and I had to leave immediately under pain of instant dismissal or words to that effect . . . It made me distinctly angry. After all, it shows a pretty poor opinion of one, and a good deal of lack of confidence, to send that sort of message at a time like this.' It was decided that Trevor Gough should go to Valverde straightway to see Alexander Hall, the general manager, who had just arrived there from England but could not get permission from the miliatry to proceed any further. The meeting took place and the deputy manager explained to his chief the full details of the situation at the mines but afterwards found that his permit to return was ignored by the troops who would not let him pass back to Rio Tinto. As a result, Laurence Hill became the only member of the British staff left at the mines.

The following morning (Thursday 20 August) the mines' area was subjected for the first time to aerial attack, bombs being dropped on el Campillo, where three people were killed, and Nerva where more than twenty died. Hill saw the local leaders of the defence committee to arrange for the women and children to take refuge in the underground workings. The Spaniards were anxious to get them all into Bella Vista village and he tried to explain to them the advantages of dispersal of people as opposed to their concentration when under air-attack, but he made little impression. 'I expect when the salvation army does eventually arrive we shall have a hell of a riot of terrified women and children in B.V. and I shall probably beat it for Atalaya open-cast!' During the next few days air-raids became a common occurrence.

On Sunday (23 August) Hill attended a meeting of the defence committee in the town hall, and tried to persuade its twenty members to do their fighting in the fields away from the villages since this would help save the lives of women and children. He was grimly amused at the indignation which some of the committee expressed at Queipo de Llano for attacking Rio Tinto. Since the place was unimportant strategically they thought he should leave them alone and concentrate on Madrid! Laurence Hill tried to get them to release their prisoners but without success.

That afternoon an aeroplane dropped leaflets giving the republicans twenty-four hours in which to surrender, and the next day another aircraft flew over looking for white flags. But there were none. The miners' leaders had decided 'to fight to the last gasp'. There was no doubt but that the militia at the mines represented a considerable problem for the attacking army. There were an estimated 10,000 men waiting for the assault, well armed and with plenty of ammunition,' and the roads into the area were mined with dynamite.[28]

The attack began at last on Tuesday 25 August at 8 a.m. Three columns of nationalist infantry moved in with bombers preparing the way ahead of them. The carefully prepared armoured vehicles which Hill had seen being equipped at the very beginning of the crisis went into action on the road between el Campillo and Zalamea, but were soon knocked out of action by the nationalists' artillery and mortars. By 2 p.m. the advance had reached the edge of el Campillo. The fighting continued throughout the day and at 11 p.m. Laurence Hill was asked to meet a deputation from the miners.

They told him that the men were handing in their arms at the town hall, that they had no option but to surrender, and asked him if he would join a party to meet the nationalist commander to treat for terms. Hill told them that they should try and telephone Zalamea immediately to tell the army of their decision, and that he would see the commander on his arrival and do his best for them. The right-wing prisoners held at Rio Tinto were, to Hill's relief, released unharmed that night, and he was assured that those at Nerva would not be hurt though, to ensure their protection, it was thought best to leave them locked-up overnight and to release them next morning.

The men defending el Campillo, however, refused to release the right-wing captives being held in the church. At dawn the next day, as the nationalist column advanced, the miners set fire to the building and some middle-class houses. Twenty-two people died in the flames

*El Campillo.*

or were shot trying to escape. The atrocity marred the reputation for moderate behaviour which the Rio Tinto miners had up till then enjoyed, for though the most barbaric acts had become common in other regions they were eschewed at Rio Tinto until that morning. And the army was to exact a terrible revenge.

At 9.30 a.m. the first of the nationalist columns arrived, followed an hour later by the second. The third arrived at lunch-time. In all there were about 1,000 men and 50 officers, a surprisingly small force.[29] The rounding-up of the republican leaders commenced at once. The nationalist commander had arrived with a list of over 300 names written out on an omnibus death warrant and in a few days all the men who had held senior positions in the militia or trade unions, and some of those who had assisted in the atrocities at el Campillo, were under arrest and being tried by court martial. If found guilty (and the vast majority were) they were immediately executed. The most prominent prisoner taken by the troops was Saturnino Delicado, who had been one of the leaders of the communist party in Seville and had fled to Badajoz when Queipo de Llano carried out his coup. He had again eluded capture when Badajoz was stormed and had taken refuge at Rio Tinto where he took command of the defence of Nerva.[30]

It is not possible to calculate the numbers of those executed in the days after the capture; the records are not full enough. But many men, and some women, were shot. Day after day, when work was once more resumed at the mines, it was a common happening for soldiers to arrive and to arrest a man at the work-face or at his bench. On one occasion a woman rushed into the smelter to find her husband, having heard that he was to be arrested. She begged him to flee but he refused, being afraid that if it was known that he had escaped because she had warned him she would be shot in his place. He returned home with her and later was taken and executed.

Normally the women who were arrested were released after their heads had been shaved. 'Miss America' suffered this indignity and in addition, as punishment for her vitriolic tongue, was forced to drink a litre of castor oil. On finishing it she laughed in the faces of her captors to show her contempt. They made her drink another litre.[31]

All the men of Atalaya, the small village on the edge of the great

*The Atalaya open-cast pit with the 'village of the widows' on its western edge.*

open-cast pit, who had taken part in the defence of el Campillo, died; many of them shot one evening after their day's work in the vast quarry. Atalaya became known as 'the village of widows'. The board of directors cabled to Queipo de Llano asking for all possible leniency to be shown; and the intercessions of the British staff on the spot (for those evacuated to Punta returned to the mines on 27 August) certainly saved the lives of some.[32]

When work was resumed on 28 August it was with a much reduced work force: 5,400 men compared with 8,500 on the day on which the war broke out; and armed soldiers supervised them as they worked.[33] It was all in strong contrast with the disorder which had reigned for so long. A large number of the young men, who a few days before had been in the workmen's militia, were now conscripted into the nationalist army.

There were other, even stranger, changes. Some of the most demagogic of the workmen started to appear in the blue shirts of the *falange*, the Spanish fascist organization and one of the political parties of the right which had given its support to the military revolt. Leonard Salkield recalls a worker at the concentrator, an anarchist who had caused much trouble as a CNT delegate, coming to work one day in a blue shirt and demanding time off to hunt for republican fugitives in the hills: 'I looked straight at him and said, "So you intend to shoot your brothers and sisters?" He looked at me rather sheepishly and replied "You understand". I did. In a civil war people must be for the side in power for their own safety.'

As the position at the mines returned a little more to normal, Laurence Hill returned to England on leave to join his wife, Nan, and his two sons. As soon as news of his safety had reached London following the army advance into Rio Tinto, Richard Preston had written to Mrs Hill on behalf of the directors:

> 'I want to congratulate you doubly; first because your husband is safe and on his way to rejoin you in England; and secondly on his brave and generous action in staying on at the mines in order to try and save the lives of the Spanish staff. You must be a proud woman today.'

*Partial view of Rio Tinto from el Valle about the time of the civil war. In the left-middle-distance is the third hospital, and in the right-middle-distance Bella Vista village. Behind Bella Vista is the Atalaya open-cast pit.*

# 18

# *The threadbare years*

## Problems of war and of peace: 1936–1945

The Spanish civil war dragged on until 1 April 1939, and though after the fall of Rio Tinto on 26 August 1936 the mines were far from the fighting, its effects continued to be felt by the miners and their families. Within days of the mines' capture the workers' organizations were refashioned and nationalists put in charge of their administration. Similarly with the *ayuntamientos*: new mayors and councillors were appointed to help in the work of reconstruction and the senior local government officers were replaced.

Inevitably (for the nationalists were traditionalists who believed not only in unity within the state but also in religion) religious conformity was enforced and the protestants of the area were no longer permitted their own separate forms of worship.* A great open-air mass was celebrated at Rio Tinto attended by almost everyone save those still in prison. Money was collected to restore the damage to the local parish churches and the fabric of the church at Zalamea la Real, set on fire by hot-heads at the very start of the war, was skilfully restored.[1]

The management of the Rio Tinto Company was committed to an attitude of neutrality and the avoidance of any form of political involvement. The staff at the mines, of course, had to accept the orders of the *de facto* government for the area whether this was republican, as it had been earlier, or nationalist.[2] The directors saw it as their responsibility to both the shareholders and the workmen to get the mines working and in production again as quickly as possible. The fact that this implied the company's acceptance of the administration set up by General Francisco Franco (who on 1 October 1936 was elected head of state and commander-in-chief by representatives of the armed forces and of those political parties which supported the nationalist cause) led to considerable criticism by republican sympathizers in Britain. Auckland Geddes, in public normally one of the most imperturbable of men, was moved to exclaim at a shareholders' meeting that newspaper reports gave the impression that the board of directors was 'composed of violent fascists actively participating on the side of General Franco.'[3]

*By the law of religious freedom the protestants are now once more allowed their own churches.

Not unnaturally, after the long turmoil and confusion of the second republic all businessmen operating in Spain welcomed the re-establishment of order. The company had co-operated with the popular front administration and accepted its instructions, taking on extra men at the government's bidding in an effort to reduce unemployment, paying full arrears of wages to men (many of whom had been in prison) laid off earlier with the consent of the then ministry, and accepting wage increases of 15 per cent ordered by the government; yet that same popular front administration had been completely unable to help the company in its endeavours to keep the mines working normally. Added to the effects of the general economic depression, this had meant that the mines had been kept running only with difficulty; while it had begun to look as if the disruptive tactics of the Spanish anarchists and the communists might lead to power passing to an extremist left-wing régime under which the company would not be able to operate in Spain at all. Consequently it would hardly have been surprising if the directors had believed that a nationalist victory would be in the company's best interests. Nonetheless, and whatever the thoughts of individual directors might have been, the company was not a combatant and the board took the view that its proper concern was to operate the mines in the best possible way in any given circumstances.[4]

In September the nationalist régime authorized employers within the area it controlled to dismiss men recruited surplus to needs on the orders of the republican administration since the previous 15 February and to peg wages at what they had been at the same date. As a result of the conflict at the mines, however, the company found itself with fewer employees at the time this decree was issued than on 15 February. Consequently, there was considerable discussion with the military authorities of Huelva province as to whether the company was obliged to find men to make up the deficiency in numbers, and finally to resolve the matter José Sánchez-Mora was despatched to Burgos, where the nationalist civil government was being set up, in order to get an answer. He was asked to wait and see the secretary general of the administration, Nicolás Franco the general's brother. It was well after midnight before the interview could take place but Sánchez-Mora was impressed by Nicolás Franco's clear understanding of the needs of the Rio Tinto mines and his concise analysis of the problems the company was facing. The secretary general agreed that the company could continue to operate with its then work-force and without needing to recruit additional labour, although he stipulated that it would have to re-employ workmen held as prisoners when their time came to be released.[5]

At this meeting Nicolás Franco spoke of the necessity to get exports moving in order to earn the foreign currency which was so desperately needed by Spain. The economy was clearly going to be in a bad way by the end of the war and it could not be too soon to start re-laying its foundations. He demonstrated his concern, and his detailed knowledge of events in the nationalist-controlled zone, in an amusing exchange with José Sánchez-Mora. The latter had just refused an invitation from the new governor of Huelva to join the Huelva port committee which was destined to play an important role in organizing the shipping of exports from southern Spain. The secretary general wanted to know why Sánchez-Mora had declined and the latter retorted that he understood it involved joining the *falange* and wearing the movement's blue shirt. Though he had no objection to army uniform (he was indeed an officer in the military reserve) he could see no point in dressing up like a Rio Tinto boy scout! Franco smilingly told him that he would be exempt from any need to become a *falangista* and Sánchez-Mora thereupon agreed to serve.[6]

By the end of 1936, with the release of some of the left-wingers who had been detained, the number of workmen at the mines had risen to 6,000; still 2,500 men fewer than at the outbreak of the war five months earlier. Yet in the month of December they produced 143,000 tons of ore, some 30,000 tons more than in the previous May, the last month during which there had been uninterrupted working.[7]

Although the shortage of food which prevailed at Rio Tinto during the month of August was ameliorated somewhat after the capture of the mines, it continued to persist. Bread was strictly rationed and most foodstuffs were in short supply. The nationalist administration ordered a stabilization of food prices and consequently charges in the company stores remained pegged, but the general lack of supplies meant that these shops were frequently empty of essentials and the women-folk were compelled to buy wherever they could, normally from street and market vendors. Troops were put on duty in the Rio Tinto street markets to prevent profiteering but the attempt was only partially successful and rising food costs thus became a real problem.

In 1937 General Franco's government decreed new wage-scales for Huelva province to meet the higher cost of living. These gave the Rio Tinto miners increases in pay of from 10 to 15 per cent and introduced a new concept: allowances for all children after the first child, provided by contributions levied upon employers of 5 per cent of their wages' bill and upon workers of 1 per cent of their earnings. Another government levy, bearing upon both employers and employees, was introduced to provide free meals for all women and children left without means of support.[8]

The Rio Tinto Company still found a need, however, to continue its own relief work among its employees at the mines and Huelva, granting pensions to widows, and providing food and clothing for the families of employees and former employees in the greatest need. The directors also made a gift of 4,000 square yards of land to the Huelva authorities for building houses for the war-disabled.[9] When an association of businessmen approached the company asking it to subscribe to nationalist army funds the board explained that as a foreign concern it had to observe strict neutrality in such matters. The company, however, was prepared to give generously to succour the victims of the conflict, providing in the first year of the war alone 1,500,000 *pesetas* (some £36,000) for distribution through recognized relief agencies and earmarked specifically for assisting the aged and the war-wounded, and additional to what it was spending on alleviating distress amongst its own employees.[10]

The company's relations with the nationalist government during the rest of the war were not easy; inevitably so for the priorities of a nation at war and the controls which have to be imposed often conflict with the needs or desires of business. Copper, pyrites and sulphur were essential war-products, and, though the company had overseas contracts to fulfil, half of the entire output of the mines was requisitioned by the military during the duration of hostilities. Produce worth well over £3 million was taken in this way against promissory notes. Though these were eventually honoured the delay in payment caused the company considerable financial difficulties and led to sharp exchanges between the British government and the Spanish nationalists.

For political reasons the Burgos administration completely forbade the export of pyrites to France and Czechoslovakia. It reasoned that in a few weeks those two countries would be so short of sulphuric acid, an important element in the manufacture of high explosive and therefore vital to their national defence, that their governments would be compelled to moderate their hostility to the Spanish nationalists. For similar reasons the flow of pyrites to Belgium, Holland, Denmark and Poland was at times interrupted, and the company's officers had to spend much time and energy doing their best to get deliveries to customers resumed.[11]

As early as 28 August 1936 Burgos decree number 70 authorised the requisitioning of the products of the Rio Tinto mines. Colonel Trujillo (a member of the Seville military staff) explained to mines' manager Alexander Hall that 'the tremendous amount of help from Germany had to be paid partly in Rio Tinto copper and pyrites, and the company has to hand over the sterling proceeds of its deliveries to other countries'. On 1 January 1937 the company was informed that it would have to supply 50,000 tons of pyrites with the highest possible copper content each month for shipment to Germany; one half of the total amount the Germans wanted to receive from Spain. A few days later the German government indicated it would prefer to obtain these shipments by voluntary agreement with the Rio Tinto Company and the other producers, holding out the inducement of using its influence with the nationalist administration to arrange a return of some of the sterling which the British companies had been obliged to surrender in order to get any Spanish export permits at all.

This was not a proposal which appealed to Geddes, for the tonnages demanded not only made it impossible to meet commitments to other customers but threatened to cut into the 20,000 tons of ore which he had promised to deliver to the British government each month for its own war-reserve stock. On 22 January the military governor of Huelva, *Comandante* de Haro, at the request of Louis Claus, the German consul in the town, issued orders to the company that preference must be given to German vessels so as to permit their rapid despatch, a 'flagrant case of flag discrimination and breach of mercantile usage' wrote Geddes to Sir Robert Vansittart, the permanent under-secretary at the foreign office, adding 'So far we have ignored this order'.

Geddes demanded that the British cabinet consider the problem and issue the company with instructions. It was plain that Germany was seizing the opportunity to stockpile pyrites against possible future hostilities and he told Vansittart, in a long letter reinforced by an appendix containing detailed statistics, that not only was this 'plain robbery' which was placing Britain's own supplies of essential raw materials at risk, but was also having an adverse effect on British industry. The Port Talbot copper smelter had been forced to close in October 1936 creating further unemployment in an already depressed area and the loss of valuable contracts throughout the world. At a time of world shortage of iron ore, the cutting off of Rio Tinto supplies to the Ford works at Dagenham and to the Carron Company in Scotland would also have serious repercussions. Complaining that Germany 'is practically master in our house' Geddes demanded that the British cabinet give the fullest protection to his company's interests in Spain:

'I cannot state too clearly that his Majesty's government will not get the pyrites and other products that our national safety requires as long as Germany is permitted to seize for her own benefit British property. A change of heart in a German is only to be expected when he realizes that force greater than he can command is opposed to him. If, therefore, his

Majesty's government requires pyrites from the Rio Tinto Company it must be prepared before it embarks on diplomatic action to resort if necessary to naval action to prevent Germany seizing our pyrites . . . It is, therefore, in my view essential that before anything is done there should be a cabinet decision to proceed, if necessary, to the application of naked force.'

On 1 March (1937) Anthony Eden the foreign secretary sent a copy of Geddes's letter with a memorandum of his own to the other members of the cabinet, setting out the legal position and his own recommendations. He felt that the treatment of the Rio Tinto Company was one 'which no government of this country can afford to acquiesce in'. It was true that normally under international law requisitions such as had taken place in Spain were not forbidden provided that they were for reasons of public interest and also that adequate compensation was paid. However, by the Anglo-Spanish commercial treaty of 1922, article 4, British subjects (including companies) were exempt from requisitions of a military character. Consequently this treaty had been broken. In normal times it would have been sufficient to take retaliatory trade measures but these would scarcely be effective in this case, and Anthony Eden declared:

'If, therefore, there is to be any prospect of remedying this intolerable state of affairs, I am convinced that stronger measures are required . . . By making use, if need be, of our warships to intercept any ship which leaves Huelva carrying any such products to Germany or to Italy we have it in our power to end this state of affairs . . . We should not, however, embark upon this course without clear determination to go through with it to the end.'

A couple of days later the cabinet met and Anthony Eden informed his colleagues, meeting under the chairmanship of the prime minister, Stanley Baldwin, that he wanted immediate authorization to make 'very strong representations' and to demand financial redress for the Rio Tinto Company and other British companies concerned, and that he did not 'exclude the possibility later on of consideration being given to the proposal in his memorandum.' In the discussion it was suggested by one minister that despite the 1922 treaty the Burgos government was legally entitled to requisition supplies, and that if compensation was inadequate then that government should be asked to appear with the British government before an international tribunal to determine the matter. The difficulty with this proposal, it was pointed out, was that this would be tantamount to recognizing the Burgos government which the British did not wish to do until it was certain that General Franco and his companions were assured of success.

The perplexity of the British government as to its course of action was well illustrated by the discussion on another suggestion: that the United Kingdom should threaten to repudiate a recently negotiated commercial agreement with the Burgos administration, and possibly suspend the importation into Britain of Spanish oranges. Despite the endorsement of the president of the board of trade, Walter Runciman, who believed that in this event 'the prospects for Spain would be somewhat grim', Anthony Eden pointed out that if the agreement were terminated it 'would damage this country more than the Spanish insurgents'.

The chancellor of the exchequer, Neville Chamberlain, commented that the foreign secretary's suggestion for the possible use of the navy 'had caused him some concern'. He thought the proposal 'was not only useless but must lead to a very serious situation with Germany and Italy.' The cabinet agreed with Chamberlain and authorized the foreign secretary only to make representations to General Franco's government after further consultation with Walter Runciman and Stanley Baldwin. Geddes's request that the government should give him 'any counsel they can afford me in this deplorable situation' was ignored. In the circumstances all he could do was to accept the position the company found itself in.

The Spanish civil war aroused in Britain and many other countries a violent partizanship, much of it based on illusory views derived from certain newpapers and politicians, who depicted the conflict as an heroic struggle for liberty by an oppressed people against a small band of blood-thirsty military adventurers. This grotesque caricature served to conceal the origins and the true nature of the conflict and also to disguise the real tragedy of a nation almost evenly divided within itself. It also produced in the minds of many non-Spaniards an over-simplified view as to the course of events in Spain and led to considerable criticism of companies trading within the nationalist zone.[12] The Rio Tinto Company's directors and staff found themselves frequently accused of nurturing the 'oppressors'. Newspaper stories alleged that the company had given £500,000 to the nationalist army and that it had voluntarily delivered vast quantities of copper to Germany and Italy in part-payment for the assistance those two countries were giving to the Burgos government.

The reports of the gift of money may have had their origin in the fact that in the published company accounts for 1937 there was a total of 22 million *pesetas* (approximately £500,000) which was shown as either being owed to the company for requisitioned produce or else standing to the company's credit in Spanish banks, yet given a book-value of only £3. This low valuation was due simply to the fact that the company's auditors were uncertain as to whether the debts, or the credits in the banks, would ever be recovered and so felt obliged to ascribe notional values to them. In fact almost all the money concerned was eventually retrieved. (The only losses which had to be written off were relatively minor sums in republican territory.)

The stories of copper deliveries to Germany and Italy were equally inaccurate though they were widely accepted; the commonest version was that during a twelve-month period in 1937–8 the company had sent 300,000 tons of copper to the former country and 65,000 tons to the latter. This would have required the production of 1,000 tons of copper for each day of the year but in fact during the whole of that time the company produced slightly less than 12,000 tons, or about 40 tons for each working-day. The German government was certainly trying to get the Burgos administration to pay for war supplies by the shipment of raw materials; and Guglielmo Danzi, the Italian fascist representative to the Franco government, seems to have believed that Rio Tinto copper was being shipped to the Germans on a large scale. In fact between 12 November 1936 and 23 February 1937 the military requisitioned 2,756 tons for shipment to Germany, 700 tons for Italy and a little under 4,000 tons for their own use within Spain. But shortly after this the Burgos government assured the company that the requisitioned metal was now all being used by manufacturers within the territory it controlled. Even so, because of the persistent rumours, the board set up a subcommittee to make an investigation; its conclusions appear to have satisfied the directors that the nationalists' assurances represented the true situation. Large quantities of requisitioned pyrites were, however, still sent at that time to Germany and Italy despite the protests of the company.[13]

Many of the economic controls imposed by the Franco government stemmed from the fact that all the country's gold reserves were, at the start of the war, in Madrid which was the principal stronghold of the popular front ministry; and the nationalists, therefore, were at a considerable financial disadvantage compared with their opponents. There was some £225 million in gold in the vaults of the Bank of Spain and in the first September of the war the republicans shipped this abroad. Bullion and specie worth £50 million were sent to France, which returned them to Spain after the nationalists' victory. The remainder, worth £175 million, was sent to Russia and none of it was ever recovered. The loss of this vast wealth was a cause of many of the economic difficulties which faced the country during and after the civil war.[14]

Their lack of funds meant that the nationalists had to adopt measures which made the Rio Tinto Company's normal business highly complicated. The company was required to deposit with banks in nationalist territory, pounds sterling to the value of all products it shipped out of the country. The rate of conversion for the sterling value of goods requisitioned by the army was fixed at between 40 and 42 *pesetas* to the pound, a figure which the company considered abnormally low. The fact that payment by the military for the requisitioned produce was always well in arrears meant that there was a constant struggle to find the money to meet the company's own financial obligations within Spain. By the end of the civil war there was a sum of 36,000,000 *pesetas* (then estimated to be worth about (£700,000) owing. The Burgos régime, moreover, made its payments in *pesetas* that were not permitted to be transferred outside Spain, while the spectre of inflation meant that it was difficult to judge what their real value would be by the time that they were actually paid over.

The need to resume pyrites shipments to customers, with whom there were contractual obligations, also provided a source of intense frustration to company staff. Export licences took a long time to obtain and frequently arrived too late for the company to charter ships in the normal way, and often were obtained only after such ships as had been chartered had been compelled by their schedules to sail away empty. By the company's sales agreements some 200,000 tons of pyrites should have been leaving Huelva every month; but in 1937 it was possible to export on average only 119,000 tons per month. The shipping companies and the mineral producers had to operate on a day by day basis and there was considerable competition between the mining companies to get the available vessels. Consequently confusion abounded.[15]

Daily life for the British staff at the mines was highly uncomfortable. The London office did its best to ensure their food supplies but there were inevitable shortages. There were also problems connected with money. In the absence of official exchange quotations the British employees could at first cash their cheques only with speculators, who were intent on making inordinately large profits. The directors then authorized the general manager to provide this facility at what he himself thought was a reasonable rate of exchange. The men had refunded to them by the company the cost of their families' passages from Gibraltar back to Britain and, when it became plain that the separations were likely to be lengthy, the board instituted allowances of £3 a week for a wife and 15s for each child in the United Kingdom.[16]

By March 1937 so many of the wives were worrying the directors for permission to rejoin their husbands that it was decided they could return any time after Easter at the company's charge. On their arrival they found their menfolk still working frantically for long hours to repair the damage to the mines' economy and in most cases were able to ease their husband's burdens by remaking their homes and providing them with support during the difficult and uncertain months which followed.

Many of the British women were deeply moved by the plight of the Spaniards at the mines. By careful household management they were able to save some food from the supplies sent to them from England and pass these on to the families of their domestics and to some of the women and children left without the support of a wage-earner. Some of the British women wished to do still more to ameliorate the problem of hunger facing the vast majority of Spanish families in the mining villages. A group of wives therefore came together and made sallies into the countryside around and across the Portuguese border to obtain supplies for distribution to yet more needy Rio Tinto families. They made sure that their husbands never knew of their activities; for what they were doing, though fairly common, was technically illegal. They exchanged luxury items from their own larders, such as tea, coffee and sugar on the Spanish black market for flour, beans, goats' meat, rice and olive oil; or else purchased the same vital foodstuffs on the other side of the frontier and brought them back to the mines in their cars.[17]

The concern shown by the British women for the innocent victims of the war, and the endeavours' of the British staff both to co-operate with the *de facto* authorities and to assist their workmen, by interceding for prisoners, arranging pensions, and finding new jobs wherever possible for men dismissed as a result of government decrees, served to bring the British and Spanish communities at Rio Tinto closer together than they had ever been since the British first isolated themselves in their Bella Vista village.[18]

The British government afforded formal recognition to General Franco's government in February 1939 and the civil war itself came to a close in the next month. By then the whole country had suffered a terrible devastation. Because at an early stage in the fighting the area of the Rio Tinto mines had ceased to be part of the combat zone it was not as badly ravaged as many parts of Spain. But the miners, because they had been crushed in the opening stages of the conflict, had endured early the indignity and penalties of defeat; and for the next thirty months of the war they had to watch the continuous collapse of the cause for which they had fought. Thus they suffered a devastation of the spirit which for some was more terrible than any physical deprivation.

However, a new element had come into the picture. The old atmosphere of turbulence had gone and law and order was once more established. The nationalist government declared its intention of ruling for the benefit of the whole nation, not just one section of it; and, although it was committed to punishing those of its opponents who had infringed the criminal code during the war, it was anxious for the scars to be healed. With human resilience asserting itself it looked as if it would again be possible at Rio Tinto, where the workmen were still one of the highest paid groups of workers in Spain, to rebuild a community in which people would be able to lead reasonable and contented lives. What was not foreseen by most people was that the coming summer would see the beginning of a world-wide conflict of such magnitude that even though Spain stayed neutral it would suffer economic upheavals as great as those which had faced the country in the time of the civil war.

Among those who did anticipate the forthcoming outbreak of hostilities between Nazi Germany and Britain and her allies were the directors of the Rio Tinto Company. As early as April 1939 they ordered the suspension of pyrites deliveries to Germany in view of their importance to that country's armaments industry, only to be instructed by the British foreign office to resume them since their action was an unfriendly act against a technically friendly country. However, there was little they could do at Rio Tinto in the way of preparation to protect the mines' community from the war they feared would come. The declaration of hostilities on 3 September in that same year meant that with scarcely a breathing space the British staff in Spain had to cope with war-time conditions for nearly eight continuous years.

In the first few months following the outbreak of the European war the British staff who were of fighting age returned to the United Kingdom with the company's permission to volunteer for service in the armed forces. The British at the mines dwindled in numbers until eventually there were fewer than thirty. Life at the mines continued fairly uneventfully for these few, however, until the fall of France to the German army in the summer of 1940; they were all working hard, keeping the mines running and coping with the new regulations which the government was introducing as part of its programme of social and economic reform.[19]

The civil war over, the Spanish ministry of industry and commerce set about establishing a regulating commission for the pyrites industry, intended to hold the ring between the producers and purchasers, and to regulate prices and agreements between the two. Consequently the company was required to submit its contracts for approval to the ministry; and though at first Sir Auckland Geddes refused, quoting the 1873 deed of sale which gave the company rights freely to exploit and export the minerals of Rio Tinto, he had to give way when the minister declared that no exports at all would be allowed until the company co-operated.[20]

This, like other changes decreed by the government, was seen as an important element in its policy of ending the struggles between capital and labour which previously had dominated Spanish industrial history. The official view was that, in the future, conflict could be avoided and living standards raised by controlling the earnings of both sides of industry, by providing minimum standards with regard to work conditions and social welfare, and by exercising control over economic development and establishing a system of priorities. The methods adopted were an elaboration of those instituted earlier by the administration of Primo de Rivera, though more complex and efficient. Thus the charter of labour made dismissals of workmen virtually impossible without official consent and minimum wages for each industry were decided by juries. The series of welfare enactments, which began in 1939 with children's subsidies, went on to provide for holidays with pay and payment for Sundays and other official holidays, sick-pay, old-age pensions, and various bonuses tied in with the cost of living, the cost of these having to be met largely by employers.

*Alexander ('Noah') Hall.*

This immense enterprise of remaking Spain was reflected in the extra administrative work and the additional strains imposed on the greatly reduced British staff at the mines, and made for considerable difficulties in the operation of the business. Alexander Hall, appointed general manager at the mines in 1935, had one of the most frustrating tenures of that post, reflected in his nickname of 'Noah', for almost all requests from his staff which involved the outlay of funds had to be rejected, their memoranda being returned to them marked 'No. A.H.'*

After the fall of France in June 1940 the mines' staff and their families had to be prepared for immediate evacuation at any time. It was assumed that Spain would stay neutral but in those critical days it seemed as if there was considerable risk of a German army crossing the Pyrenees, either with or without Spain's permission, in order to take Gibraltar. General Franco, having made plain to Adolf Hitler that German troops could not enter Spain by agreement, had then to convince the German *führer* of the unwisdom of risking direct action against the Spaniards. He succeeded and his policy later won warm praise from Winston Churchill: 'If Spain had yielded to German blandishments and pressure our burden would have been much

*Alex Hall retired in 1941, his successor being Trevor Gough.

heavier. The straits of Gibraltar would have been closed and all access to Malta would have been cut off, from the west . . . [But] the critical moment passed.'[21]

During those anxious days a special train was kept ready at the mines, and suitcases kept packed in readiness, in case it proved necessary for the British families to travel quickly to Huelva and board a boat.[22] A subcommittee of the board of directors was formed and made alternative arrangements for evacuation in the event of no boat being available at Huelva, obtaining frontier passes which would allow the British to pass into Portugal.[23]

Then came the winter of 1940–41, the most terrible in the history of the mines. The civil war had affected Spanish agriculture disastrously and had completely dislocated the country's transport system, destroying half of all the rolling-stock on the railways. The collapse of European commerce which accompanied the second world war made it impossible to relieve and repair this damage. The food shortages worsened and the distribution difficulties increased suddenly with the onset of unusually bad weather in the November of 1940. At Rio Tinto the cold was particularly intense and sustained, arriving early and lasting longer than usual. As a result of the combination of these factors, famine struck. The ordinary food of the mining families – wholemeal bread, dried cod, eggs, beans, cheese, rice, olive oil, goat-meat, fat-bacon, sardines and all the vegetables – completely disappeared save for those people able to deal with the black-marketeers.

The situation was bad throughout Spain, and the supplies of flour and chick-peas which were rushed to the mines by the government were sufficient to give each person only 8 oz (later reduced to 6 oz) of bread and thirty or forty chick-peas a day. Some of the local officials and bakers made even this diet worse by mixing the flour with adulterants such as potato-meal and ground chick-peas. Herbert Potts saw 'many hundreds of men and women die through sheer starvation'. The appearance of the people changed: 'All had swollen limbs and feet' and 'a dreadful colour – more like walking corpses than living men. They were all very weak and naturally incapable of much physical effort, so that the standard of efficiency in work fell off tremendously.' Conditions became so bad that many of the younger men applied for government permission to go to Germany to work. Over 1,000 went. Most of these went to the factories of Stuttgart and many of them were killed in the devastating air-raids on that city by the bombers of the allied air force.[24]

The members of the British staff and their families, too, had difficulties in getting enough to eat that winter, though they had friends and individual contacts throughout Spain and Portugal which enabled them to get occasional supplies of nourishing food. However, even for them there were long periods when bread was unobtainable and they had to substitute for it potato-cake and pressed paste of figs. Their situation was greatly improved when the United Kingdom government was able to arrange special monthly deliveries of flour, tea, tinned meat, sardines and other foods to British subjects in Spain.

Even with the advent of spring and summer in 1941 the work-people still suffered from a shortage of supplies. The company's officials urged the government in Madrid to improve the rations allotted to Rio Tinto employees if it did not want the mines' production (essential to the country's industry) to drop.[25]

The first year of the war had produced (as in 1914) a loss of most of the company's markets. In Europe, all that was left were the Spanish and United Kingdom sales. The company's United States contracts could not be fulfilled because the British ministry of shipping, trying to cope with the German blockade of the United Kingdom and to maintain essential supplies for the war effort, refused to direct to Huelva any ships *en route* to the U.S.A. to pick up a cargo of pyrites for American customers.[26]

Yet (as also happened at the start of the first world war) the shortage of minerals produced a rise in prices. In 1939 the company's net return before tax from sales of Rio Tinto pyrites and copper had been £289,000. As a result of wartime scarcities, and with the aid of a British government contract for copper refined at the company's Port Talbot works, the net return in 1940 rose to £528,000. However, it then gradually declined until in 1944, the last full year of the war, it was £278,000. During the same years dividends from subsidiary and associated companies provided, in relative terms, an important source of income: in 1939, £284,000; in 1940 £242,000; though in 1944 this declined to £163,000. By then the company's total income was £130,000 less before tax than it had been in 1939, and all the while the tax burden in Spain and Britain had been growing progressively greater.[27]

Moreover, the Spanish need to obtain foreign currency for government needs meant that the onerous exchange-control restrictions had to be continued and the company was still unable to draw on its Spanish earnings save as a part-contribution towards the cost of importing equipment needed at the mines.[28] Consequently the ordinary shareholders who, owing to the troubles in Spain had not received any income from their shares since 1931, had to continue without dividends throughout the war and for two years after.*[29]

*An 8 per cent dividend was declared for 1947; and even that was possible only because of the growing return from other investments outside Spain.

As well as declining revenues and rising taxes the company had to face increasing costs in operating the mines due to continuous inflation within Spain. Most of the war years were marked by summer water-shortages at Rio Tinto which seriously affected day to day working. Coal and coke for the production of electricity at the mines' generator and of copper at the smelter were also in short supply, and even when solid fuel was obtainable the run-down state of the railways and the shortage of ships made it difficult to transport it from the north of Spain to the south. Requests were made to the United Kingdom government for British coal to ship to the mines but it was virtually impossible to get the necessary permits.[30] Decrees made by the Spanish government, awarding pay increases to the workmen and levying additional contributions from employers towards the cost of welfare services, added to the problems.

In October 1940, when the Tharsis company and Piritas de Huelva received government permission to introduce short-time working, the Rio Tinto Company applied for and obtained the same concession. There existed a legal obligation to replace all workers who left the mines but eventually the government conceded that the problem of surplus labour was too great a burden, and permitted the company to postpone the filling of vacancies caused by employees retiring or going to Germany. But because most of those going to work in German factories were young men, the company was left with an ageing, as well as a largely unfit, labour force. At times work in many departments had to be restricted to a three-day week and even then many men had to be employed on such tasks as building roads and planting trees. The beginning of 1944 saw a complete shut-down of the mines and railway when coal stocks were temporarily exhausted.[31]

As early as 1941 the directors were highly apprehensive about the future of the mines and the heavy and continuous drain on the company's *peseta* reserves.* In July 1941, when the minister of industry suggested to Ulick Charles, the company representative in Madrid, that a Spanish concern should take over the production of copper at Rio Tinto and run the smelter, the directors felt it was impracticable to split the company's activities in this way; but as a result of considering this suggestion they decided to offer a lease of the mines, their equipment and attached properties to the Spanish government. Colonel Preston was sent to Madrid but was unable to get the minister to agree on terms. Preston, however, was able to convince the minister of the seriousness of the situation facing the company and was able to get his consent to an increase in Spanish prices for copper, sulphur and pyrites.[32]

The effects of the war on neutral Spain were such that its government felt itself justified in taking all possible steps to maintain over-

*Which were, in fact, completely exhausted in 1943.

seas trade in order to mitigate the damage being done to the country's economy. Moreover, at the outbreak of the European war there were still debts outstanding to Germany and Italy for assistance which those two countries had given to the nationalist cause between 1936 and 1939, and these were offset by the despatch of pyrites to the axis powers. For example, the French-owned company, the Société Française de Pyrites de Huelva, sent 39,000 tons to Germany in 1941 and 13,600 tons in 1942, while a Spanish company, Castillo de las Guardas, sent 16,000 and 3,700 tons to Germany in those same years. Although such deliveries undoubtedly assisted the war efforts of Britain's enemies, and were resented by some segments of British public opinion, they were in line with Spain's policy of neutrality. Nor did that policy work to Britain's disadvantage, for the British-owned companies such as Rio Tinto were able to send much vaster quantities of their produce to the United Kingdom, and Britain was the overall gainer. (Stories circulated from time to time alleging that Rio Tinto pyrites were being exported to Germany, but the company's statistics show that not a single delivery was made to that country during the war years.)

One strange event occurred which gave the British staff in Spain an opportunity to contribute directly to their country's war effort, though they did not know the true nature of their contribution until after the cessation of hostilities.* Early in 1943 British intelligence, in an attempt to mislead the German high command as to the precise spot where the allied forces would invade Italy, persuaded Winston Churchill that the body of a dead man should be placed in the sea dressed in the uniform of a Royal Marines' major and carrying misleading despatches, with the simulated appearance of having been drowned following an aeroplane crash. Commander Ewen Montagu and his colleagues in Naval Intelligence had to create a personality and life history for this fictitious major and find a suitable body for the deception (which was code-named *operation mincemeat*).

The only corpse they could obtain was of a civilian who had died of pneumonia. Packed in ice the body was transported by submarine, and on a dark night in May 1943 was lowered into the sea near Huelva to be washed ashore by the tide, with its briefcase, containing the falsified papers, chained to it. Huelva had been selected since, though Spain was neutral, it was the headquarters of an active German agent† with such excellent contacts that he could be expected to gain access to the allegedly secret papers.

Precisely as planned the body was discovered and the dead man assumed to have died from drowning. The corpse was handed over to Hazleden, the British vice-consul in Huelva, and as previously arranged

*Much of the story is told in Ewen Montagu's book, *The man who never was* (London, 1953).

†In fact, this was Louis Claus the consul.

*The grave of 'the man who never was' in Huelva cemetery.*

immediately interred in Huelva cemetery by the Rio Tinto Company. Meanwhile, the German spy obtained photographs of the documents in the briefcase for forwarding to Berlin. (As a wry footnote it should be noticed that the Spaniard employed to do this photography was a member of the Rio Tinto Company's staff.)

It was appreciated that once German intelligence officials received the copies of the despatches they would want an examination made of the body to ascertain whether the dead man had indeed drowned after an aeroplane crash at sea. Thus the company's staff in Huelva were asked, without being told the reason, to cover the grave with a stone as soon as possible and in the meantime to heap it over with flowers from the gardens of *casa* Colón, the Rio Tinto office in Huelva, so that any attempt to surreptitiously remove the corpse would be noticeable. The amount of time and care lavished on the grave of the fictitious Major William Martin, RM, by the company staff in the days following the burial service frustrated the German agent's attempts at an informal exhumation. The Berlin high command accepted the papers as genuine and the allied invasion in Italy, when it came, took the enemy by surprise and unprepared. Thousands of lives were saved, and the British general staff were able to send a message to Winston Churchill, then in the United States, 'mincemeat swallowed whole'.

With the advent of peace in 1945 the company's staff and directors hoped for a quick return to full working at the mines. The post-war reconstruction of Europe that obviously lay ahead would provide excellent opportunities for the company if it was in a position to take advantage of them. The needs of European and American agriculture seemed certain to lead to a growth in the use of fertilizers which would expand the demand for sulphuric acid, and of vine and potato sprays which required sulphate of copper. New communications systems for the devastated mainland of Europe would need copper cables. But there was already a note of caution in the address of Lord Geddes* when he presided in 1946 at his twentieth and last meeting of the shareholders. By then it was plain that the Spanish economy still faced considerable difficulties while many of Rio Tinto's pre-war customers in Germany and eastern Europe were cut off by the iron curtain gradually being lowered across the continent.

*Sir Auckland Geddes was created a baron in 1942.

*The negotiators of the 1954 agreement are received in audience by the Spanish head of state in the Pardo palace on 7 July. On General Franco's left is the Conde de Benjumea and on General Franco's right are Lord Bessborough, Val Duncan and H. A. Mellor.*

19

# *Climacteric*

## An end and a beginning: 1945–54

At the conclusion of the second world war the Spanish economy was exhausted. Spain, the non-combatant, had not been able to avoid the economic effects of the conflict; and though the war-ravaged countries of western Europe could rely on massive aid from the United States, the Spaniards had to face their problems alone. For Spain was unloved by the governments of the world, many of which were opposed to the Spanish government on ideological grounds, while some were too apprehensive to become involved with it lest they incurred the censure of others. The British labour government which took office in 1945 under Clement Attlee was particularly unwilling to assist a country to whose régime it was implacably opposed; while the American president, Harry S. Truman, a baptist in religion, responded to the appeals of his coreligionists and refused to extend Marshall aid to Spain unless its government changed its religious policy.

The great injection of capital which Spanish industry desperately needed was denied it until the late 1950s. In the meantime General Franco's government was obliged to make use of all the resources which the country possessed in order to prevent an even greater decline in its economy, and of these resources one of the most important was Spain's mineral wealth. The part which the Rio Tinto mines had to play in helping to overcome the nation's problems was thus a major one, but the accompanying government-imposed restrictions made some formidable difficulties for the Rio Tinto Company.

Any contribution which the mines could make to the resurgence of the Spanish economy in the immediate post-war years depended in great part on what could be done to improve the technical processes, for the mines were in a badly run-down condition. Another factor which had to be borne in mind was that at Rio Tinto there was a large population of wage earners and their families who depended entirely on the company for their livelihoods; and these had to be provided for, whatever policies the company or government adopted.

As a result of the civil conflict and the second world war it had been virtually impossible to install any new equipment at the mines for fifteen years. The antique steam-shovels which had first been employed to dig the Panama canal at the beginning of the century were still in use at Rio Tinto in 1945. Much of the excavated ore continued to be manhandled: in the deep and hot underground workings, near-naked men shovelled as much as fifteen tons of broken mineral per man each day into esparto-grass baskets which was then tipped into small wagons, passed through the workings and up the shafts where it was then emptied by still more men into waiting railway wagons.

The mines' managers had been forced to become specialists in improvisation. For so long had it been impossible to bring in equipment from abroad that means had to be found of devising ways to repair every single piece of machinery. By the mid-1940s the mines were almost a museum of early-twentieth-century British engineering technology, where a Babcock and Wilcox boiler was still working after four decades of operation, where electricity plant installed before the first world war was still being coaxed into generating power and where welding machinery was often far older than the blacksmiths using it. The fact that despite frequent breakdowns such equipment could be got to function at all was an impressive tribute to the general manager Trevor Gough, to Laurence Hill the deputy manager in charge of technical matters, and to their assistants.

Thus by 1945 the Rio Tinto mines were in great need of modernization. But the desire to modernize had to be tempered by the availability of new equipment and the fact that there were 11,000 men who had to be kept in work. Even if the Spanish government had been prepared to accept large-scale redundancy (and it would not have been) as the price to be paid for improved technology, there was no desire on the part of the company to impose it. The state of the Spanish economy at that time was such that any men laid-off would have had the greatest difficulty in finding fresh employment. Equally, the company could not afford to replace men by machines and continue to pay the men full wages. Thus the technical staff at the mines were faced with a nice question of judgement as to how to keep a proper balance in deciding what and where to modernize and the rate at which to proceed.

Consequently in the years 1945–54 Trevor Gough, Laurence Hill, Charles Julian (the second deputy manager and later general manager) and Leonard Salkield (by this time the head of the smelter) had to be extraordinarily ingenious with their solutions. Drawing heavily on the support and expertise of the technical department in London headed by Jimmy Lawford and the purchasing department under Lester Andrew, they overcame, with considerable success, most of the obstacles and problems they encountered. The purchasing department in its turn had to find and transport to Rio Tinto a substantial quantity every year of British equipment and spares as replacement parts to meet the mines' requirements, in itself no easy job with the shortages being experienced by British industry.*

Moreover, before the company could increase production to fulfil its customers' orders, the mines had to get coal for running the locomotives and for generating power, and coke for smelting, both of which proved at times to be almost unobtainable. It was a constant struggle to obtain even minimal supplies. Spain could not meet her own internal demands for fuel, while in the first few years the British government would make only a little available for export to Spain. Added to this, all the time, with the fall of the internal value of the *peseta*, costs continued to rise at the mines although the international exchange rates remained pegged.[1]

Thus 1946, the first new year since the end of the war, provided not the respite to which the staff in Spain and London were entitled after so long a struggle to keep the company solvent, but a feeling of even greater despondency. A Spanish government decree raised wages another 15 per cent and increased employers' contributions to the state welfare services. The French markets, for political reasons, remained closed to Spanish pyrites. During four months of the year it was impossible to make any copper at all at the mines because of lack of fuel.[2] Consequently the feeling of buoyancy and optimism with which the company's staff, like most other people, had greeted the coming of peace in 1945, was sadly brief. By 1947 it was obvious that the essential conditions for operating the Rio Tinto mines viably were still beyond reach and likely to remain so for some time to come.

When the young men had started to return from the British armed forces, taking up their jobs again at the mines or in the London office, they had been glad to get back; but, imbued as they were with the enthusiasms of their generation, they could not but notice a certain weariness in the topmost echelons of the company. The men who had carried the burdens for so long and so well were growing old.

Colonel Richard Preston, who had run the small head-office in the city of London throughout the war years (it had consisted of six people only, the other members of the London staff who had not served in the war being evacuated to the country to avoid the air raids) resigned in March 1946 after twenty years as a director. In February 1947 the company's Madrid representative for nineteen years, the invaluable Captain Ulick Charles, died. Lord Geddes, the chairman of the company since May 1925, who had served as the government's wartime regional commissioner for the south-east and later the north-west of England, found that his general health and

*By 1954 the annual value of this equipment exported to the mines had grown to almost £200,000.

*Lord Bessborough.*

eyesight had suffered under the strain, and in March 1947 he relinquished his chairmanship.

Of those who remained on the board in 1947, the Earl of Bessborough, GCMG,* who succeeded Geddes as chairman, had been a director since 1936; J. N. Buchanan, DSO, MC, since 1926; and Sir Francis Joseph, bart., KBE, since 1937. Etienne du Castel, DSO, who had joined the directors in 1936 as the representative of the French Rothschilds, but had been unable during the German occupation of France to act as such, had returned to the board with the end of the war.

These men had recognized the need for some new blood and had recruited to the board in 1946 (as managing director) Thomas Robbins, CB, CBE, MC, a member of the pre-war staff who had just returned, covered with distinction, from his army service; and also the Right Honourable Anthony Eden, MC, MP (later the Earl of Avon, KG) eminent conservative statesman.[3] But two years after the end of the war the board of directors still consisted, for the greater part, of men whose participation in the company had been conditioned by the problems of the inter-war years. The very experience they had gained in the company, and the service they had given to it, were likely to influence their thinking when faced with the need to devise new policies and plan the company's future.

A third of the British staff at the mines (eighteen out of fifty-two) were, by 1947, in the fifty-five to sixty-five age group, and twelve of them were over sixty-two years of age.[4] These men in particular, and their general manager, A. T. Gough, had been faced with the most exhausting problems and worries for almost twenty years by the time they came to retire in the years 1949 to 1951. They themselves recognized that, if change were to come, it would be better carried out by younger blood.

The changes in the board-room began quietly in 1947. Thomas Robbins had briefly met Mark (later Sir Mark) Turner, the under secretary in charge of German economic affairs at the control office for Germany and Austria. In the summer Mark Turner had returned to the city of London and to his job as a director of Robert Benson Lonsdale, the merchant bankers, when he received a lunch invitation from Thomas Robbins. In the course of the meal he was asked if he would join the Rio Tinto Board, and after discussing the proposal with his chairman, Sir Rex Benson, decided to accept.

From July to September Mark Turner was busy on two foreign office missions to Washington and it was not until October that he was able to attend his first executive directors' meeting of the Rio Tinto Company. By that time two other new directors had also been

*Governor-general of Canada, 1931–5.

*Mark Turner.*

recruited, Gerald E. Coke and Colonel (later Sir) Eric Gore-Browne. At his first directors' meeting, Mark Turner noted the disagreements that obviously existed between some of the longer-standing board members, and was left with the impression that further changes could not be long delayed in the board's composition. Mark Turner also found that he had to play a larger role in the company's affairs than he might have expected as a new director, for as a prominent city of London man his advice and financial expertise counted for a great deal.

In January 1948 Mark Turner, Gerald Coke, Eric Gore-Browne and Lord Bessborough were named as the company's new management committee. They made a strong group, possessing important contacts and the knowledge of how to cut through the red-tape of government bureaucrats. Bessborough's first task as chairman had been to visit the Spanish minister for industry and commerce, Juan Antonio Suanzes y Fernandez, to negotiate a higher price for pyrites and copper sold within Spain. An increase was granted and, although not so large as had been hoped for, provided useful relief, and in the breathing-space thus given the new committee was able to apply itself to other aspects of the company's affairs. Mark Turner was active in London and on the continent, attending discussions in Paris which led to increased pyrites' and cinders' sales to France; Laurence Hill, by then retired but still a consultant, was sent to the United States to investigate a new cobalt process for use on lead ores; new links were established with other producers and statistical information exchanged to improve long-term planning.[5] Old problems were tackled with renewed energy. Anthony Eden, at the management committee's request, negotiated with the board of trade to secure an increase in the amount of coal allocated for export from Britain to the mines.[6] The Spanish minister of industry and commerce was again met in Madrid and the way prepared for the easier and quicker issue of Spanish export permits.[7] But the problem of the Spanish *peseta* exchange rate seemed unamenable to improvement.

The pace of the company's administration had rapidly increased: the new management committee began to meet weekly (fortnightly in July and August) compared with the former committee's eight times a year. Much time and energy was expended on trying to revive and reinspire the company, but it was hard work, and, because it was operating within the framework of the company's traditional activities (the mining of pyrites, their sale and the making of copper), the committee's success was circumscribed by the controls imposed by the Spanish government. Then in the December of 1947 came an intense argument at board level. Thomas Robbins and J. N. Buchanan resigned and Mark Turner was asked by Lord Bessborough to become managing director. Mark Turner discussed the offer with Sir Rex Benson, and finally agreed to take the appointment on the understanding that he would hold it for a maximum of two and a half years, devote three days a week to Rio Tinto business and find and nominate his own successor in the job.

The new managing director faced considerable difficulties. There had been some improvement in the situation in 1947 when production and exports of pyrites rose as a result of contracts which the Rio Tinto Company had obtained to supply shipments to the British and American zones of Germany where they were needed for fertilizer production. Exports climbed again the following year (1948), but they were then profitless because rising costs in Spain offset the anticipated gains.[8]

This period of world shortage meant that the company could sell everything it could export. However, under Spanish regulations (necessary because of the state of Spain's economy) prices were fixed by the government, and in addition all foreign currency earned by the sales, save for a very small proportion, had to be transferred into Spain for conversion into *pesetas* at the various rates of exchange prescribed by the Spanish authorities. The company's staff had to cope with no fewer than twenty-three different rates under this differential exchange rates system. In 1950, for example, for every £1 sterling, or its equivalent in other currencies, the company received in Madrid 49 *pesetas*. But when it wished to pay its overseas bills it had to purchase the pounds sterling back from the government at a much higher rate. In that same year it had to pay 90 *pesetas* for every £1 intended for the purchase of railway equipment. For working expenses in England, or for purchasing plant or general stores, it had to pay 110 *pesetas* for every £1 bought from the Spanish authorities.

This had the effect of imposing a heavy discount on the company's receipts. In 1950, when for the first time in many years the company made a profit on its sales within Spain, losses on exchange were still suffered when some of this profit was remitted to London to make essential purchases.[9] However, as a result of long discussions in Madrid the company was permitted in 1951 to remit a proportion of its profits in sterling to the United Kingdom; and that year, too, the price of pyrites on the mainland of Europe rose so greatly that the company, despite the adverse exchange rates, actually made a *peseta* profit on its transactions. Thus in 1951, for the first time since 1930, Spanish profits contributed to an ordinary dividend payment.[10] (35 per cent was paid as the dividend for that year compared with 12½ per cent for

*Val Duncan.*

1950 and 10 per cent for 1949 and 1948.)

But in 1952 fresh Spanish restrictions once again cut the transfer of sterling remittances to Britain almost to nothing; and although the profits from the company's other investments continued to hold up well, and the ordinary dividend was still a respectable 20 per cent, which was maintained in 1953, it began to look as if there was to be no end to the financial difficulties besetting the company's Spanish operations.[11]

This was the context in which the company was operating when Mark Turner took over the managing directorship in 1948, and set about analysing the problems facing him. He came to the conclusion that his first essential was to dismantle the authoritarian régime that had been established by Lord Geddes and his predecessors so that the senior staff would no longer be inhibited from talking freely to the directors but would state their views frankly and fully. The second essential was to find and to train the right man as his successor: a man with the abilities needed to deal with the problems which he had analysed. The man he determined on was Val Duncan.*

At the end of the war Val Duncan had been a colonel 'Q movements' on the staff of General Montgomery, and earlier had been on the staffs of Generals Eisenhower, for the invasion of north Africa, and Alexander, for that of Sicily and Italy, before joining the team planning the invasion of France. Following the allied invasion of the continent he had collaborated closely with civil officials of the Dutch and Belgian governments to establish a system of priorities at a time when army requirements for road, rail and air movements were impinging on the needs of the civil authorities. Subsequently he had joined the allied control commission in Germany, where he obtained a reputation as a man capable of successfully resolving conflicting interests and making major policy decisions. At one stage he had gone to Essen to study the coal situation – of vital importance to the German population and economy – and had found that the efforts to distribute coal were being frustrated due to a shortage of railway wagons. This had arisen because of the export of coal as reparations to France and the refusal of certain French officials to return the empty wagons. He issued an ultimatum: if the wagons were not returned the coal exports to France would be discontinued. The deadline had expired with no response from the French and he had cut off the coal deliveries. In the ensuing controversy he was summoned to the foreign office in London where British diplomats were trying to placate irate French officials (and informed him privately that they were grateful the action had been taken by a soldier and not themselves). However, the French returned the railway wagons and Val

*Now Sir Val Duncan, OBE, and chairman and chief executive of the Rio Tinto-Zinc Corporation.

*Frank Byers.*

Duncan acquired a certain fame.

He had also met, for briefing purposes, Mark Turner, then under secretary at the control office (later to be renamed the German section) of the foreign office, and in April 1946 he was invited to join the control office as an assistant secretary. The two men worked well together and became close friends. Eventually, with an eye to the future, Mark Turner had suggested that Val Duncan should get some commercial experience. Sir Arthur Street, the former deputy head of the control office, had been appointed deputy chairman of the new National Coal Board established in 1948 to operate Britain's nationalized coal industry, and he invited Val Duncan to join the board as assistant director of marketing with special responsibility for organizing transport. In his new job Val Duncan recruited eight former lieutenant-colonels who had worked for him in the army to serve as the transport heads in the regions. He also travelled to Geneva to attend the frequent international meetings concerned with European coal production where, because of his past experience in Germany, he was able to argue the needs and problems of the German coal industry in addition to the British case. It was at this stage of his career, in October 1948, that Mark Turner invited him to join the Rio Tinto Company, with the prospect, if things went well, of succeeding to the managing directorship.

Val Duncan became the head of the Rio Tinto Company's commercial department in November, and attended the meetings of the management committee together with a handful of other senior departmental chiefs, and with effect from 1 January 1951 was chosen to become managing director of the company in succession to Sir Mark.[12] Shortly afterwards the board was strengthened by the appointment of Frank Byers, OBE (later the Right Honourable Lord Byers, PC) as a director.

The way in which Frank Byers was recruited is indicative of the informal and old-fashioned way in which the company still ran its affairs at board level as late as 1951. He had recently lost his seat as a liberal member of parliament, but during his period as the party's chief whip he, like his predecessors, had held weekly meetings with James de Rothschild of the international banking family, and a former liberal MP and minister in the wartime coalition government, to discuss the political scene. One day he told Frank Byers to expect an invitation to lunch at merchant bankers N. M. Rothschild and that he was to accept it no matter what other engagements he might have. Though the lunch seemed no different from any other given by merchant bankers, being attended by a varied selection of people from different walks of life, Frank Byers, without knowing it, was being vetted. A month later, Jimmy de Rothschild invited Frank Byers to his home in Saint James's Place to have lunch with him and an elderly gentleman whose name Frank Byers did not catch. At the end of the meal, James de Rothschild uncharacteristically left the room and the other guest addressed Frank Byers saying, 'Jimmy has a high regard for your ability. Would you care to join our board?' At that stage, Frank Byers did not even know the name of the company they were talking about, and when told it was Rio Tinto had to admit he did not even know what it did. The other guest, who was in fact Lord Bessborough, told him that the board had recently appointed a bright young managing director called Val Duncan, and wanted to lower the average age of the board members by recruiting one or two younger men.

The mention of the name aroused Frank Byers's immediate interest, for during the war, when he had been staff duties' lieutenant-colonel in the headquarters of Field Marshal Montgomery's eighth army and twenty-first army group, he had worked closely with Val Duncan, though neither had seen the other since 1945. Consequently Frank Byers agreed to join the company, though when his appointment was made he did not even know what fees he would receive, and not until later still did he discover that to qualify he had to hold £2,000 worth of Rio Tinto shares, to acquire which he had to borrow the money.

At first he found he was involved in a routine operation, attending once a month to sign cheques and share certificates. At that time the office in Plantation House in the city of London held about forty employees, and was run on very efficient but still traditionally authoritarian lines by H. A. Mellor, the powerful company secretary (who once asked the new managing director not to be in the office before 10.15 a.m. as otherwise the post would not have been opened!). During the war years Mellor had run the company's affairs together with Richard Preston, and working in the evacuated office while Preston remained in London he had carried the major responsibility for decision making. At the sacrifice of personal popularity he had exercised rigorous administrative control and the strictest financial economy at a time when an unwise decision could have put the company's future at risk.

He had built up a good staff and it seemed to Frank Byers that the time was ripe for someone like Val Duncan to take advantage of Mellor's work and 'to make the company tick and take off'. The new managing director told Frank Byers that something had to be done to make full use of the company's potential; he was not content 'to sit back and try to get money out of Spain for the next thirty years'.[13]

*Roy Wright.*

At the beginning of 1952 Val Duncan offered Roy Wright* the job of overseas' manager. Roy Wright replied that as far as he knew the Rio Tinto Company had no overseas business to manage. 'That's quite true,' came the reply. 'But we shall have.'[14]

The arrival of these younger men saw the emergence of a new style and pattern of management within the company, at first gradual and then increasing rapidly in tempo. A new sense of energy and enthusiasm began to permeate through the London office and even the most junior members of the staff sensed that exciting changes were on their way. There was a change, too, in the company's dealings with Spain. Val Duncan, Mark Turner and Frank Byers all visited Spain with greater frequency than the company's directors had ever done before. They got to know the mines, the staff and their problems more thoroughly than anyone since the time of Hugh Matheson. They desired a better relationship and a greater involvement with Spain than had previously existed and Val Duncan in particular spent a great deal of time getting to know the country's leading men of industry and commerce. Moreover, there was a desire to prove to the Spaniards that the company no longer merited the distrust with which it had been traditionally regarded in that country. Despite H. A. Mellor's protest that 'We don't tell the Spaniards what we do', Val Duncan insisted that the Spanish authorities were to be told everything about the Rio Tinto Company that they wished to know. Spanish officials were welcomed at the mines, doubtless remembering that even as late as Walter Browning's time as general manager, when the only direct means of access to the mines' area had been the Rio Tinto railway under the company's control, they had often been unwelcome and refused permission to make the journey. Val Duncan also demanded that the international meetings of the industry concerned with pyrites' residues be opened to Spanish representatives. Gradually it became apparent to Spanish government and industry that the men running the Rio Tinto Company were trying quite seriously to do their best for Spain and accepted the obligations that went with operating in that country.

These years, too, saw a widening of the company's activities outside Spain. It was in fact the year 1949 that saw the first cautious discussions about possible expansion in southern Africa, and by 1953 the Rio Tinto Company was active not only there but also in Australia and Canada (see appendix two). In this respect, too, the new directors were reflecting the enthusiasm in British post-war society for a new order. At a time when the British were turning from a concept of empire to one of commonwealth, with independent nations linked in ties of friendship and mutual aid, the Rio Tinto Company was formulating

*CBE. Now deputy chairman and deputy chief executive of RTZ.

Gerald Coke.

a policy of expansion in terms of the creation of independent overseas' companies, controlled from within the country of operation, linked by personal friendships and looking to London for technical assistance. In Val Duncan's case the desire to do something of this sort had originated while he was still in the army. When visiting A. W. Sowtham, an economic adviser in the control commission in Austria, and talking through the night, Bill Sowtham had spoken of how he would like to do something constructive for the commonwealth. Val Duncan had agreed with him and they discussed in general terms how the commonwealth countries could be linked together. Later, when the Rio Tinto Company participated in the creation of the British Newfoundland Corporation, Val Duncan remembered the conversation and suggested that Bill Sowtham be put in charge.

The reaction of the long-serving board members in London to these proposals for change was, on the whole, favourable. Not unnaturally, there were some reservations about the new dynamism, but Lord Bessborough provided a stabilizing force, while Gerald Coke, with his penetrating questions, and Etienne du Castel with his superb administrative skills (acquired as head of the French railways) were able to ensure that the enthusiasm of the younger directors did not get out of hand. Baron Guy de Rothschild, too, though not a director at the time, was frequently present in London and provided useful advice and help when it came to discussing possible projects. Consequently the work proceeded without any of the bitter disputes that had occurred in the late 1940s.

Events were to prove that the new team was a strong one, the different members complementing each other. But, until he handed over the managing-directorship to Val Duncan, the key figure was undoubtedly Mark Turner. A busy merchant banker with many directorships, and commitments to many major companies as well as a variety of public interests, it took some courage to take on the job of managing director, even on a temporary basis. His capacity to contribute to the development of the Rio Tinto Company earned him the admiration of his colleagues. Despite his many activities he was somehow always able to make a dash to Canada, or any other part of the world where the company was at work, in order to deal with a crisis. His ability to pick up and assimilate situations without tedious preliminaries provided many short cuts to major decisions. The fact that he was not a full-time executive of the company was also of value. With Val Duncan and Roy Wright employed on a full-time basis he was able to bring to bear a more detached and financially-orientated outside viewpoint. In addition, his expertise as a merchant banker produced a situation in which the Rio Tinto Company itself became virtually a merchant bank, with the overwhelming proportion of the large funds spent in developing the resources of various parts of the world being raised directly by the company with the relevant institutions.

In Spain, meanwhile, the stringent regulations which inhibited the company were continuing to make it difficult for the mines and their resources to be used so as to give the utmost benefit. By that time Spain's own industrial and commercial companies were beginning to thrive, for General Franco's government had been highly successful in giving the country the political stability it needed for fostering the right conditions for the country's economic development.

Frequently in the past various Spanish politicians had expressed concern at one of Spain's major industrial enterprises being in the hands of a foreign company, and General Franco had himself called Rio Tinto 'Spain's economic Gibraltar': a reference to the fact that unlike most mines, which are the subject of temporary concessions, Rio Tinto was a British-owned freehold property, and to the unusually wide powers possessed by the company in the mines' area by virtue of the rights conferred by the deed of sale in 1873.

Because of the increasingly healthy state of the Spanish economy in this period, the possibility of a reversion of the mines to a Spanish concern came to be thought of as a practical proposition given that, for the first time since 1873, Spanish resources themselves could provide the large capital sums needed to finance the operations. The directors considered the situation. A considerable quantity of capital was tied up in the mines, capital which was showing little return and which was potentially valuable if it could be withdrawn from Spain and applied to the diversification of the company's interests. Yet the ties with Spain were long-standing and close, and there was a strong sentimental attachment, felt not only by the long-serving members of the staff but also by the new directors as a result of frequent visits to Rio Tinto and familiarity with its affairs arising from the time and energy they had expended on its problems. They were not unaware that at Rio Tinto their company had created an enterprise unique in mining history, even though they may have viewed the mines rather more romantically than those who had to work there permanently.

After a number of preliminary approaches serious discussions began early in 1954 with a Spanish financial institution, with a view to its acquiring a major interest in the ownership of the mines. The talks reached an advanced stage when the Spanish government indicated that some of the proposals which the would-be purchaser was making to the ministry of commerce (particularly with regard to concessions in the currency regulations) would not be acceptable. The Spanish government also had doubts about the wisdom of allowing control of the mines to pass to a single institution.

At this time the under secretary at the ministry of commerce was General Antonio de Torres y Espinosa, a distinguished soldier who had acquired a good working knowledge of the pyrites industry. In particular he was closely concerned with 'operation P', a working party, consisting of ministry officials and representatives of the pyrites' producers, set up to deal with the problems of the Spanish pyrites industry. Thus when another round of negotiations began in connexion with a possible purchase of the mines by Spanish interests, Antonio de Torres was the government official charged with the task of organizing the talks.

This second series of discussions had its origins in a conversation between Val Duncan, Carlos Sundheim Díaz Truchuelo and Javier Benjumea Puigcerver, all three of whom had met as participants in 'operation P'. Moreover, Carlos Sundheim and Javier Benjumea were first cousins, and also grandsons of the Sundheim who had been a founder of the Rio Tinto Company. They expressed the hope to Val Duncan that they could be instrumental in recovering for Spain the mines which their grandfather had helped the company to acquire eighty years before.

An approach was then made to the ministry of commerce by the Banco de España, the Banco Español de Credito and the Banco Hispano Americano, indicating a willingness to hold fresh talks with the Rio Tinto Company, organized by the general manager of the Banco Hispano Americano, Andrés Moreno. Although a Spaniard, he had made a career for himself in England with the Midland Bank, which had been quick to recognize his talents and had eventually suggested him as a possible general manager of the Spanish bank when asked to help find someone to fill the vacant post. His knowledge both of British industry and British businessmen was to be of considerable help when the new discussions got under way, for he was able to help both sides to understand the other's position, while his own personality contributed greatly to making the talks easy and friendly.

The ministry as sponsor for the negotiations decided to enlarge the talks to include five other principal Spanish banks: the *bancos* de Bilbao, de Vizcaya, Central, Exterior and Urquijo, and these (together with the *bancos* Español de Credito and Hispano Americano) formed the group which was to acquire the majority interest in the mines of Rio Tinto. Also invited to the discussions were Carlos Sundheim and Javier Benjumea.

The *Conde* de Benjumea was nominated as chairman of the talks. Governor of the Banco de España, and three times minister of finance, he also had an intimate knowledge of the mining industry, and possessed two other exceptional qualifications: he had the complete confidence and respect of Spain's head of state, General Franco, and a high reputation for fair-dealing and business integrity. The fact that he was the uncle of Javier Benjumea was coincidental, though it was yet another factor in promoting informal and easy discussions.

Before the talks actually began, Val Duncan saw the Spanish minister of commerce, Sr Aberoa. Saying that he wanted this 'to be an elegant negotiation', the minister added that as a result of the recent Barcelona Traction case (in which an inferior-court decision had transferred control of a foreign-owned company operating in Spain to the Spaniard controlling the company's preference shares, a verdict which had produced an outcry abroad), Spanish integrity had suffered and foreign businesses were regarding Spain with suspicion. 'If we reach a fair agreement with the Rio Tinto Company, which is the biggest foreign interest in Spain,' commented the minister, 'you will be able to say so.'

When the British and the Spanish teams met the talks progressed so smoothly and rapidly that both sides were a little surprised. Unusually for business negotiations of this sort there was an atmosphere of complete mutual confidence. Those of the Spanish team who did not previously know Val Duncan were impressed by his grasp of the facts and his understanding of Spain's economy. They found him an easy mixer and, although they thought him unmistakably British in his characteristics, his sense of humour and relaxed manner, and his willingness to trust the Spaniards fully, were the cause of much good will. Even the normally stern H. A. Mellor, the secretary of the British company, was found congenial by the Spanish contingent; and Lord Bessborough (as chairman of the Rio Tinto Company the titular leader of the British team), with his affable manner, inspired a lot of friendship. The resulting trust of each side in the other was epitomized by the reaction of the *Conde* de Benjumea when asked by a Spanish official if there were certified figures to verify the reserves' position as stated by the Rio Tinto Company: 'We do not need that sort of thing,' he commented. 'We are dealing with the English.'

The series of talks was marked by an increasing intimacy between the two sides and the forming of lasting individual friendships such as that between Val Duncan and Antonio de Torres. Lord Bessborough, impressed by the efficiency with which Javier Benjumea and Carlos Sundheim worked together as one, expressed his opinion of their essentiality by dubbing them 'Brandy and Soda' (from their initials B and S). That this very English joke was enjoyed by the Spaniards themselves is an indication of the relaxed way things were going.

The Spanish banks represented at the talks made plain to their government in advance that if a deal was struck with the British they would not be asking for any special concessions or any advantages not extended to other pyrites' producers. Thus they knew that when agreement was reached there would be no governmental obstacles to overcome.[15] From the Rio Tinto Company's point of view, given its operating problems and need for capital to undertake new activities in other parts of the globe, the Spanish offer when finally worked out was almost entirely what had been hoped for: the Spanish assets of the Rio Tinto Company to be transferred to a new Spanish company in exchange for a one-third interest in the new concern, £7,666,665 in seven annual instalments (the major portion being paid in the first three years and carrying interest of £200,000) and £333,333 representing accumulated profits. The *bancos* Español de Credito and Hispano Americano were each to take, approximately, a 14 per cent interest in the equity of the new company, and the other five banks roughly a 7 per cent interest each.[16] The terms combined the three required elements: the provision of much-needed capital, escape from the Spanish government's economic regulations and the maintenance of the company's connexion with the mines.

At the final meeting on 20 June 1954 there was a new arrival, a large well-built man whom Val Duncan had not met before and who was introduced to him as Pablo Garnica, the chairman of one of the banks participating in the negotiations. To Val Duncan's surprise he declared aggressively, 'This price is ridiculous! It is too much.' With a slightly sinking feeling, and wondering if all the effort that had been put into the talks was about to be rendered superfluous, Val Duncan protested, 'The minister told me that this was to be an elegant negotiation.'

The Spanish banker looked at the Englishman thoughtfully. 'Won't you accept anything less?' 'No,' retorted Val Duncan. An expansive grin spread over Pablo Garnica's face. 'Done!' he exclaimed, extending his hand. The agreement, in fact, met the requirements of the Spaniards as well as the British; and the fact that both sides were well satisfied was in great part a tribute to the policy of absolute frankness and trust in dealings with the Spaniards which had been initiated three years before and had set the tone for the recent discussions. Such a satisfactory result would have seemed impossible only a short time earlier, and indeed when Val Duncan broke the news to Lord Bessborough, telling him, 'We've done the deal!' even his chairman exclaimed, 'It can't be true!'

The signing of the agreement was an impressive event. It took place in the Banco de España under the brilliant chandeliers of a sumptuously furnished room, the walls of which were hung with gilt-framed Goyas. Some weeks later, to mark the event, there was a ceremonial audience with the Spanish head of state. Meanwhile the new arrangements had to be explained, as a matter of priority, to the employees at the mines. Charles Julian, the manager at the mines, had joined the later stages of the negotiations, and so was fully briefed, but on the evening of his arrival at Rio Tinto Val Duncan recounted the terms to the company's senior Spanish employee, José Sánchez-Mora the legal adviser in Huelva.[17]

The next day in the club Val Duncan explained the details of the agreement to the British staff, and also the Rio Tinto Company's obligation to provide the new company with advice and technical services, and four British directors. Charles Julian, the then general manager, and Frank Spettigue, his deputy, had agreed to continue for a minimum of five years, not only because they were highly valuable men but also because it was hoped by the Spaniards that this would reassure the other British staff at the mines that they, too, were still wanted. Some of the staff, however, were alarmed at the prospect of the change. They were promised by the new company that most jobs and salaries would not suffer as a result of the sale, and Val Duncan promised financial help as compensation for any of the British staff who might not be needed by the new company, or felt their own interests made it necessary to resign. Nevertheless, their apprehension was natural, for they were exchanging a known for an unknown employer and in the event some decided to look for fresh jobs. Others remained at the mines in the new Compañia Española de Minas de Río-Tinto, S.A.*

On 12 August 1954 the shareholders of the Rio Tinto Company, meeting in London, authorized the sale of the mines and their dependent properties with effect from the following 1 January. Thus the 'noble corpse' purchased by Matheson's consortium eighty-one years earlier returned to Spanish control, while the company he founded went on to expand its interests in other parts of the world.

The acquisition of the Spanish interest in the mines was widely welcomed in Spain when it was revealed. Although by 1954 responsible public opinion accepted that the British company had achieved great results at Rio Tinto, the fact that a foreign business organization had owned the freehold of all the land and minerals in the mines' area was as much a cause of public concern then as it had been at any time during the previous eighty-one years. Rightly or wrongly, it seemed to most Spaniards that a corner of their country had been turned, by the sale of 1873, into what was virtually a colony of the British empire. It was true that in 1954 most Spaniards no longer felt the indignation

*At the time of writing (1972) retirement, death and other departures have reduced their numbers and only one Englishman from the old company, Joe Jowers, now lives at Rio Tinto and he has recently transferred to Río Tinto Patiño S.A.

or anger which the mere name of the company had been sufficient to evoke in the early part of the century. The improvement in both official relationships between the company and the government, and in personal relationships between the British staff and individual Spaniards, meant that the comment of the two deputy managers in 1921, that the name of the company was detested throughout Spain, no longer applied. Thus the 1954 sale, though generally approved in Spain, was not seen as a reason for a *fête nationale*, as would have been the case only twenty years earlier.

The reaction of those Spaniards employed by the company on hearing the news was mixed. During the previous ten years, working and living conditions had improved, at first slowly and then at a growing pace. By 1954 the workmen had become reasonably contented. Food and clothes were again plentiful and, though these commodities were dear, as in the whole of Europe at that time, wages were still high compared with the region around. The basic rates plus indirect payments by the company gave a labourer a daily average earning of 47.2 *pesetas*, while men such as drillers, fitters, boilermakers and blacksmiths earned as much as 50 per cent above that. Only the old men could remember, and perhaps sigh for, the days when 3.50 *pesetas* a day seemed good money and the man that was earning it could put both meat and wine on his table.[18]

However, even though the majority of the workmen had developed a *modus vivendi* with the British company, many of them expected that the new Spanish company would be able to provide them with higher rates of pay, since it would be freed from some of the economic problems which had burdened the Rio Tinto Company. Moreover, a good many of them could look back to four generations of discontent and conflict with the British concern, and it would hardly have been surprising if some of them had not thought 'good riddance'.

Amongst the Spanish staff, however, feelings were rather different. Since civil war days social contacts had increased with the British, national characteristics no longer proving impassable barriers. It had been accepted, long before, that the senior positions at the mines should be made available to Spaniards with the proper qualifications. Thus when a Spaniard had taken a position formerly filled by an Englishman he had also moved into Bella Vista where by that time houses were allocated irrespective of nationality; and the Spanish language had become as familiar in Bella Vista and the *club inglés* as English. Some of the Spanish staff even played a respectable game of cricket, and others had developed an affection for the music of Gilbert and Sullivan as a result of performing in Bella Vista operettas with their British colleagues.

Consequently, many members of the Spanish staff realized that the change in ownership might accelerate their promotion to the topmost jobs but that the same process would mean losing friends for whom many had come to have a deep affection. Perhaps, too, they were aware, consciously or unconsciously, that just as the British who went to Rio Tinto had absorbed something of Spain, so they in their turn had acquired some of the British company's ethos, and had developed attitudes to work and to life which reflected the characteristics of the people of a country which most of them had never even visited; and that, in a hardly definable way, a part of their lives had become English. It was, perhaps, not entirely inappropriate that the mass of the work people referred to the Spanish managers as *los inglesados*.

In 1873 Hugh Matheson had bought a seemingly unworkable mine; in 1954 Val Duncan had been the principal mover in disposing of what was apparently the company's *raison d'être*. In some respects the two decisions were similar, carrying an element of risk which to many made them seem a gamble. It is interesting, though, to reflect that both men were of Scottish origin, cool-headed, long-sighted and prepared to play for big stakes when most other men would have stayed their hands; that both had that type of personality which enables men to win the support essential for realizing their visions. The comparison, of course, must stop there. In most respects they would have found each other incompatible. But it seems not unreasonable to suppose that Hugh Matheson might have approved of the later decision; might well, perhaps, have taken it himself in the same circumstances.

At the end of this investigation it is possible to see that the eighty-one years during which the British company ruled the mines were not without importance in Rio Tinto's long history. The changes in the circumstances of life there for the Spanish miners and their families, between the start of our period and its completion, are obvious. The evidences of material progress abound: the inherited problem of smoke from the *teleras* had been solved; living standards had risen; there had been a great improvement in the housing of all classes. The expansion of the mines, the construction of the railway, the building of the pier at Huelva and the great increase in the number of ships entering and clearing that port all contributed to the advance. One result was that there was more to eat, and a greater variety of consumer goods available to the people of the mines and the region around. Life, despite the horrors of starvation experienced during the civil and second world wars, became progressively less hard for ordinary people than it had been in 1873. Much of Spain, of course, advanced economically in those years, but a great deal of the progress in the area of Rio Tinto had its origins in the operations of the British company.

As important, perhaps, as the material improvements was the change that is discernible in things of the spirit. During those eighty-one years a greater value came to be set upon human life and the avoidance of suffering (paradoxically maybe in a time which saw so much bloodshed and starvation); and though accidents to life and limb are inevitable in a mining community, by 1954 the safety record had become a source of pride. Not the least of the company's achievements were the great successes of the medical service in the fields of surgery and preventive medicine, which provided no small contribution to the sum total of human happiness in the villages around. In addition, the company schools opened up new vistas for children at a time when there was little stimulation of the intellect for the vast majority, and they equipped many young people with the basic skills to break out of the vicious circle of poverty that had circumscribed the lives of their parents. Moreover, they provided a sound foundation for the expansion of the educational system which has taken place at Rio Tinto since 1954 and which has seen the introduction of technical education at a superior level.

It is interesting to observe in the records how, during this period, men begin to emerge more distinctly as individual personalities from the background of their social groups. And though there were marked changes in the life of the local communities between the time of the arrival of the Rio Tinto Company and its departure, there was also considerable continuity and many of the old communal traditions survived and served to strengthen the greatly enlarged Spanish villages.

The political and industrial conflicts of the period, tragic and bloody though these often were, still provided a precious gain; for they contributed to the process by which the ordinary people awoke to the need for self-consciousness and self-questioning, and thus achieved a greater degree of maturity. For the Spaniards, therefore, those years, though they contained much suffering, represented a most significant period in the progression of their society.

As for the British colony at Rio Tinto, it was always small and at its largest was no more than about 120 families. But a society as pocket-sized as that, which in the course of less than a century had Hugh Matheson as its far-sighted suzerain, Walter Browning as its courageous governor, Russell Ross as its dedicated physician and Laurence Hill as its pattern for gallantry, may well bear comparison with any community of similar size in British history.

When one thinks of Bella Vista, a hundred and more images and memories come crowding in upon the mind: the British families raising their children in the Victorian villas; the church where they worshipped and sang the hymns of home; the club which was the focus of social life; the small gaily-flowered cemetery where many of them now lie sleeping. Or perhaps one thinks of the ruins of the old hospital on the Mesa de los Pinos where nurse Ferrier became a legend; or of a now-deserted little schoolhouse of yellow brick where a young Scottish schoolmistress presided over a class of lively Spanish girls. Perhaps it is Huelva which comes into mind and one sees again the flower-beds and palm trees of the *casa* Colón and the women strolling along the paths in the large hats sent specially from London and shaded by their parasols; or maybe the great curved pier which terminates the railway and serves as an impressive monument to George Bruce who designed both.

The visible memorials of these people, the homes they loved, the photographs they cherished, the letters they wrote and sent across the seas, the patterns they imposed so dramatically upon the Rio Tinto landscape, and much more besides, survive for our examination. But there are other reminders of them, less tangible but no less real: the traditions which they created and the attitudes they developed. For the best of these, in transmuted form, go on both in the great international mining group which has its origins in the company they fashioned and in the Spanish companies which now work the mines of Rio Tinto: a concern for the physical well-being of their workpeoples evidenced by their housing programmes and their medical and educational services; and a readiness to think of mining in heroic terms, to tackle the technical problems with originality and zest, even enjoyment. For those who have succeeded them in the business, their British and Spanish heirs, and who are prepared consciously to look back with an intellectual awareness to the past, the old exploits and triumphs can never be wholly without interest, and the period of the Rio Tinto Company in Spain will always have upon it the dew of morning.

Nor, for those who are detached observers, unconnected with Rio Tinto and its mines, is the tale without significance. For it forms a chapter in the long story of the relations between Britain and Spain; an episode less celebrated but just as material, for those who seek to understand how the present has grown out of the past, as the intervals of martial glory. The great armada of 1588, the war of the Spanish succession of 1701 to 1713 and Wellington's peninsular campaign are all rightly remembered. But peace has her triumphs and disasters too, and, in the nature of things, their effect tends to be more enduring.

# 20 *Epilogue*

The creation of the Compañía Española de Minas de Río-Tinto in 1954 marked the beginning of a new phase in the history of both Rio Tinto and of the British company. This is not the place to tell the story of either in the twenty years since then, though both stories will eventually be told and found not without interest.

Briefly, however, it may be noted that the Rio Tinto Company expanded into a highly successful international mining organization. On the death of Lord Bessborough in 1956 Gerald Coke became chairman. In 1962 there was a merger with the Consolidated Zinc Corporation to form the Rio Tinto-Zinc Corporation (of which Sir Val Duncan is now the chairman and chief executive).*

The new Spanish company came into existence under the chairmanship of the *Conde* de Benjumea, proposed for the post by the Banco de España on account of his experience and reputation. Similarly, the bank proposed Antonio de Torres as a member of the board and as general manager of the company, both because of his knowledge of the pyrites industry and his independence of any of the participating banks. Javier Benjumea (who was later to succeed his uncle as the company's second chairman) and Carlos Sundheim were also selected for the new board in recognition both of their expertise and of the services they had voluntarily rendered their country in contributing to the success of the negotiations. Of a total of sixteen members of the new board of directors, four were British: Lord Bessborough, Val Duncan, Charles Julian and H. A. Mellor.†

Charles Julian remained in charge at the mines and of the four assistant managers appointed to work with Antonio de Torres, one was Leonard Salkield, who has figured from time to time in the pages of this book.

The sale of the two-thirds interest in the ownership of the mines to the Spanish group had an importance beyond giving Spain control over one of its most important natural resources. It pointed the way to a new type of partnership which has since proved of great value to Spanish economic growth. As a result of the 1954 agreement, Spain acquired not only control over the mines but the opportunity (of which every advantage was taken) of collaboration with a group of foreigners who had both an understanding of Spain and also ex-

*The first chairman of RTZ was Alfred Baer of Consolidated Zinc. On the merger Val Duncan became managing director.

†Gerald Coke, Percy Truscott (who had worked at the mines as administrative deputy manager and then had become secretary of the Rio Tinto Company in succession to H. A. Mellor and later still became secretary of RTZ) and Laurence Hill subsequently succeeded Bessborough, Mellor and Julian respectively. Robert Lethbridge, another member of the staff of the British company well known to the Spaniards, later succeeded Laurence Hill.

perience gained outside that country. In the case of the Rio Tinto mines it was a collaboration based on personal friendships, providing the opportunity for specialist technical assistance and the chance to draw upon the know-how of the British company's directors and staff throughout the world. In Spain until that time partnerships of this nature had not existed save in the sherry industry, but the industrial partnership which resulted from the 1954 agreement possessed a wider significance. For it served as a model for the collaboration with foreign companies which began to invest in Spain in the late 1950s, and today there are many such partnerships in Spanish industrial and commercial life. Thus this rearrangement of the shareholdings in the mines of Rio Tinto has proved a significant milestone in the growth of the Spanish economy since the late 1950s, the development of which is now earning Spain an important place in the society of nations.

The friendship between the British company and the Spanish continued to grow deeper. It was based on such things as Val Duncan presenting the Spanish point of view at the international 'tea-parties' of the pyrites industry, when producers from many countries met to discuss world demand and production; on the work of the Rio Tinto Company's subsidiary, the Pyrites Company, in marketing the products of the Rio Tinto mines outside of Spain; on the successful efforts of the purchasing department in London in advising on and providing the mines with new equipment, and seeing that it arrived in good time; and on the short-term visits to the mines of many technical experts. In return the Spanish officials looked after the British company's interests in Spain. Val Duncan had early come to the view that Antonio de Torres was 'a totally honest and honourable man', and was soon able to tell him that the British company was happy, in the event of any sudden emergency in connexion with its Spanish interests, for Antonio de Torres to take whatever action he thought fit on its behalf.

Antonio de Torres and his Spanish colleagues in the Compañía Española de Minas de Río-Tinto did not see themselves as serving a completely new commercial organization, but as the heirs in Spain of the British company. They saw their history beginning not in 1954 but in 1873, their traditions not to be made but already existing. And this view was true in a real sense, for the employees at the mines were largely the same people after the sale as before, with the same attitudes and the same customs.

Change, of course, was bound to occur, for a business which does not change soon dies, but there was a desire for continuity and to avoid upheaval. Immediately after the new agreement was signed, for example, Antonio de Torres declared that the British church at Bella

Vista would remain as a chapel of the church of England as long as it was required as such by the British community, and at a meeting with the Roman catholic bishop of Huelva he arranged for the diocesan authorities to agree to what was then an unusual procedure: the maintenance of an Anglican church by a Spanish company in a country where (at that time) the only religion officially permitted was the Roman catholic faith.*

In Spain, as in England, the Rio Tinto Company's system of administration by the early 1950s, though honest and in its own way efficient, was remarkably old-fashioned. Just as Mark Turner and Val Duncan had to make changes in the London office, so Antonio de Torres and his assistants had to introduce new methods at the mines. They ensured, however, that it was done gradually. Not unnaturally, some of the older members of the staff, both British and Spanish, were not enthusiastic about the new techniques but care was taken to avoid too much upset. The death of Charles Julian from a heart attack and of Frank Spettigue in a car crash, both within a few weeks of each other in the summer of 1955, was a minor tragedy for the new company, which lost not only two skilled and loyal men but, as a result of the shock experienced by the British community, a number of other senior men who decided with their wives that if they were to return home then this was the time to do so. There was, however, no difficulty in finding skilled Spaniards to replace them and the years 1955–6 saw the start of the modernization of the mines.

At that time began a large exodus of Spanish workers from Spain to Germany and many went from Rio Tinto. This continuous reduction in the labour force gave the management the chance it needed to speed up mechanization and thus solve the problem of surplus labour which had for so long shackled the British company. At the time of the sale over 10,000 workmen were employed at the mines. Migration and retirement reduced this to 7,000 by 1962 and 5,000 by 1966. But at the same time production increased. The tonnage of pyrites produced rose from 860,000 in 1954 to 887,000 in 1962 and 1,129,000 in 1966. Copper manufactured at the mines went from under 6,000 tons in 1954 to almost 12,000 tons during the same period. Production of sulphuric acid was increased from under 10,000 tons at the earlier date to 180,000 tons by 1966. This achievement, carried through by the Compañía de Minas de Río-Tinto with the support and advice of the British company was one of the first fruits of the new partnership.

In 1966 the Compañía Española de Minas de Río-Tinto joined with the Patiño Mining Corporation and the Rio Tinto-Zinc Corporation to form a new company called Río Tinto Patiño, with respective shareholdings of 55 per cent, 40 per cent and 5 per cent, to make a new mine

*When the British community eventually disappeared save for a couple of Roman catholic families General de Torres arranged with the Anglican bishop of Gibraltar for the fittings of the church to go there.

at Rio Tinto on the *cerro* Colorado and to construct a new copper smelter on the coast at Huelva. Under the chairmanship of Antonio de Torres the new mine has already proved successful and once again the British company and the international group centred on it has contributed in the way of technical advice.

By 1970 pyrites producers throughout the world were facing sales' difficulties and the Compañía Española de Minas de Río-Tinto was no exception. But by then it had proved itself and demonstrated its importance in the Spanish economy. Rationalization suggested a merger with the large Spanish chemical combine, the Unión Española de Explosivos. The company resulting from this is Unión Explosivos Río Tinto, one of the great companies of Europe. Interestingly, although the RTZ shareholding in this new company is very small, and though the Spanish company has wide-ranging interests, the old friendships have been carried into it. Thus the negotiations of 1954 proved far more than just a commercial agreement between a group of businessmen and government officials sitting around a table and talking money. They were the basis for a co-operation and an understanding which has benefited Spain and its people, and replaced mistrust and even dislike with confidence and friendship: a fitting end, it may be thought, to the story of the British at Rio Tinto.

# APPENDIX ONE *The Archaeology of Rio Tinto*

Since the rediscovery of the mines in 1556 their subsequent history has been recorded in an ever-increasing quantity of documents. Before that date they are effectively part of prehistory with their story recoverable only with the aid of archaeological techniques, dependent on discovery by the archaeologist's trowel or chance exposure by fortuitous blast or the excavating machinery of the mining engineer. For the greater part the early history of Rio Tinto is still unknown. However, there have been discoveries (most of them in recent years) which are now making it possible to build up, piece by piece like a mosaic, a picture of the life of the communities which in ancient times inhabited the area around the mines, and gradually to place them in the context of the old Mediterranean culture to which they belonged.

## *The pre-Roman period*

Rio Tinto lies in the Iberian pyrite field, extending eighty miles from Aznacollar near Seville to Aljustrel in Portugal, and forming one of the greatest mineralized belts in Europe where from earliest times men have mined a variety of minerals and produced a number of metals, but especially copper, silver and gold. This area of southern Spain lies between the river Guadalquivir on the east, which runs through the cities of Cordova and Seville to enter the Atlantic near Cadiz, and the river Guadiana on the west which provides part of the frontier with Portugal. Between the two run the lesser rivers, the Tinto and the Odiel, which originate in the very heart of the pyrite lodes and converge into the same estuary near Huelva on the gulf of Cadiz.

At certain places this mining field originally showed surface signs of two highly coloured copper minerals, bright-green malachite and deep-blue azurite. At Rio Tinto their outcrops were to be found on the sides of the hill called *cerro* Colorado. These minerals could be easily mined by ancient men and copper produced by smelting with charcoal. Since the rich silver ore lay immediately above the malachite and azurite ores, it would have been only a question of time before the ancient miners, attracted to the spot by the green and blue outcrops, discovered the more precious ore overlaying them; although it would seem from the interpretation of the archaeo-metallurgical evidence that, as we measure time, the discovery took many centuries to make.

During the bronze age (from about 2500 to 1000 B.C.) this region was controlled by the Tartessians, whose culture was based on tilling the rich soils of the river plains and extracting minerals from the hills. Their prosperity eventually drew to the area the foremost traders of the ancient world, the Phoenicians from the cities of Tyre and Sidon. They arrived as the bronze age was being succeeded by the iron age (approximately 1000 B.C.) and soon their commerce was financed with ever increasing quantities of Tartessian silver.

From 750 B.C. to 580 B.C. the Phoenicians had an effective monopoly of the south Iberian metal market. According to Diodorus Siculus they carried bullion away from Tartessos in such quantities that when their ships were freighted to the full they threw away the anchors and substituted new ones of silver.

Their connexion with Tartessos seems to have been severed as a result of Nebuchadnezzar's thirteen-year siege of Tyre, after which the Phocaeans from Greece replaced them. It was the Phocaean traders who introduced wine and olives to the area, bartering them for the local metals. Their interest in the region, however, was short-lived, for in 535 B.C. the Phocaeans were defeated in battle by the Carthaginians, who set about creating an empire for themselves in southern Iberia, in the course of which, in 510 B.C., they destroyed the city of Tartessos. The Carthaginians in their turn were conquered by the Romans about the year 205 B.C.

The archaeological evidence proves that the Rio Tinto mines were the source of some of the silver which figured so prominently in the economy of the later Phoenicians, those of the eighth and seventh centuries before Christ. It is possible that the mines were being exploited even before then, in the time of the earlier Phoenicians, but so far no proof has been found that would completely satisfy the archaeologists of this. The folklore of the district, which must be treated with caution, assumes a history which goes back to the earlier period, for the traditional belief is that, during the first centuries of Phoenician influence in southern Iberia, the mines of Rio Tinto were in production and were indeed the mines of the celebrated King Solomon who ruled the tribes of Israel between 971 and 931 B.C. (or, according to some historians, between 961 and 920 B.C.)

Certainly the fleets of Hiram, king of Tyre and Solomon's great ally, took to the Jewish king cedars from Lebanon, gold from Ophir and silver and gold from distant Tarshish. (So distant that the round journey took the fleet three years.)[1] When Jonah quit Joppa to flee 'from the presence of the Lord' (to escape, that is, from the lands where the god Yahweh was supreme) he fled to far-off Tarshish where other gods held sway, and eventually prophesied its destruction: 'Howl ye ships of Tarshish for it is laid waste'.[2] And the Tarshish of Hiram and Jonah was fairly certainly the Tartessos of the Phoenicians, that city which, with its hinterland, lay beyond the western limits of the Mediterranean at the very edge of the known earth.[3] And if this identification is correct then the precious metals shipped to Solomon certainly had their origin in the many small and large mines which existed in the area of the modern province of Huelva, but plainly it is difficult to prove that any specific mine ever supplied King Solomon.

However, the folklore of this part of Spain is full of stories of those times which have impressed themselves vividly on the minds of the common people. It is a rich tradition, though, like most folklore, much of it cannot be proved and much of it is based on demonstrably false premises. In the mid-nineteenth century the French entrepreneur Ernest Deligny discovered some ancient mines not far from Rio Tinto and situated in a range of hills which the local people told him was called the *sierra* Tharse. He was convinced that the little village nearby marked the site of Tartessos and so called his discovery the mines of Tharsis, claiming for them a pre-eminence in antiquity which is improbable, for that great trading city was fairly certainly situated on, or close to, the coast.[4]

Many other stories have plainly arisen merely to explain place names. At el Cerro near Tharsis is a hill called Andévalo (the *v* being pronounced as a *b*) which is alleged in local folklore to have been a sacred place of the Phoenician deity Baal.[5] The imposing *cerro* Salomón at Rio Tinto, the ruins at its summit of the so-called *castillo de* Salomón (almost certainly nothing more than a Roman look-out post or store-house) and the town of Zalamea la Real are all associated in popular legend with the biblical king. But while the *cerro* Salomón and the town of Zalamea la Real clearly possess cognate names they seem more likely to have been derived from the arabic *salama* meaning 'peace' and are hardly likely to be earlier in date than the Moorish occupation of the middle ages.[6] In fact close to *cerro* Salomón was Zalamea la Vieja ('old Zalamea', now known as Nerva) and it seems likely that the hill originally took its name from this *pueblo* as Zalamea la Real might well have done, too, for the names of the two townships suggest a parent and daughter settlement.

At Alosno the people consider that they are descended from the Phoenicians and claim that their music has an eastern origin.[7] The *esquila* of Rio Tinto, a strange and beautiful music sung from house to house during the novena of Our Lady of the rosary, may also have originated in the east; though one suspects the influence here and at Alosno is more likely to be Moorish and mediaeval than Phoenician and the first millenium B.C.

The great hills of ancient slag which are so impressive for visitors to Rio Tinto also present problems of historical interpretation. There is some contrast between the heaps. Some are obviously the result of planned tipping and lay-out and tend to reach great heights, indicative of well-regulated dumping operations, while others seem to have been created less methodically; but from appearance alone it would be unwise to attribute (as has been done) the former to the Romans and the latter to their predecessors. The Rio Tinto ancient slag amounted, at a conservative estimate, to 16½ million tons which is almost two-thirds of the 25 million tons to be found in the whole of modern Huelva province and adjoining eastern Portugal.[8] Some tenth of this total used to be attributed to pre-Roman workings[9] but the late J. C. Allan, a mining engineer, and L. U. Salkield, a metallurgist with long experience of Spanish mining, came to the conclusion that a far higher proportion than this resulted from the pre-Roman smelting for silver.[10] Leonard Salkield has demonstrated that the ancient slag was produced from the smelting of jarositic earths to extract silver and from the smelting of cupreous pyrites for copper, that they are in the ratio of 15 to 1 respectively, and that most of these jarositic earths were mined and smelted before the arrival of the Romans.[11] The Romans, when eventually they went to Rio Tinto, resmelted these ancient heaps, recovering still more copper from the copper slags and silver from the very high silica content of the silver slags. (Why the Phoenician slags contained so much silica, the white pieces of which can be seen throughout the slag, is still not known.)

Between 1880 and 1950 there were a number of finds of the Phoenician period, including a great deal of pottery, two heads with horns carved in stone which may represent Ashtaroth (but may possibly be crude work of a later date), Phoenician coins of the seventh century and stone implements.[12] All save the coins are undatable and none was recovered during scientific excavations which alone would have given them historical significance. The amiable and eccentric English archaeologist, Mrs E. M. Whishaw, claimed in the 1920s that a pre-Roman furnace had been found at Rio Tinto consisting of a half-burned heap of alternate layers of ore and charred twigs,[13] but by modern standards this discovery (potentially of some importance) is not properly documented. Three stone axes were also found on a hill-track near Rio Tinto from which it has been argued, though not very convincingly, that the track was a pre-historic road to the mines.[14] These various finds, though unsatisfactory as archaeological evidence, taken together with the slag analyses made by Leonard Salkield, showed that there had certainly been pre-Roman mining at Rio Tinto. But it was still difficult fully to comprehend the significance of what had been discovered and to speak in definite terms about the pre-Roman miners. Then the picture was radically transformed by the excavations conducted at Rio Tinto in 1966–7 by Dr A. Blanco and Dr J. M. Luzón, two archaeologists of Seville university.

On *cerro* Salomón, which still dominates the Rio Tinto landscape, though its lower slopes have disappeared as a result of open-cast mining, they found the remains of a town which had been occupied in the eighth and seventh centuries B.C. by a community of miners with houses and equipment very similar to those known a little earlier in Palestine.[15] The soil covering the hill is largely stony and thin, and a shallow clearing of the ground by the archaeologists was enough to reveal almost on the surface the remains of the buildings of the pre-historic settlement which had once extended all over the hill and was over 900 yards in length from east to west.

The walls consisted of two or three courses of undressed stones with no trace of mortar. Each house had been low-standing and probably roofed with thatch, with a varying number of small rectangular rooms with mud floors, some of which were covered with slate slabs, and stepped pavements where the incline of the hill made it necessary. The entrances were protected on the outside by projecting stone walls, a detail not previously found in prehistoric Spanish architecture though it was a feature which has been found in mining villages of approximately the same date in the near east.

The hammers left behind by the inhabitants of this town were made of porphyry, diorite or other hard stone. Their metallurgical implements included granite pestles and mortars. Many of those found in the houses had been considerably worn from rotary grinding, and were similar in shape to those found in mining communities of the tenth century B.C. in Palestine. They had been used either for crushing slag, so that the metal pellets could be extracted by hand or, more likely, to pulverize the minerals before reduction in the furnaces.

The slag left from smelting the metal was found scattered throughout the houses together with charcoal and drops of lead, and mixed up with the inevitable pottery and bones, showing that for these people metal founding was a domestic activity and not concentrated in large foundries or factories. Analyses of this slag by Leonard Salkield showed that the copper content was almost nil but that the silver content was surprisingly high: an average of 600 grammes per metric ton, a quantity comparable with the richest of silver ores. Since these ancient metal-workers were not capable of extracting metals so well as to leave the ores free of them, it is plain that these slags were the result not of copper but of silver mining. (Nonetheless, at about this time there would appear to have been some copper being mined from the malachite and azurite found in the neighbouring peak of the *cerro* Colorado.)

From their discoveries, Dr Blanco and Dr Luzón were able to reconstruct the techniques by which the silver had been extracted by the inhabitants of the settlement. It had been done by digging small holes in the ground, lining them with clay and then filling them with the pulverised ore, adding a silica flux and charcoal. The last was ignited and air from a primitive bellows forced through a clay pipe near the bottom of the bowl-furnace until the contents melted. On cooling, the lead and the silver in the ore formed an ingot underneath the slag and it was necessary to separate the two metals by the process of cupellation, in which the ingot was melted and its lead contents oxidised in a strong current of air, leaving the pure silver behind in its solid state. The origins of this process are obscure but the fact that these early men knew of it indicates how advanced was their technology.

The large quantities of pottery found in and around the houses of the settlement are all demonstrably of Phoenician origin of the eighth and seventh centuries B.C., showing that these people were either Phoenicians or natives using Phoenician goods. They received a number of their supplies in globular shaped, two-handled neckless jars a foot or more in height. They lit their homes with saucer-shaped lamps with one or two open spouts and kept the oil for these in jugs with ovoid-shaped bodies and pointed bottoms.

It has long been known that in the eighth and seventh centuries before Christ the coast from Villaricos to Cadiz was lined with Phoenician trading stations (of which Huelva on the mouth of the river Tinto, as a result of excavations conducted by Dr Luzón and Professor Juan Pedro Garrido Roiz, is proving to be a notable example). The new evidence from Rio Tinto shows that at the same time the Phoenicians were also penetrating some distance inland and benefiting from the silver mining in the hinterland either as traders or as miners and metallurgists.

The settlement on *cerro* Salomón was abandoned sometime in the seventh century B.C., at much the same time as the trading centres on the coast, for reasons which are not obvious: perhaps there was an anti-Phoenician outburst by the native inhabitants, or a regrouping of Phoenician communities in the western Mediterranean area as a result of commercial and economic change. But the north-east corner of the settlement probably experienced a second stage of occupation about 400 years later, and before the Roman occupation, for this part of the site has yielded a quantity of red-figured Attic wear of that time.[16]

## *The Roman period*

The period of the Roman occupation, following the defeat of the Carthaginians, does not present such great difficulties of interpretation. The inhabitants of the area around the upper part of the river Tinto, in the first centuries at least of Roman rule, were a Celtic tribe called the Turdetani which had intermingled with the northern Tartessians, and these people must have provided much of the labour force at the mines when the Romans began to exploit them. But the mining activities of the Romans at Rio Tinto in the first two centuries of their occupation were clearly limited in extent.

Under the Roman republic, the state exercised ownership of all mines but let out claims to be worked on licence, normally by small prospectors. These men would have had little difficulty in rediscovering the mines even if they had been abandoned following the departure of the Phoenicians, for the dark-red water of the river would lead them up into the *sierra*, the bright-red capping of iron oxide on the hill tops would show them where the minerals lay, and the old workings of their predecessors would serve as sign posts to the ore bodies. The more sophisticated of them might have observed the pattern of the fractured rocks and noticed the taste of the drainage waters, and sunk new shafts to explore below the surface. But most were probably content to follow in other men's footsteps and to work seams already uncovered.

During the Roman empire, which replaced the republic in the first century B.C., cartels of bankers took over mining operations and the small prospectors gradually disappeared. For by that time to work mines such as those of Rio Tinto in an economically sound way it was necessary to have considerable capital investment in order to get below the exhausted veins close to the surface and down to the deeper deposits. As early as the reign of Vespasian, between A.D. 69 and 96, the mining industry of the empire was dominated by syndicates. These either operated mines on leases or else loaned the capital for state-run enterprises, but control of all mines by the state was fairly strict, with output regulated to maintain prices.

Towards the end of the first century of the Christian era Rio Tinto was certainly well on the way to being developed as a large enterprise and was probably state-administered, for it was then in the charge of a senior civil servant, the *procurator* Pudens. By that time, the Spanish copper mines had been well developed and as a result of their enormous production surpassed in importance those elsewhere in the empire. It therefore seems likely that this was the time that the mining of sulphide copper ores began at Rio Tinto. Certainly within a short time the mines there became of great significance, apparently serving as a control centre for all the mines in the area around, and acting as a school of mines, training engineers and despatching experts to solve the problems of mines elsewhere.[17]

It is noteworthy that no Roman remains have been found at Rio Tinto which date from before the birth of Christ when mining would have been in the hands of small prospectors; and that what has been discovered all comes from between the end of the first century and the beginning of the fifth, with the greater part of the many coins and other datable objects falling into the second and third centuries, suggesting that this was the time of maximum exploitation.[18] The earliest Roman coin in the Rio Tinto museum is of the emperor Nerva who reigned from A.D. 96 to 98, and a statue of the same emperor was found at Rio Tinto in 1772. From that short reign, too, dates a bronze plate, in size 3 feet by 2 feet, discovered by Francisco Sanz.[19] It reads:

IMP. NERVAE CAESARI AUG
PONTIFICI MAXIMO TR.
POTEST. P.P. COS.III
DESIG.IIII. PUDENS AUG. LIB.
PROCURATOR
PAT. SUO POSUIT.

In expanded Latin this runs: *Imperatori Nervae Caesari Augusto, Pontifici Maximo, Tribunitie Potestatis, Patri Patriae, Consuli tertium, designato quartum. Pudens augusti libertus procurator patrono suo posuit.* Translated it reads: 'To the emperor Nerva Caesar Augustus, supreme pontiff, wielder of the tribunican power, three times consul, consul designate for a fourth term. Erected by the *procurator* Pudens, imperial freedman, to his patron.' The tablet, now severely damaged, is in the Museo Arqueológico Nacional, Madrid.

Pudens, the freed slave and mines' manager, was doubtless intending to record the completion of the long gallery, cut straight into the east side of *cerro* Salomón, from which the plate was recovered; and it may well be that it was the brief reign of Nerva which marked the start of the great Roman expansion of the Rio Tinto mines.

The Romans at that time must still have been mining silver in large quantities at Rio Tinto, despite their interest in copper, and it was the precious metal which caught the imagination of men. Most of the Roman writers who dealt with the subject concentrated on it to the exclusion of other metals. They referred to the Iberian mines often enough, though the information they give about them is never as precise as modern historians would like. And they recorded the names of many of the ancient mining towns of the Baetica region. One of these was the city of Urium which took its name from the river known as the Urbero or *ur-yero*: the Phoenician for 'river of fire'.[20] Traditionally this city is identified with Rio Tinto. Dr Luzón's study of the archaeological evidence, and particularly the coins, of Andalusia has made the traditional identification as certain as it is ever likely to be and pin-pointed the site of Urium as that which is now partially occupied by the Rio Tinto village of la Dehesa.[21]

It is from the time of the emperor Nerva onwards that the vast majority of the ancient shafts recorded at Rio Tinto probably date, for though it was possible for veins close to the surface to be worked by the open-cast method, those at any depth involved the making of vertical shafts down into the earth. The Rio Tinto shafts were normally about 30 inches across. Most of them were rectangular and given linings of wooden planks, though some were circular. The former were in the softer soil, which carried the danger of collapsing sides; the latter in the harder ground. Some of the deep shafts would have required ladders inside them but in most there were small holes in the sides to provide the hand and toe-holds by which the miners, backs pressed against the opposite wall, descended or climbed to the top. Many of these shafts existed simply as vertical openings with no galleries leading off them: doubtless these revealed no minerals worth the mining.

But if an exploration was successful and the mineral so deep that open-cast working was not appropriate, then galleries were excavated, radiating out from the bottom of the shaft and following the seam of ore until it petered out or became too hard to work. Sometimes, perhaps because trial shafts showed the location of a particularly rich body of mineral, the Romans penetrated the hillside with horizontal galleries driven straight at the heart of the seam from the external slopes of the hill. Even though these entrances could reach over 200 yards in length, considerations of economy still required that they were kept as low and as narrow as possible, and they were normally about 5 feet high, with floors just broad enough to let a man walk, and with the walls widening out towards the top to allow sufficient space for a man bearing a loaded basket on his back.

The internal galleries were seldom more than 2 feet across and 3 feet high, with slightly-arched tops and a small drainage channel of about a foot deep cut in the centre of the floor. Along these the miners must have crawled on their way to the work-face, which at times took the form of large caverns, and the mined ores must have been dragged in grass baskets by the men, or possibly boys, employed to do the haulage work. Really rich seams required more workers and made it worthwhile to open up larger tunnels, many of which have been found at Rio Tinto, often 5 feet high and 4 feet wide. One achieved both a height and a width of 9 feet suggesting a considerable amount of two-way traffic. Whether high or low, all the galleries occasionally widened out into larger chambers; a sign that at these particular spots enriched deposits were found, and some of those discovered in the course of open-cast mining at Rio Tinto since the 1870s have been remarkable for their size.

These galleries and chambers have caused major problems to more recent miners. The Romans, despite their advanced engineering skills, seem to have been weak on their survey work, sinking shafts and carving out galleries without keeping proper maps and records. Thus at times tunnels were cut too close to earlier workings and the effects of oxidization on the minerals in the rock eventually weakened the walls, often causing them to cave in. In 1875 an extensive network of tunnels with numerous side-galleries dating from the Roman period was found situated immediately underneath similar tunnels on the next level, with only a foot of rock between the roofs below and the floors above. There had been numerous falls and the whole was in a highly dangerous condition which made the modern work extremely hazardous.

Working underground, often at depths greater than fifty feet, the Romans had to solve five basic mining problems: the transportation of the ores, the provision of lighting, the support of the roofs of the galleries, ventilation and the drainage of the mines. At Rio Tinto the underground workers of the Roman period packed the ores and debris at the working face into small baskets of esparto grass. (The remains of many of these baskets have been found and resemble the baskets called *espuertas* which were used on a large scale in the early days of the Rio Tinto Company and are still in use today in parts of Spain.) When full, the baskets were dragged through the low tunnels by their handles, or carried on backs through those larger galleries where it was possible to walk almost upright, to the bottoms of the shafts. Then they were either carried up ladders to the surface or, in the case of the deeper shafts, hauled by rope-winch (the remains of one of these has been recovered). The many circular copper rims dating from this time and found at Rio Tinto were probably inserted into the baskets as an aid either to hold them open during loading or to hold them whilst they were raised by winch to the surface. Once on the top the grass baskets would have been carried to a nearby dumping-ground situated close to the head of the shaft and made by cutting a clearing in the brushwood on the hillside.

The galleries were dimly lighted at intervals by small terracotta lamps which were placed on shelves cut in the rock and which have survived in vast numbers, and these must also have provided the illumination at the working face. The roofs, and often the walls, were supported by props made of oak obtained from the nearby forests and these, too, have been found in great numbers, preserved into modern times as a result of being saturated with water containing copper salts in solution, and impregnated with flecks of native copper resulting from the reduction of the salts.

Ventilation in the deeper galleries was a considerable problem, for as air and dampness came into contact with the sulphur in the rock, oxidization produced dangerous sulphur-dioxide gas and the equally dangerous hydrogen sulphide. The Romans therefore dug airshafts between some of the tunnels and the surface, and probably increased the circulation of air by lighting fires at the bottom of these to cause strong upward draughts.

Drainage was the greatest difficulty of all. When mining was being done inside the upper portion of a hill, then a water adit could be made, sloping gently downwards and emerging out of the hillside. The most impressive of these was at the east end of *cerro* Salomón and survived until this century. Some 264 yards long, it was pierced by twenty shafts from the hillside above, the innermost being 65 yards deep. It was made by sinking shafts at regular intervals 12 yards apart and then cutting a gallery into the rock from the bottom of each shaft to the bottom of the next. Where the adit emerged from the hillside it was continued down to the river

Tinto in the form of a stone trough six feet deep. Plainly it was designed to carry a great volume of drainage water. Thirteen such adits are recorded as being found at Rio Tinto, having been constructed by the Romans to drain the upper levels. In the smaller galleries, and during the preliminary excavations for the larger tunnels, the small channels cut in the floor drained the water to the bottom of the entrance shaft where it was scooped into buckets of esparto grass lined with pitch and passed to the surface, either from hand to hand or by means of ropes. But once workings reached any size or depth the volume of water became too great to be taken out of the mines in this way. If the workings were buried too deep inside the hill to be reached by an external adit, then the Romans coped with the problems by installing elaborate systems of water-wheels. The drainage water was carried in a trough to a wheel 14 ft. 8 ins. in diameter, the scoops of which raised the water to another trough above, where it was again lifted by another wheel. These wheels were made of wood with bronze axles and over thirty have been discovered at Rio Tinto, including a nest of sixteen arranged in sets of two at eight different levels and providing a lift of over 100 feet.[22]

The Romans treated the ores in furnaces, a number of which still survive at Rio Tinto. J. F. Allan, who worked at Rio Tinto (and father of J. C. Allan who collaborated with Leonard Salkield) described them in 1887: 'From the few remaining furnaces in existence, one side appears to be formed by excavation of the solid rock, with a semi-circular wall about 2.1 metres high built in front of it, leaving a circular section 0.67 metres in diameter, with an opening below for the outflow of slag and the admission of blast.[28] (Dr B. Rothenberg of Tel Aviv University located about twenty surviving Roman furnaces at Rio Tinto in April 1973.) The charge of ore these smelted was small, probably no more than three or four hundred pounds at a time, so that given the extent of the operations and the resulting slag they must have used numerous smelting furnaces. On the stone base were placed alternate layers of charcoal and ore (mixed with flinty rocks containing silica to aid melting) which were then heaped over with clay and earth. When ignited, air was blasted into the furnace through leather bellows and eventually a mass of fused metal sank to the bottom of the heap. Further refinement was needed and the silver was recovered by cupellation with lead.* These same furnaces were also used to smelt the sulphide ores containing the copper. This process was complicated and lengthy, requiring the sulphide minerals and the products from the smelting to be roasted a number of times before

*Lead was imported to the mines together with most other supplies. Two blocks of lead marked *Carthago Nova* have been discovered at Rio Tinto.

the copper was obtained. Charcoal was used in the process because it was the only known fuel that could produce the high temperature needed, but it was an expensive commodity, for in order to produce a pound of charcoal, at least four times that quantity of wood had to be burned.

The miners of that period had few aids in their work. Some of their tools have been found: single and double-headed picks, levers, hammers weighing between five and ten pounds, shovels and tongs. Since they did not have explosives for blasting, work at the rock-face had to be done by splitting the rock, which involved heating it with fire and cooling it rapidly with water, and then driving in wooden wedges which were then wetted to make them expand and so open the split in the rock still more. If there was a fall of rock between the miners and the shaft then rescue was highly unlikely, and during the nineteenth and twentieth-century operations at Rio Tinto many skeletons, with their tools alongside them, have been found entombed in the ancient galleries.

The mine workers under Roman rule were of three classes: free citizens, slaves and criminals. It must have been the still-shackled bones of a criminal that were uncovered in a grave in the Roman necropolis near la Dehesa village at Rio Tinto. Working alongside the criminals in the heavy labouring jobs were the slaves. Pliny claimed that there were 20,000 of them in his day working in the mines of the Roman province of Boetica (which covered western Andalusia). Since over half of the ancient slag in that area is at Rio Tinto, his statement might imply that some 10,000 or more of these slaves might have been employed at the mines there. They would have been despatched from the various parts of the empire as Rome's boundaries were expanded following new territorial conquests, though the ending of the wars and then the easing of the lot of slaves by the emperor Constantine (A.D. 306–37) following his adoption of Christianity must have led eventually to a decrease in their numbers. Citizens were also employed and possibly local hillsmen from the area.[24] It seems likely that most of the administration and the skilled work would probably have been in the hands of free, or freed, men. It was argued convincingly by Ernest Deligny that the miners, who worked alone and in constant danger at the underground rock face, required both the skills and the self-discipline that could only be expected of free men, and that this *élite* would do no useful work unless they did it freely.[25]

Remains of Roman buildings survive at Rio Tinto under later slag heaps and some of the stonework recovered from them shows a high standard of craftsmanship. Stone sarcophagi have been found on the slopes of *cerro* Salomón. Close to la Dehesa village is an extensive burial ground, and though many of the monuments naming Roman miners and their families have been taken by those using the necropolis as a quarry for building-stone over the centuries, a number have been preserved in the Rio Tinto museum. From these graves have come many glass tear-bottles, so much used by upper-class Roman women, and household items such as jars, dishes and bowls, showing that the women-folk of the senior employees at least were present at the mines and despite the inaccessibility of them, had some domestic comforts. Such as these were, they had to be brought in from outside, over rough tracks and on muleback, together with the vital necessities for the rest of the community: the food, the iron tools, the lead for the cupellation of silver, the oil for the lamps, the esparto grass for the buckets and the pitch for their water-proofing, and the hemp for the ropes.

The organisation of the community at the mines must have been highly complicated and well-regulated. In 1876 in a slag heap at the Roman Vispasca mines at Aljustrel in Portugal, were found two bronze tablets, obviously part of a series that contained the public code of that community in the early days of the empire; regulations which would have applied equally to Rio Tinto and to other mining communities. In minute detail the code covered such things as the monopoly affecting the sale of boots and the taxes to be paid by the shoemakers; regulated the employment of the public barbers while permitting slave-owners to have their hair cut by their domestic slaves without penalty; conferred on the chief mines' administrator a monopoly in the supply of clothing; provided for the keeping of a roll of the names of the labourers at the mines; exempted schoolmasters from taxation; and provided for the free use of the public baths by freemen, soldiers, public officials and children while other men paid half an *as* and women customers one *as*.[26] One tablet detailed the tax that had to be paid for the men employed in collecting, cleaning and smelting the ancient copper and silver slags. There can be no doubt that for the upper-classes at Rio Tinto some of the comforts of Roman civilization were available and that some must have enjoyed a high standard of living. The bronze statuette (perhaps representing the goddess Minerva or one of Juno's hand-maidens) of a crowned female figure, contemplative and elegantly robed, that has been recovered from a slag heap, must have graced the home of a wealthy household. The bronze model of a wild boar which has also been found (the whereabouts of which is unfortunately now unknown) shows that there was an indigenous culture flourishing at the same time, for it probably represents one of the Celtic gods and is modelled on the long-legged boars which roamed the *sierra de* Aracena.

Roman trade and commerce reached its peak during the reign of Trajan (A.D. 97–117); it was then that conditions were most favourable for the development of the Iberian copper mines, and the evidence from Rio Tinto suggests that it was indeed at this time of boom for the Roman economy that the mines there underwent their great expansion and development. When during the third century the economy of the empire ran into difficulties, there was a severe spell of inflation which greatly increased the demand for copper coinage and this also probably served to stimulate copper production, although the drop in the slave population of the empire by the early fourth century must have caused considerable manpower problems. However, as the century progressed the different parts of the empire tended to fragment politically and economically and this presumably had an adverse effect on the copper mines. Coins have been found in considerable numbers at Rio Tinto for the reigns of the emperors Theodosius (A.D. 379–95) and Honorius (A.D. 395–425) which suggests that the mines continued to be worked throughout this period of economic difficulty until a comparatively late date. But in 410 the Visigoths sacked Rome and the Romans started to withdraw from the Iberian peninsula.

The Roman withdrawal marked an end of underground mining at Rio Tinto for thirteen centuries. But the scale of their operations and those of their predecessors had been such that the evidences of their work survived the centuries to impress the minds of later men. The Spaniards who rediscovered the workings in 1556 were not the last to have their imaginations stirred by the sight of the ancient remains. Many mining men and even poets (including Lord Byron, who visited Rio Tinto on a journey from Lisbon to Seville [27]) have been moved and their fancies excited on seeing the prodigious survivals from that distant time.

## *After the Romans*

Following the sack of Rome, the barbarians went on to ravage Gaul, and then spread southward into Spain, consolidating their control over the south by about 475. They had only a small coinage and no need to mine copper on a large scale. But even if they had wished to continue working the mines of Rio Tinto it is hardly likely that they had the necessary skill to do so. For, once the Romans had withdrawn, the water-wheels would have stopped operating and the drainage adits become blocked through lack of maintenance. The mines would soon have been flooded by the rising waters and the galleries progressively collapsed. To remove the accumulation of water and to start the submerged water-wheels would have required an even more sophisticated system than that employed by the Romans, which, at best, had enabled them only to keep abreast of the waters, not to remove large scale inundation.

In their turn the Visigoths were conquered in the course of the eighth century by the Moors from Northern Africa. Caliph adb-ar-Rahman the first made Cordova his capital and the centre of an independent caliphate. For 600 years, during an occupation as long as that of the Romans, and in many ways far more creative, the caliphs ruled the area that included the Rio Tinto mines. In that time copper was mined at Cordova and Toledo, but there is no archaeological evidence to indicate that they exploited the mines of the Rio Tinto district. Yet the Moors were certainly settled at the mines and throughout the area around. Their coins have been found there, and to them is probably owed the name of Zalamea. On the north-east side of *cerro* Salomón were a number of Moorish wells and a water conduit 140 yards long at the site known known as *huerta de la Cana.** It may be that although the Moors did not operate beneath the ground like the Romans, they perhaps mined on the surface, and that some of the old shallow open-cast pits represent small operations carried out during their period. Perhaps, too, they produced copper by precipitation, discovering, as did later men, that if the rivulets of water coming from the mines containing copper sulphate were allowed to run over iron, pure copper was precipitated as the iron was dissolved. However, the Moors have left little in the way of survivals behind them at Rio Tinto compared with the abundant evidence of the Roman and earlier periods, so their work, if any, is more likely to have been small in scale.

The Moors in their turn were finally driven from Spain in 1492, in the joint reign of the *reyos católicos*, Ferdinand and Isabella. And in that year, too, Christopher Columbus sailed on the first of his momentous voyages of discovery from Palos, a small port on the river Tinto estuary, close to Huelva, taking his little fleet consisting of the *Santa María*, the *Pinta* and the *Nina*, across uncharted seas and half the world. Within fifty years of his journey, Spain had won a south American empire of great immensity and vast riches. But the inflow of wealth from the Americas was not enough to sustain the ambitions of Spain's sixteenth-century sovereigns, and a search began to locate the old Roman mines within the country itself. It

*These disappeared early in the twentieth century when they were buried under overburden removed from the open-cast pits.

was as a result of this that Mendoza's party went to Rio Tinto in 1556 and, recognizing the abundant survivals from the Roman period, realized that they had found one of the great mines of the ancient world.

# APPENDIX TWO *The Rio Tinto Company outside Spain, 1927-54*

When the Rio Tinto Company began life in 1873 its founders saw it simply as a mining and copper producing company. By the time of its last few years of independent existence, after it had sold the majority of its interest in the mines of Rio Tinto and before it merged with the Consolidated Zinc Corporation (that is, in the years 1954 to 1962) it had endeavoured, with varying degrees of success, to diversify its interests.

The beginnings of this change in the nature of the company's operations are to be found far back in the early years of Auckland Geddes's chairmanship, though until the post-1954 period its significance was not generally appreciated and so, unspectacular as it was, never excited the attention of the share-buying public as did the company's work at Rio Tinto.

What basically lay behind the new policy developed by Geddes was a desire to insure the company's future. He saw that the Rio Tinto mines, despite their immense reserves of ore, would one day (however distant) be exhausted; and appreciated that the uncertain nature of the Spanish economy and politics in that period meant that there was always the risk that the mines would become unprofitable, or that the company might even be compelled to withdraw from Spain. It seemed to Geddes, and later events proved him right, that the shareholders would best be served by the acquisition of interests in what appeared to be politically stable areas of the world.

The developments that took place as a result of this decision appear at first sight to be related only indirectly to the story of the mines of Rio Tinto. Yet in the long run both the failures and the successes of the new policy had a direct bearing on the company's Spanish operations. Shortcomings in the investment policy followed in the U.S.A. tended to make some of the directors and senior employees excessively cautious and to prefer the devil they knew to the one they did not, producing in the late 1940s a board-room split on the whole question of the company's future in Spain. On the other hand, the success of the Rhodesian copperbelt investment helped in the long term to keep the Rio Tinto mines in operation (for the operating losses which were incurred there in the 1940s were offset by the investment income), and then eventually pointed the way to a reduction of the company's interests in Spain combined with expansion in other parts of the world.

The programme began relatively cautiously. At the end of 1927 Geddes acquired complete control of the Pyrites Company and its American subsidiary, the Pyrites Company Incorporated, and made it the distributor for all Rio Tinto ores in the United States. This last company also owned at Wilmington in Delaware the second largest extraction plant in the world which produced from pyrites cinders both copper bars and sintered iron ore of a good quality, for which there was a heavy demand in north America. (In 1929, new zinc and lead extraction plants were added at Wilmington to permit the recovery of those metals.) At the same time the Rio Tinto Company purchased a substantial minority holding in the Davison Chemical Company of Baltimore in Maryland, a large buyer of pyrites and one of the largest American manufacturers of fertilizers.[1]

This last company had as a subsidiary the Silica Gel Corporation which, by treating sodium silicate with sulphuric acid, produced the product from which the company took its name. Chemically resembling common beach sand, but capable of a high degree of adsorption (that is, of attracting to its surface molecules of gases and liquids) the commercial use of silica gel was being rapidly expanded within the United States where it was employed in air conditioning plants, the refrigeration equipment of railway freight wagons and in the refining of benzol. In February 1929 the Rio Tinto Company entered a partnership with the Silica Gel Corporation to develop the manufacture and use of the product throughout the world outside of north America. By this agreement a company, Silica Gel Holdings, S.A., was set up in Geneva with Geddes as president and a British subsidiary, Silica Gel Ltd, in London with Geddes as its chairman and other members of the Rio Tinto board and staff as its directors. Additional subsidiaries were soon established in France, Germany, Japan and India.[2] In a little over a year the Rio Tinto Company invested £1 million in these subsidiary and associated companies.

At the very beginning of 1929, too, Geddes and his fellow directors began to take an interest in Northern Rhodesia where a great cupriferous area had been discovered. It looked as if the copperbelt, as it became known, was likely to provide an important part of the world's copper supplies and it was decided that the Rio Tinto Company should use some of its resources to buy its way into companies at work there. A purchase of shares in Minerals Separation Ltd made Rio Tinto its largest single shareholder, and R. M. Preston and W. G. Gray joined its board as Rio Tinto representatives. Minsep, as it was commonly called, held a number of patents for extracting metals from ore concentrates and had agreements for the use of these at some of the copperbelt mines. It also owned a 50 per cent interest in Copper Ventures Ltd, a company specifically created to apply one of the Minsep processes in the treatment of copper ores in Northern Rhodesia.

Simultaneously, a direct interest in the copperbelt was acquired by purchasing a large shareholding in Rhodesian Congo Border Concession Ltd and arranging for Auckland Geddes and J. M. Buchanan to join its board. This company was closely associated with Minerals Separation Ltd, and had begun life in 1923 with a major concession covering 50,000 square miles along the Rhodesian-Congo border, rapidly discovering there three major ore bodies at Nchanga, Mufulira, and Bancroft.

At the time of this Rio Tinto investment in Rhodesian Congo Border Concessions it appeared as though the American copper producers might obtain control of the copperbelt and oust British and local interests there. Fears on this score were intensified when on 28 January 1929 shareholders in the Nchanga mine received a letter setting out proposals for obtaining finance for the development of Nchanga from the American Smelting and Refining Company on terms which would have given that company control.

The *Financial Times* immediately attacked the proposal:

> 'By what can only be considered a clever coup the famous American Smelting and Refining Company has secured, provisionally, the technical control and the reversion of the financial control for immediate payment of under £210,000 (118,750 shares at 35s each against a current market price of well over £4) . . . what are our Kaffir houses doing to let the A. S. & R. get a throttlehold on one of the best of the coming Rhodesian copper producers on such easy terms? . . . Is it too late for one of the others [i.e. a South African mining company] to make a better bid than the Guggenheims have put forward?'

In fact a counter-offer came from Sir Ernest Oppenheimer's Anglo-American Corporation, with the backing of Geddes, which led to the withdrawal of the American Smelting and Refining offer. Geddes had been persuaded of the need to ensure the exclusion of American participation then or in the future if the British and the (then) imperial interest was to be served. He also saw an opportunity to advance his new policy of diversification for the Rio Tinto Company.

With enthusiasm he started to organize financial support for Rhodesian Congo Border Concession and for Nchanga. This campaign, like all those he organized, was one of shuffling papers and talking across tables. It was not the sort of fight that would have appealed to a man like Walter Browning, but it was one for which the intellectual and austere Auckland Geddes was superbly suited. He was also well placed to manage it.

He brought together Rio Tinto and N. M. Rothschild and Sons (who had a large shareholding in the Rio Tinto Company) in an alliance with Oppenheimer's Anglo-American. The consulting geologist of the latter company, J. Austen Bancroft, had been a professor at McGill when Geddes had himself been a professor and then the principal there, and Geddes knew him to be a man worth backing. The British South Africa Company, the British Metal Corporation, the Johannesburg Consolidated Investment Company, the Union Corporation and the Anglo-Metal Company were all gradually brought into the consortium.

The agreement they hammered out provided for the group to take up 300,000 shares at £5 in Rhodesian Congo Border Concession. From the funds which this share issue provided, that company was to subscribe for 100,000 Nchanga shares at £3, with a further 200,000 shares being offered to the Nchanga shareholders. As a result of this influx of capital, while the Nchanga company could proceed to develop and equip its own property, the Rhodesian Congo Border Concession was free to undertake the intensive prospecting that the remainder of its concession merited.[3]

It would seem as though at the beginning of the outcry against the growing extent of American participation in the copperbelt, Geddes could see no objection to it in principle. But 'those in high places' (presumably he meant those in government) urged him to prevent American dominance since 'They'll be there as overlords, not as partners'. Thus encouraged he decided to act. After the victory he wrote: 'I came to the board of this company and I went to see several friends and . . . a group was formed of which the Rio Tinto Company is a member . . . The agreement [with American Smelting and Refining] was not consummated by the Nchanga company, and when the dust of conflict settled we found ourselves in a position to control Minerals Separation Ltd, the Rhodesian Congo Border Concession and the Nchanga companies.'[4]

Rio Tinto, as the largest of the shareholders in Rhodesian Congo

Border Concession and Nchanga, appointed four directors to their boards, and took over the administration of their technical committees. This was followed up by acquiring ordinary shareholdings in the Roan Antelope and Nkana mines (the latter through the Bwana Mkubwa Copper Mining Company), and in the Rhodesian Selection Trust which owned the highly profitable Mufulira mine. As a modest celebration of the victory over the Americans, Auckland Geddes, Austen Bancroft and Vibert Douglas, the Rio Tinto Company's principal geologist (also a graduate of McGill where he had studied under Bancroft), brought together a collection of specimens illustrating the geology of Northern Rhodesia and the southern Congo which was presented to the Natural History Museum at South Kensington.[5]

The money for the Rio Tinto Company's incursion into the north Rhodesian copperbelt was raised by the issue of 50,000 ordinary shares of £5 each, offered to shareholders on a one-for-eight basis at £50 a share. These yielded £2,250,000.[6]

In 1930 Oppenheimer proposed the amalgamation of the Nchanga company, the Bwana Mkuba Copper Mining Company and the Rhodesian Congo Border Concession. Such a move would make borrowing easier in view of the greater amount of security this would provide and thus help in the financing of technical development. Expenditure of between £8 and £9 million was envisaged, with a development period of four years before there would be any return on these copperbelt investments, but by 1935 (Oppenheimer told Geddes in a letter) the combined annual production should reach 195,000 tons, equivalent to 10 per cent of what world production had been in 1929.[7]

The merger took place in 1931, when the Rhodesian Congo Border Concession increased its capital from £750,000 to £2,000,000 and acquired the other two companies by an exchange of shares. It adopted the name of the Rhokana Corporation Ltd (an amalgam of Rhodesia and Nkana) and Auckland Geddes became its first chairman.[8] An issue of convertible debentures followed, largely taken up by the Rio Tinto Company, which in turn had to raise £2 million by an issue of redeemable debenture stock.[9] These heavy investments in Northern Rhodesia, made at the height of economic depression and the collapse of the copper market, were prompted more by optimism for the future rather than hope of quick gains. By the end of 1931, of Rio Tinto's investments totalling £4 million, £2 million was tied up in the copperbelt.

The Rhodesian investments also caused Geddes to change one aspect of his policy. In October 1926 the Rio Tinto Company had joined a newly established combine being set up in America under the name of Copper Exporters Inc., which consisted of producers of 90 per cent of the world's copper. The intention of the combine was to eliminate the price-fixing function of the London Metal Exchange by restricting supplies. However, the higher prices thus brought about led to consumer resistance and to the substitution of other metals and materials for copper, as well as stimulating operations by low-cost producers in Northern Rhodesia and Katanga.

It may have been this American alliance which at first made Geddes reluctant to go into the copperbelt, but once Rio Tinto was committed to participate in Northern Rhodesia it was plain that its membership of Copper Exporters Inc. would be an embarrassment and would have to end. In fact the combine collapsed in 1932 when the United States imposed tariffs on copper imports in the interests of its own producers and Britain imposed a similar levy on copper coming from non-empire sources. In consequence copper prices once again came to be determined largely by the law of supply and demand, operating via the London Metal Exchange.

1932 saw the world economic outlook become progressively darker. In that year the Rio Tinto Company was selling its copper at a loss, while the Davison Chemical Company, in which it had such an important shareholding, was having a difficult time due to the depressed state of American agriculture, resulting in its having a receiver appointed, and in its subsidiary the Silica Gel Corporation Inc., also being unable to meet its debts. When U.S. government inspectors alleged grave irregularities in the conduct of Davison Chemicals, the Rio Tinto Company decided to write off its investment in the American company and the Silica Gel subsidiaries. In addition, it became necessary for the Pyrites Company to close down temporarily both its mines in Cyprus and its subsidiary's plant at Wilmington.[10]

In 1935 Rhokana paid its first dividend, which in turn allowed Minerals Separation to pay an increased dividend due to the additional revenue provided by its own holding in Rhokana. Rio Tinto in turn benefited, its income from investments rising from £31,000 in 1934 to £91,000.[11] In the next year Rio Tinto took up 425,000 shares in a new company, Nchanga Consolidated Copper Mines, formed to take over the Nchanga and other properties from the Rhokana Corporation, while its income from investments rose yet further to £108,000, soaring still higher in 1937 to £316,000.[12]

In that year some shareholders expressed considerable discontent with the extent of the company's commitments in Rhodesia and there were some demands, which Geddes rejected very firmly, to sell out the copperbelt holdings in order to pay off the debenture holders.[13] It was as well that he did stand firm, for these investments (which Rio Tinto still held at the time of its merger with Consolidated Zinc in 1962) were the mainstay of the company during the second world war.

The debate within the company, however, about its future role as either a mining house or as an investment holding company (or as a combination of both) was left unresolved by these events and the experience of the between-war years. The discussion was at times acrimonious, contributing in 1948 to the resignation of two senior directors. It was at this time and against this background that the new men, who were in a short space of time to lead the company into a fresh round of diversification of company interests, were recruited. The appointments in particular of Mark Turner, Gerald Coke, Val Duncan and Frank Byers as directors, and Roy Wright as overseas' manager can be seen in retrospect as a crucial element in the policy which was soon to be evolved so successfully.

Yet this group of young men were not taken into the company to promote any specific policy. Each was wanted for his specific abilities and to a great extent remained unaware of the past arguments. Consequently the policy with which they became so closely identified was evolved largely through their own appreciation of the situation facing the Rio Tinto Company. In part the desire to diversify interests still further arose from a conscious assessment of the best ways to make the company more profitable and to utilize its capital (largely tied up in Spain) and the expertise of its senior staff. But there was also in this an element of frustration with the position confronting them once they had settled into their new jobs. It was not only Val Duncan who was unwilling to sit at a desk for the next thirty years trying to get *pesetas* out of Spain, though he expressed this view more forthrightly than the others. The Rio Tinto Company provided a vehicle for their talents, and in any case it was hardly likely that they would be immune to the feeling of buoyancy which was gradually spreading through British commerce in the early 1950s after the long and sterile war and post-war years.

The first cautious discussions about possible expansion in southern Africa took place in 1949. Towards the end of that year Mark Turner reported on his discussions with the Canadian Frobisher Company about the Kilembe copper deposits in Uganda. Soon after, Rio Tinto acquired an interest in the New Eland diamond mine and the Tororo apatite deposit, and then an interest in Vermiprop, the vermiculite mine at Palabora. By July 1952, however, the Kilembe mine clearly needed refinancing and the Rio Tinto Company began to withdraw from the operation. Similarly with New Eland Diamonds, to which Rio Tinto had advanced £62,000. By the spring of 1952 it was plain that there was no hope of attracting new partners into the venture or of Rio Tinto being able to buy out the other shareholders, and it was decided to place the mine on a care and maintenance basis.

Relations between Rio Tinto and its principal partner in Vermiprop, Dowson and Dobson, deteriorated, and by the spring of 1952 it had been decided that though Rio Tinto should act as mining consultants to the mine, and its subsidiary company Mandoval should have the exclusive world selling rights of vermiculite, the association with Dowson and Dobson should be virtually terminated. Another prospect in South Africa in which Rio Tinto became interested, in 1952, was Mallin Diamonds, but when complications arose over land titles it was necessary for Rio Tinto to back out. Disagreements with its partners in the management of the Tororo apatite deposit led to Rio Tinto withdrawing from that particular venture in 1953.

Mark Turner made an extended visit to South Africa in February 1952, and his report on his discussions with that country's leading mining companies served to mould Rio Tinto's thinking on its future there. He concluded that any Rio Tinto company formed to operate outside the United Kingdom should not necessarily seek a 51 per cent stake in a particular operation, but should bring in financial and other partners who would be content to let Rio Tinto manage the operation. He recommended, too, that the company should not set its sights too high as far as the maximum size of orebodies was concerned, but aim for a good spread of risk and the shortest possible period of non-profitability. He envisaged operations in countries other than those in southern Africa, and proposed that American and Canadian interests should be engaged, so that their dollar contributions would be available to finance any operations in north America.

While on the visit that led to this report, Mark Turner met in Johannesburg Oscar Weiss, a geophysicist of considerable reputation in mining circles. He seemed an ideal man to lead the exploration teams that Mark Turner and Val Duncan (by then managing director) were anxious to set up, and even before contracts were signed he had put forward a number of promising ideas.

At this time Val Duncan submitted a paper for consideration by the board outlining his views on the company's future. He referred to the fact that in the recent past the company's practice had been to pursue active mining on the one hand and on the other to hold investments, while refusing, for example, opportunities to participate in industries (such as fertilizer production) which consumed Rio Tinto pyrites. He now proposed the adoption of a vertical structure which would permit participation in businesses ancillary to, or associated with, active mining in Europe, Africa and north America. After much discussion, and a great deal of heart-searching, the new proposals were accepted and the exploration teams set to work in Rhodesia, South Africa and Canada. A feasibility study was even made into the possibility of reopening an abandoned mine in Scotland.

In 1953 Roy Wright, the new overseas' manager, made a reconnaissance visit to Australia and reported so enthusiastically on the possibilities there that Val Duncan and he made a lengthy tour together shortly afterwards. In consequence a field team was set up to work in Northern Territory searching for uranium.

Roy Wright's reports were often written in the course of his continuous aeroplane flights around Australia. On arrival at Plantation House, so quickly were they needed that there was no time for one typist to copy a report for circulation; one page apiece would be distributed to several secretaries for typing and the typed sheets and carbon copies then stapled together and rushed to those who were waiting for them. (It was the need to get these reports from Roy Wright into circulation as quickly as possible that led to the company buying its first mechanical copying machine).

Following a meeting organised by Rothschilds in London for Joe Smallwood, the prime minister of Newfoundland, the Rio Tinto Company became a founder member of the British Newfoundland Corporation, commonly known as Brinco, which consisted of a number of Canadian and British companies whose aim it was to develop the potential of Labrador and Newfoundland. It also set up its own exploration programme in Canada to prospect for minerals under the umbrella of Ownamin Limited.*

Then in 1954 came the sale of the Rio Tinto mines, providing a large capital sum for investment and freeing the directors and senior staff from the time-consuming problems that had been a heavy handicap for the company since the beginning of the 1930s. The sale thus provided the opportunity for the concentration of effort and money on the evolving programme of mining and investment throughout the world and pointed the way to the eventual merger with the Consolidated Zinc Corporation.

*The name was derived from 'Oscar Weiss North American Minerals', after the consultant geologist for Rio Tinto.

# APPENDIX THREE *A note on Spanish names*

## *Place names*

The following names have been given in this book in their English forms:

| | |
|---|---|
| *Andalucía* | Andalusia |
| *Castilla* | Castile |
| *Cataluña* | Catalonia |
| *Córdoba* | Cordova |
| *La Coruña* | Corunna |
| *Extremadura* | Estramadura |
| *Lisboa* | Lisbon |
| *Menorca* | Minorca |
| *río Tinto* | river Tinto |
| *Ríotinto* | Rio Tinto |
| *Sevilla* | Seville |
| *Vizcaya* | Biscay |
| *Zaragoza* | Saragossa |

## *Personal names*

Names of Spanish monarchs have been given in their more familiar English equivalents. Thus Felipe *segundo* is given as Philip the second and Carlos *tercero* as Charles the third. The names of all other Spaniards, however, have been kept in their original form, e.g. Francisco Tomás Sanz.

Normally, non-Spaniards have their names given according to the usage in their country of origin (even though in the Spanish source materials these were often Spanicised): Heinrich Schnell rather than Enrique Schnell, Franz Peringer rather than Francisco Peringer, etc. However, where these individuals themselves are known to have used the Spanish equivalents of their foreign names, these have been retained, e.g. Juan Wert, Felipe Prieto (for Philip Black) etc.

# APPENDIX FOUR *Glossary*

adit — tunnelled entrance to a mine and at Rio Tinto frequently a large underground drainage channel emerging on the upper slopes of a hill.

*administrador asentista* — administrator-contractor.

*alcalde* — mayor or chairman of local authority possessing judicial powers.

*alcova* — recess.

*andaluz* — dialect spoken in the region of Andalusia in southern Spain.

*arroba* — customary unit of weight, at Rio Tinto equivalent to 25 lbs.

*asistente* — formerly the chief officer of justice in the city of Seville.

*ayuntamiento* — municipal council; also the town hall.

Bella Vista — British village at Rio Tinto (literally 'fine view').

*caballero gran cruz* — knight grand cross.

*cacique* — formerly political boss of a district.

*caciquismo* — former system of government, exercised through local *caciques*.

calcination — reduction to friable substances by roasting.

*capitán de caballería* — cavalry captain.

*carabineros* — force of soldiers on frontier-duty.

*casa grande* — 'the big house' of a district and at Rio Tinto, in the time of the British company, the home of the general manager (now called the *casa conserjo*).

*castillo* — castle.

cementation — a process of salvaging copper from its solution in mine drainage-water, etc., by precipitating on scrap-iron.

*cerro* — hill or ridge.

*club inglés* — British (literally 'English') club.

CNT (Confederación Nacional del Trabajo) — the anarchist 'national confederation of labour'.

*colonia inglesa* — British (literally 'English') colony.

*comisario de guerra* — former army rank denoting second-in-command of a military district.

| | |
|---|---|
| *cobre negro* | crude ('black') copper. |
| *conde* | count. |
| *conflicto de carbón* | 'coal war'. |
| *consejo* | council of state or (in a commercial concern) board of directors. |
| *consejo de hacienda* | royal council of finance. |
| *copperas* | sulphate of iron; green vitriol. |
| *corta* | open-cast pit. |
| *cortes* | Spanish parliament. |
| *cueva* | cave. |
| cupriferous | yielding copper. |
| *de gracia* | free of payment; by grace. |
| *de justicia* | by right |
| *dentro de un ratito* | in a moment. |
| *direccion general de fincas* | directorate of state property. |
| *dollar* | disused coin. |
| *don* | sir. |
| *dos de mayo* | second of May, 1808 (uprising against French army of occupation in Madrid). |
| doubloon | the former *doblon sencillo* was valued at 12s. 6d in the mid-eighteenth century. (The gold doubloon was valued at 35s.) |
| *ducado* | ducat, a former coin valued at 10s. 6d. |
| *duque* | duke. |
| *ejercito salvador de España* | the saviour-army of Spain. |
| el Valle | 'The Valley', one of the Rio Tinto villages. |
| *empleados* | staff, senior employees. |
| *escoria* | slag. |
| España | Spain. |
| *esquila* | the folk music of Rio Tinto (derived from the word for sheep-bell). |
| *excelentisimo* | 'most excellent'. |
| *'fiat lux et facta est'* | 'let there be light, and there was light'. |

| | |
|---|---|
| *filón norte/sur* | north/south lode. |
| flux | substance mixed with metal ores to promote fusion during smelting. |
| Frasch process | recovery of sulphur from geological structures known as 'domes' by smelting the sulphur *in situ*, bringing it to the surface in molten form and then using it to produce acid. This elemental sulphur contains nearly 100 per cent sulphur compared with the 45 or 50 per cent normally found in pyrites. |
| *gaceta* | gazette. |
| *guardia civil* | civil guard. |
| *guardia de las minas* | the corps of mines' watchmen. |
| *horno* | oven, furnace. |
| igneous | produced by volcanic agency. |
| *juez conservador* | 'judge-conservator', a legal official with special responsibility for protecting the rights of heirs and others in property. |
| *juez sub-delegado* | assistant deputy-judge. |
| *junta* | board or committee. |
| *lago* | lake. |
| la Mina | the modern name for the eighteenth-century village on *cerro* Salomón, known in the later nineteenth century as Rio Tinto. |
| leaching | a method of removing soluble matter from ores by percolating liquid through them. |
| lixiviation | separation of substances into soluble and insoluble constituents by treatment with water. |
| *madrileño* | inhabitant of Madrid. |
| *maravedís* | former coin equivalent to an English farthing. |
| *marqués* | marquis. |
| *mesa* | plateau of land. |
| Mesa de los Pinos | 'plateau of pine trees', the name of one of the Rio Tinto villages. |
| *mina* | mine. |
| *mohinos* | the nickname of the inhabitants of Rio Tinto. |
| *pagares* | promissory notes. |
| *peseta* | monetary unit (equivalent to 4 *reales*) and through most of the period covered by this book (i.e. up to about 1914) with a value of 10d. |

| | |
|---|---|
| *plaza* | main square. |
| *plaza de toros* | bull-ring. |
| *pocos minutos* | a few minutes. |
| *posada* | inn. |
| *pueblo* | village or town. |
| pyrites | ores containing a combination of sulphur with iron and copper. |
| *quintal* | hundred-weight. |
| *real* | former coin, equivalent to a fourth of a *peseta*, and through most of the period covered by this book worth 2½d. |
| *reina* | queen. |
| *rey* | king. |
| *río* | river. |
| *san* (feminine *santa*) | saint. |
| *sierra* | mountain range. |
| *telera* | a pile of ore for open-air roasting. |
| *terremoto* | earthquake (a nickname of Walter Browning). |
| *tinto* | wine red. |
| ton | as used in this book is a short ton, i.e. 2000 lbs. |
| U.G.T. (Unión General de Trabajadores) | the socialist 'general union of workers'. |
| Vista Alegre | apartment block for British families at Rio Tinto (literally 'lively view'). |

APPENDIX FIVE

# *A note on Source Materials*

## a. *Printed sources*

1. *Rio Tinto.* A large number of publications contain allusions to Rio Tinto and most of these are to be found in the references section of this work. Three books dealing specifically with the history of Rio Tinto have previously appeared in print. These are:

RUA FIGUEROA, Ramón, *Ensayo sobre la historia de las minas de Río-Tinto* (Madrid, 1859).
ALDANA, Lucas de, *Las minas de Río-Tinto* (Madrid, 1875).
NASH, William G., *The Rio Tinto mine: its history and romance* (London, 1904).

Half of Rua Figueroa's eight chapters are devoted to the Phoenician and Roman periods at Rio Tinto; and since he wrote at a time before archaeology became a science these have little value for the modern reader. He has also one chapter on the sixteenth and seventeenth centuries, and three on the period 1724 to 1859.

Aldana's book is the most useful of the three and deals exclusively and in great detail with the period 1725 to 1873, and corrects many errors in Rua Figueroa's work. Both Rua Figueroa and Aldana were senior engineers at Rio Tinto and made considerable use of documentary material kept in the mines' offices.

William Nash was an Australian who was head of the mines' lands and sanitary department from 1886 to 1917. His book ranges from the Phoenician period to 1873 and though it borrows extensively from the works of Aldana and Rua Figueroa contains numerous mistakes due to his unhistorical approach and his tendency to prefer a romantic story to a factual one.

A great deal of technical information about mining at Rio Tinto is to be found in I. Pinedo Vara's comprehensive work *Piritas de Huelva* (Madrid, 1963).

2. *General works.* The books on which I have relied for background information on Spanish history are:

BRENAN, Gerald, *The Spanish labyrinth* (Cambridge, 1943).
BOLIN, Luís, *Spain: the vital years* (London, 1967).
CARR, Raymond, *The Spanish civil war* (London, 1961).
HENNESSY, C. A. M., *The federal republic in Spain, 1868–74* (Oxford, 1962).

JACKSON, Gabriel, *The Spanish republic and the civil war, 1931–1939* (Princeton, 1965).
MARVAUD, A., *L'Espagne au XIX siècle* (Paris, 1915).
NETTLAU, M., *Miguel Bakunin, la internacional y la alianza en España, 1868–73* (Buenos Aires, 1925).
PETRIE, Sir Charles, *King Charles III of Spain* (London, 1970).
RAMOS OLIVIERA, A., *Politics, economics and men of modern Spain* (London, 1946).
ROBINSON, R. A. H., *The origins of Franco's Spain* (Newton Abbot, 1970).
TREND, J. B., *Origins of modern Spain* (Cambridge, 1934).
VICENS VIVES, Jaime, *Aproximación a la historia de España* (Barcelona, 1952).
VICENS VIVES, Jaime, *An economic history of Spain* (Princeton, 1965).
VILLAR, Pierre, *Histoire de l'Espagne* (Paris, 1947).

## b. *Documentary sources*

For the years 1556 to 1873 I have drawn extensively on records acquired by the former Rio Tinto Company in 1873–4 for use in its London office. Some of these are originals but for the greater part they are transcriptions of documents which were then to be found either at the mines (and which now no longer exist) or in government offices.

For the period 1873 to 1954, when the mines were controlled by the Rio Tinto Company, the source materials consist largely of minute books, reports, letter books and files kept at the London office. There are still many records relating to the day to day administration of the various mines' departments which are retained at Rio Tinto, having passed to the Compañía Española de Minas de Río Tinto in 1954.

## c. *Eye-witness testimony*

I have interviewed a number of people associated with Rio Tinto and their information has been extremely helpful in amplifying the documentary materials for the period 1905–54. The post-1954 information contained in the epilogue (chapter 20) was almost exclusively obtained in this way. I have also been able to use notes made by people who are now no longer alive: Alexander Hill's notes of his early years at Rio Tinto, Laurence Hill's civil war diary (both kindly made available by Mrs L. Hill), and the late E. Potts's recollections (made available to me, with his usual generosity, by Mr L. U. Salkield). In addition Mr Salkield has let me draw extensively on the considerable notes which he himself has compiled.

# REFERENCES

The prefix 'RTC' denotes a document in the archives of the former Rio Tinto Company (now in the London headquarters of the Rio Tinto-Zinc Corporation).

The prefix 'RTE' denotes a record to be found at the mines' administration office in Rio Tinto.

The abbreviation 'BMB' indicates 'board minute book', 'LB' indicates 'letter book' and 'SB' indicates 'staff book'.


## Chapter 2
### The hill of Solomon

1. RTC 100/B/26(a), ff.1–9; R. Rua Figueroa, *Ensayo sobre la historia de las minas de Río-Tinto* (Madrid) 1859, pp. 118–9.
2. RTC 100/B/26(a), ff.9–10; Rua Figueroa, pp.118–23.
3. RTC 100/B/32(a), pp.55, 61–4; Rua Figueroa, pp.123–5.
4. Rua Figueroa, pp.125–8.
5. RTC 100/B/32(a), 70pp.–2; Rua Figueroa, pp.134–40.
6. Rua Figueroa, p.140.
7. RTC 100/B/32(a), pp.73–4; 100/B/26(a), f.10; Rua Figueroa, pp.140–1.
8. RTC 100/B/26(a), f.10; L. de Aldana, *Las minas de Río-Tinto* (Madrid) 1875, p.18.
9. RTC 100/B/26(a) f.11; Aldana, p.18.
10. *ibid.*

## Chapter 3
### Congenial souls

1. RTC 100/B/26(a), f.11; Aldana, p.19; information of Mr L. U. Salkield.
2. Rua Figueroa, pp.144–5, quoting González, *Registro de minas de la corona de Castilla* (Madrid) 1832.
3. RTC 100/B/32(b), *Manifesto por D. Lieberto Wolters von Siohielm* (Madrid) 1725.
4. RTC 100/B/27(d), ff.11–12; Rua Figueroa, pp.147–9; Aldana, pp.437–538 reprints a number of these pamphlets.
5. RTC100/B/27(d), f.14; Aldana, p.27.
6. RTC 100/B/27(d), f.15.
7. *ibid.* f.14.
8. *ibid.* ff.14–16; Rua Figueroa, pp.149–55.
9. RTC 100/B/27(d), ff.11–12, which contains a list of the original shareholders; Aldana, pp.28–9, 134–6.
10. RTC 100/B/26(a), f.16.
11. *ibid.*; Aldana, p.29.
12. RTC 100/B/32(a), p.94.
13. RTC 100/B/26(a), ff.17–18; Aldana, p.29.
14. RTC 100/B/26(a), ff.18–19; Aldana, p.31.
15. RTC 100/B/26(a), f.19; Aldana, p.31.
16. RTC 100/B/26(a), f.20; Aldana, pp.31–2.
17. RTC 100/B/26(a), f.21.
18. Aldana, p.32.
19. RTC 100/B/32(a), p.99.
20. Rua Figueroa, pp.157–9; Aldana, pp.33–4, where he incorrectly calls Lady Maria Herbert '*duquesa* de Powis'.
21. Rua Figueroa, p.160; Aldana, pp.33–4.
22. RTC 100/B/26(a), ff.21–4; Aldana, p.33.
23. RTC 100/B/26(a), f.25; Aldana, p.34.
24. Aldana, pp.35–8.
25. *ibid.*; RTC 100/B/26(a), f.28.
26. RTC 100/B/26(a), f.25.
27. *ibid.* ff.25–6.
28. *ibid.* ff.26–7; 100/B/27(d), f.13
29. RTC 100/B/26(a), f.28.
30. Aldana, p.39.
31. *ibid.* pp.49–50.
32. RTC 100/B/27(f), ff.18–22.
33. RTC 100/B/26(a), f.32.
34. *ibid.* f.25; Aldana, pp.59–61.
35. RTC 100/B/27(d), f.15.

## Chapter 4
### The tailor from Valencia

1. RTC 100/B/26(a), f.28.
2. *ibid.* f.29.
3. *ibid.* ff.29–30.
4. *ibid.* ff.30–1; Aldana, p.74.
5. RTC 100/B/26(a), f.31.
6. *ibid.* f.32.
7. RTC 100/B/27(c), ff.1–7, 21–2; 100/B/27(e), ff.9–10; Aldana, pp.83–9.
8. RTC 100/B/27(e), ff.16–19.
9. Aldana, p.126; RTC 100/B/27(f), ff.18–22.
10. Aldana, p.76; RTC 100/B/26(a), f.32.
11. RTC 100/B/26(a), ff.32–4.
12. *ibid.* f.34.
13. *ibid.* f.35.
14. RTC 100/B/27(f), f.22.
15. RTC 100/B/27(b), f.1; Aldana, p.97.

16. RTC 100/B/27(b), ff.4–6.
17. *ibid.* ff.6–17.
18. *ibid.* f.16.
19. Aldana, p.109.
20. *ibid.* pp.101–13.
21. *ibid.* p.110, n.l.
22. *ibid.* pp.115–20.
23. RTC 100/B/26(b), f.16.
24. RTC 100/B/27(i), ff.21–3.
25. RTC 100/B/27(h), ff.1–6, 11.
26. RTC 100/B/26(b), ff.12–13.
27. *ibid.* f.14.
28. *ibid.* f.15.
29. *ibid.* ff.15–16.
30. *ibid.* ff.17–19.
31. *ibid.* ff.19–21.
32. *ibid.* ff.22–5.
33. *ibid.* ff.26–33.
34. Aldana, pp.6–7, 202.

## Chapter 5

## The new broom

1. Aldana, pp.141–4
2. *ibid.* pp.144–7.
3. *ibid.* pp.148–50.
4. *ibid.* pp.150–1.
5. *ibid.* p.152.
6. *ibid.* pp.152–3.
7. *ibid.* pp.153–4.
8. *ibid.* pp. 154–5.
9. *ibid.* p.158.
10. *ibid.* pp.156–7.
11. *ibid.* pp.163–168.
12. *ibid.* pp. 168–9.
13. *ibid.* pp.171–3.
14 *ibid.* p.174.
15. *ibid.* pp.174–7.
16. *ibid.* pp.177–82.
17. *ibid.* pp.182–8.
18. *ibid.* p.179.
19. *ibid.* p.178.
29. *ibid.* p.159; Rua Figueroa, p.186.

## Chapter 6

## The devil of the mines

1. RTC 100/B/28(i), ff.32–4.
2. *ibid.* ff.1–2.
3. RTC 100/B/27(b), ff.36–9.
4. RTC 100/B/28(i), f.2.
5. RTC 100/B/27(b), f.39.
6. RTC 100/B/28(i), ff.2–3.
7. *ibid.* f.3.
8. *ibid.*
9. *ibid.* f.4.
10. *ibid.* f.5.
11. RTC 100/B/28(k), f.14.
12. Aldana, p.221; RTC 100/B/28(j), ff.8–9.
13. RTC 100/B/28(j), f.16.
14. *ibid.* ff.26–38.
15. RTC 100/B/28(i), f.5.
16. *ibid.* f.6.
17. Rua Figueroa, p.209; Aldana, p.218.
18. RTC 100/B/28(j), ff.5–7; Aldana p.219.
19. RTC 100/B/28(j), f.8.
20. *ibid.* ff.9–16.
21. *ibid.* ff.17–18; Aldana, p.226.
22. RTC 100/B/28(j), f.19; Aldana, pp.228–230.
23. RTC 100/B/28(j), ff.20–1.
24. RTC 100/B/26(b), ff.40–1.
25. RTC 100/B/28(j), ff.2–3; 100/B/28(k), f.14.
26. RTC 100/B/28(j), ff.25–6.
27. *ibid.* ff.27–37; Aldana, pp.237–46.
28. Aldana, p.248.
29. *ibid.* pp.247, 249–54; RTC 100/B/28(k), ff.1–8.
30. RTC 100/B/28(j), f.38.

## Chapter 7

## And the birds ceased to sing

1. RTC 100/B/28(k), ff.8–11.
2. *ibid.* ff.11–14; Aldana, p.261.
3. RTC 100/B/28(k), ff.14–18; Aldana, pp. 264–5.
4. RTC 100/B/28(k), ff.19–21.
5. *ibid.* f.22; Aldana, pp.271–2.
6. *ibid.* ff.23–4.
7. RTC 100/B/29(f), ff.1–22.
8. *ibid.* ff.24–5; Aldana, pp.325–39.
9. RTC 100/B/29(f), f.21; 100/B/29(m), f.1.
10. RTC 100/B/29(m), ff.2–7.
11. R. Ford. *A handbook for travellers in Spain* (London) 1845, vol.2, pp.288–90.
12. RTC 100/B/29(m), ff.9–10,
13. *ibid.* ff.7–9.
14. *ibid.* ff.10–12; Aldana, pp.340–3.
15. RTC 100/B/29(m), ff.12–13.
16. *ibid.* ff.14–15; Aldana, pp.346–8
17. RTC 100/B/29(m), ff.16–18.
18. RTC 100/B/29(p), ff.1–2.
19. RTC 100/B/29(o), ff.1–17.
20. *Revista Minera*, vol. iv (1853), pp.343, 400
21. RTC 100/B/29(o), ff.21–3.
22. *Revista Minera*, vol. v (1854), p.673.
23. RTC 100/B/29(o), ff.25–32.

## Chapter 8

## That noble corpse

1. RTC 100/B/29(p), ff.2–7.
2. *ibid.* ff.8–9; Aldana, pp.408–10.
3. RTC 100/B/29(p), ff.8–9.
4. *ibid.* ff.12–13.
5. *ibid.* ff.14–16.
6. *ibid.* ff.17–21.
7. *ibid.* ff.21–2.
8. *ibid.* f.24; *el Criterio* 16 November 1856
9. Aldana, p.424.
10. RTC 100/B/29(p), f.26; Aldana, p.425.
11. RTC 100/B/29(p), f.27.
12. Aldana, p.394; RTC 100/B/29(p), f.26.
13. J. L. Thomas, *Notes on the mines of Rio Tinto* (London) 1861
14. A copy of Thomas's report came into the possession of Hugh Matheson.

15. *The Times*, 1 October 1968; Aldana, pp.427–8.
16. RTC 100/B/31(i); Aldana, p.429.
17. Aldana, pp.429–30.
18. *ibid.* p.432.
19. *ibid.* pp.430–2.
20. *ibid.* pp.433–4; RTC 100/B/29(p), f.29.


## Chapter 9

## Don Hugo and friends


1. See Matheson's obituary in *The Times*, 10 February 1898.
2. Copy of sermon preached at Matheson's funeral, preserved at RTZ London office.
3. RTC BMB/1, p.18.
4. *ibid.* pp.2–5.
5. RTC 100/A/30, pp. 1–21, 151–8.
6. RTC 100/B/1(1); prospectus of Rio Tinto Co. Ltd., 8 July 1873.
7. RTC BMB/1, p.1; 100/B/1(d).
8. S. G. Checkland, *The mines of Tharsis* (London) 1967, pp.112, 114.
9. RTC prospectus.
10. Checkland, p.114.
11. RTC annual report for 1884.
12. *ibid.* for 1913.
13. RTC BMB/1, pp.29, 34.
14. *ibid.* p.226.
15. *ibid.* p.28.
16. *ibid.* p.29.
17. RTC 100/A/30, pp.201–7, 251–6, 257–63, 377–80; and Alexander Hill's notes on the first few years of his employment at the Rio Tinto mines, in the possession of Mrs L. C. Hill, widow of his son Laurence Hill.
18. RTC 100/B/59(a); annual report for 1879.
19. RTC annual report for 1876.
20. *ibid.* for 1881.
21. Checkland, p.144.
22. RTC BMB/1, p.258; BMB/2, pp.211, 237.
23. RTC BMB/2, p.119, 124, 127; 100/B/l(1).
24. A. Hill's notes: RTC 100/B/1(h); 100/B/59(b); 100/B/58(b) and (c).
25. RTC annual reports for years stated; BMB/10, p.286.
26. RTC annual reports for 1895 to 1909.
27. RTC annual reports for 1887, 1888, 1889; 30/A/2.
28. RTC annual reports for 1890, 1891; 30/A/2; Checkland, p.142.
29. RTC annual reports and chairman's addresses for 1889–91; 30/A/2.
30. Checkland, p.142.
31. *ibid..* p.143.


## Chapter 10

## A vision realized


1. RTC 100/A/30, pp.151–69; 100/B/1(m); 100/B/5.
2. Information supplied by the Keeper of Manuscripts, National Maritime Museum.
3. RTC 100/A/30, pp.162–72; annual reports for 1874, 1875; 100/B/15; 100/B/57(b).
4. RTC 100/A/30, pp.162–72; annual report for 1876; 100/B/15; 100/B/56.
5. RTC 100/A/30, pp.1–10.
6. *ibid.* pp.25, 31.
7. *ibid.* pp.9–14.
8. *ibid.* pp.25–6.
9. *ibid.* p.4.
10. RTC 100/A/1(e)
11. RTC 100/A/2.
12. RTC 100/A/3.
13. RTC annual reports for 1882–5; 100/A/3; 100/B/35.
14. RTC 100/A/10.
15. RTC 100/A/6.
16. RTC 100/B/20; 100/B/37; 100/B/60; Checkland, pp.135–6.
17. Information of Mr L. U. Salkield.
18. RTC 100/A/7 to 9; information of Mr L. U. Salkield.
19. RTC 100/A/30, pp.6–8.
20. *ibid.* p.28.
21. *ibid.* pp.14–15.
22. *ibid.* p.16.
23. *ibid.* pp.17–18; 100/B/33(b); 100/B/34.
24. John Allan's hitherto unpublished poem was supplied to me by Mr L. U. Salkield; RTC 100/A/2 to 4; 100/B/34 and 35; 100/B/58(f).
25. RTC annual reports for 1897 to 1913; Checkland, p.165; 100/B/58(d); 100/B/60(a)1.
26. Information of Mr L. U. Salkield; RTC 100/A/10; 100/A/14; 100/D/3 and 7


## Chapter 11

## A time to weep


1. This chapter is based largely the reports of the 1888 riots contained in RTC 100/B/8 (iii).
2. RTC BMB/1 p.18.
3. The telegrams are contained in RTC 100/B/8(iii) appendix A.
4 Information of the late D. Juan Wilkins Rincón; RTC 100/A/3.


## Chapter 12

## Not on Queen Victoria's birthday


1. Alexander Hill's notes, in the possession of Mrs L. C. Hill.
2. RTC SB/2, 'List of principal employés in Spain, 1877'.
3. RTC SB/3.
4. See T. W. Bamford, *The rise of the public schools* (London) 1967, pp.236–48.
5. Information of the late D. José Sánchez-Mora and others.
6. RTC 100/A/1 and 2.

7. Information of Mr L. U. Salkield.
8. RTC SB/2, pp.15, 20, 28, 66, 109, 120, 205.
9. *ibid.* pp.13, 31, 66, 268.
10. *ibid.* pp.5, 32.
11. *ibid.* p.91
12. RTC 100/A/2 and 3.
13. Information of Mr L. U. Salkield who provided me with a copy of the rules.
14. SB/2, p.249.
15. RTC SB/3, p.231.
16. *ibid.* p.87.
17. RTC BMB/7, p.192; BMB/8, p.198; SB/3, p.231.
18. RTC SB/2.
19. RTC 100/B/59(k).
20. *ibid.*
21. For this anecdote I am indebted to Mr Robert Wilton Brown.
22. RTC BMB/7, p.309.
23. Information of Mr E. Rowe and others.
24. RTC BMB/7, p.182; BMB/10, pp.168, 374.
25. RTC BMB/7 p.123; BMB/9, pp.221 and 244; BMB/10, p.144; BMB/11, p.157.

## Chapter 13

## Los mohinos

1. RTC 100/A/2, 9 and 11; RTE *Estadisticas*, vol. 1, ff.201v–204v.
2. RTC 100/A/2; 100/B/34.
3. RTE LB/San Dionisio *correspondencia general* (1908) f.187.
4. RTC 100/A/2; information of Mr L. U. Salkield and the late D. Juan Wilkins Rincón.
5. RTC BMB/4, pp.267–317.
6. RTC BMB/5, p.198.
7. RTE LB/Underground, vol. 1, f.433.
8. *ibid.* vol. 2, f.99.
9. *ibid.* vol. 1, f.468; vol. 2, f.205.
10. *ibid.* fol. 1, ff.368–413.
11. RTC *Compra-Venta* deed of 1873; 100/B/8; 100/B/33(iii) and (a); 100/B/44.
12. RTE LB/*correspondencia general* (1909–10), 25 May 1910.
13. *ibid.* 28 March 1910.
14. *ibid.* 6 August 1909.
15. Information of D. Francisco Martín-Pernil and others.
16. RTC BMB/7, pp.286, 329.
17. RTC BMB/10, pp.35, 37, 40, 43.
18. RTC 100/B/8(i).
19. RTC 100/B/8(iii)
20. RTC annual report and chairman's address for 1889.
21. RTC 100/B/8(i), (ii) and (iii).
22. RTE LB/1, ff.367–9.
23. RTC annual report and chairman's address for 1890.
24. Information of Mr L. U. Salkield.
25. RTC annual report and chairman's address for 1884; 100/A/4.
26. C. M. Cipolla, *Literacy and development in the west* (London) 1969, p.115.
27. *ibid.* p.128.
28. RTC SB/2, p.62; SB/3, p.42; annual reports and chairman's addresses for 1880 and 1885; 100/A/3.
29. RTC 100/B/8(e); 100/B/44; SB/3, p.146.
30. RTE LB/*Filon norte* (*varios*) 1900, f.75. LB/Underground (1904–5), ff.19, 116, 135.
31. RTE accident books, vols. 35 and 36.
32. RTE LB/San Dionisio *correspondencia general* (1909), f.421.
33. RTC BMB/7, p.235; BMB/8, p.271; BMB/10, pp.131, 305.
34. RTE LB/Underground, vol. 1, p.135.
35. RTC *Compra-Venta* deed of of 1873.
36 Copy supplied by Mr L. U. Salkield.
37. RTC SB/2, p.1; LB/Mines/1, f.18.
38. RTC SB/2, p.153.
39. RTC 100/A/19; 100/B/44.
40. RTC 100/B/8(iii).
41. RTC SB/2, pp.1, 157; 100/A/2 to 5.
42. RTC SB/2, p.1.
43. RTC SB/3, pp.18, 51, 216; 100/A/6 to 9; information of Mr L. U. Salkield and the late D. José Sánchez-Mora.
44. Information of Mr L. U. Salkield.
45. RTC 100/A/3 to 7.
46. RTC 100/A/8.
47. RTC SB/3, p.90; information of Mr L. U. Salkield.
48. RTC 100/A/5.
49. RTC 100/A/6.
50. RTC 100/A/21 and 22.
51. II Timothy 2, v. 15.
52. RTE LB/San Dionisio *correspondencia general* (April 1908), p.92.
53. RTC 100/B/8(i).
54. RTC. LB/2, pp.92–4.
55. RTE LB/San Dionisio *correspondencia general* (April 1908), ff.53, 93, 263.
56. *ibid.* ff.53, 264, 265.
57. RTC 100/B/8(ii); 100/B/44.
58. RTC 100/A/7 to 9.
59. Juan Guerrero Ruiz, 'El ayuntamiento de las Minas de Ríotinto' in *Revista de Estudios de Vida Local*, vol. 17 (Madrid) 1944, pp.724–8.

## Chapter 14

## A king in Huelva

1. RTC annual report and chairman's address for 1889.
2. RTC SB/2, pp.231, 235; SB/3, p.141.
3. RTC SB/3, p.141.
4. RTC SB/2, p.106.
5. Information of the late D. José Sánchez-Mora.

6. RTC SB/3, p.41.
7. *ibid.*
8. Information of the late D. José Sánchez-Mora.
9. Information of Mr C. Browning.
10. Information of the late D. José Sánchez-Mora.
11. RTE LB/San Dionisio *correspondencia general* (1908), pp.8, 10, 28, 77–9, 80–2; information of the late D. José Sánchez-Mora.
12. Precisely the same point was made by many people who knew Browning, including Mr E. Rowe, the late D. José Sánchez-Mora and D. Francisco Martín-Pernil.
13. Unpublished memoirs of the late H. Potts in the possession of Mr L. U. Salkield.
14. *ibid.*
15. Information of the late D. José Sánchez-Mora.
16. *ibid.* and of D. Francisco Martín-Pernil and others.
17. RTC LB/Private official, nos. 2025, 2013, 100/B/59(g); copy of the 'Report by Robert Palmer on the San Dionisio disaster' made available by Mr L. U. Salkield; information of the late D. Juan Wilkins Rincón.
18. Robert Palmer's report. Mr L. U. Salkield explained its technicalities to me.
19. RTC 100/A/10.
20. RTC 100/A/12 to 25.
21. RTC 100/B/19(c).
22. RTC 100/A/6 to 13.
23. RTC 100/A/10; 100/B/19(b); information of Mr L. U. Salkield.
24. RTC SB/3, p.52.
25. Copy of Caddick's note supplied by Mr L. U. Salkield.
26. RTC 100/B/19(a).
27. RTC annual report and chairman's address for 1906.
28. Information of Mr L. U. Salkield.
29. RTC 100/A/11, 12, 16 and 28.
30. Information of the late D. José Sánchez-Mora.
31. RTC BMB/14, pp.56, 63, 69–92, 206, 221–9; information of the late D. José Sánchez-Mora and the late D. JuanWilkins Rincón.
32. Information of the late D. José Sánchez-Mora.
33. Information of D. Francisco Martín-Pernil.
34. Information of the late D. José Sánchez-Mora and others.
35. H. Potts, memoirs.
36. RTC BMB/14, pp.235–255; information of the late D. José Sánchez-Mora.
37. RTC BMB/14, pp.255–7.
38. RTC 100/B/8(i).
39. RTC 100/B/8(ii).
40. RTC 100/B/8(v); BMB/14, pp.255–7.
41. RTC SB/3, pp.66 and 92.
42. RTC BMB/14, pp.257, 268–9, 323.
43. Information of Mr L. U. Salkield.
44. RTC SB/3, pp.82, 182; BMB/14, p.296.
45. H. Potts, memoirs.
46. RTC 100/B/14; H. Potts, memoirs.
47. RTC SB/3, pp.182, 192.
48. *ibid.* p.82.
49. Information D. Francisco Martín-Pernil and others.
50. H. Potts, memoirs.
51. *ibid.* and information of D. Francisco Martín-Pernil.
52. Information of the late D. José Sánchez-Mora.
53. *ibid.* and H. Potts, memoirs.
54. Information of the late D. José Sánchez-Mora; RTC 100/B/43 and 45; 100/D/4.
55. RTC SB/3, p.33; BMB/15, pp.79, 102, 129.
56. RTC Add. Letters/1.
57. RTC BMB/15, pp.132, 135; SB/3, p.33.
58. RTC Add. Letters/1; BMB/15, p.246; accounts ledgers November 1928; SB/3, p.33.
59. H. Potts, memoirs; information of the late D. José Sánchez-Mora and D. Francisco Martín-Pernil.


## Chapter 15
## The labyrinth


1. *El Liberal* 19 November 1899.
2. RTC 100/B/8(ii)
3. RTC BMB/1 pp.2, 10, 49–50; SB/2, p.58; *La Epoca*, 30 December 1889.
4. RTC 100/D/3 and 7; 100/B/8(iii).
5. RTC annual reports and chairman's addresses for 1888 and 1889.
6. RTC 100/B/58(f) and (g); 100/D/3; 100/D/7.
7. RTC SB/2, p.58; SB/3, p.23; annual reports and chairman's addresses for 1889 to 1891.
8. RTC 100/B/60(g); annual report and chairman's address for 1882.
9. RTC 100/B/60(g); annual report and chairman's address for 1883
10. RTC 100/B/60(f); BMB/7, p.108.
11. RTC BMB/4, p.144.
12. RTC SB/3, p.23.
13. *ibid.* BMB/4, p.191.
14. RTC BMB/4, p.193.
15. RTC SB/3, p.222.
16. *ibid.* and p.263; SB/4, p.275.
17. RTC BMB/10, pp.29, 62.
18. *ibid.* pp.43 and 63.
19. *ibid.* pp.35, 37, 40, 43.
20. *ibid.* p.63.

21. RTC annual reports and chairman's addresses for 1899–1905; 30/C/1, 2, 3, 8, 9, 27 and 28.
22. *ibid.* for 1900; 100/A/6; BMB/10, p.348
23. RTC BMB/10, pp.348, 354, 356.
24. RTE LB/Underground, vol. 2, pp.420, 431.
25. *ibid.* vol. 1, pp.63 and 64.
26. *ibid.* p.198.
27. RTE LB/San Dionisio *correspondencia general* (1908), pp. 182, 237.
28. RTC 100/B/8(i).
29. *ibid.*
30. *ibid.* LB/Private Official, ff.2013 and 2025; Robert Palmer's report on San Dionisio disaster; 100/B/59(g).
31. RTC 100/B/8(i); 100/B/21; information of the late D. Juan Wilkins Rincón.
32. *ibid.* and 100/A/13 and 14.
33. RTC annual reports and chairman's addresses for 1914 to 1918; information of the late D. Juan Wilkins Rincón.
34. RTC 100/B/8(i).
35. *ibid.* RTC 100/A/27, 28 and 29; annual reports and chairman's addresses for 1915 to 1918.
36. RTC 100/B/8(i); 100/A/28; information of the late D. Juan Wilkins Rincón.
37. *ibid:*
38. RTC 100/B/8(i); 100/B/48.
39. *ibid.* and annual reports and chairman's addresses for 1919 and 1920.
40. RTC 100/B/8(i).
41. *ibid.*
42. *ibid.* and 100/B/49.
43. *ibid.*
44. *ibid.*
45. *ibid.*
46. RTC 100/B/8(i).
47. *ibid.* 100/B/49; 100/B/51.
48. RTC 100/B/8(i); 100/B/51.
49. *ibid.*
50. *ibid.*
51. *ibid.*
52. RTC 100/B/8(i) and (ii); annual report and chairman's address for 1921.
53. *ABC* 24 November and 29 December 1920; RTC 100/B/49.
54. RTC 100/B/8(i).
55. *ibid.*
56. *ABC* 28 November 1920; RTC 100/B/49.
57. *ibid.* RTC 100/B/8(i).
58. RTC 100/B/8(i); information of the late D. Juan Wilkins Rincón.
59. RTC 100/B/8(ii).
60. *ibid.*; information of the late D. José Sánchez-Mora.
61. RTC 100/B/8(ii); annual report and chairman's address for 1920.
62. RTC 100/B/8(ii).
62. *ibid.* 100/A/27.
64. RTC 100/B/8(ii).
65. RTC 100/B/8(v).
66. Hugh Thomas, *The Spanish civil war* (London) 1968 edition, p.37, quoting Tomanes, *Estructura ecónomica de España*, p.203.
67. RTC 100/B/25; 100/B/47; SB/3; pp.48, 54.
68. RTC annual reports and chairman's addresses for 1924 to 1929.
69. *ibid.* for 1928; *The Times* 5 April 1928.
70. RTC annual reports and chairman's addresses for 1928 to 1930; 100/D/8; interim report to shareholders for January to June 1931; *The Times* 5 April 1928, 25 April 1929; information of the late D. José Sánchez-Mora.
71. Information of the late D. Juan Wilkins Rincón and the late D. José Sánchez-Mora.

## Chapter 16

## Fight the good fight

1. RTC annual reports and chairman's addresses, 1914–18; 100/A/27 to 29; BMB/14, pp.166, 168, 172, 224; 100/B/59(h).
2. RTC BMB/14, pp.175, 179; 100/B/12(c); W. G. Checkland *The mines of Tharsis* (London) 1967, p.199.
3. RTC annual reports and chairman's addresses 1919 and 1920; BMB/14, pp.267, 271; Checkland, pp.197–8.
4. RTC BMB/14, p.279; Checkland, p.198
5. RTC BMB/14; p.281; Checkland, pp.198–9.
6. RTC BMB/14, p.283; Checkland, p.199.
7. RTC BMB/14, p.284; Checkland p.199.
8. RTC BMB/14, pp.296, 323; Checkland, p.199.
9. RTC BMB/14, pp.323–5; annual report and chairman's address for 1923; Checkland, p.200.
10. RTC annual report and chairman's address for 1924.
11. *ibid.* for 1925; BMB/14, pp.367, 373.
12. Checkland, p.200; RTC 80/A/1.
13. RTC annual report and chairman's address for 1926; BMB/15, pp.7, 8, 11, 29, 70.
14. RTC annual reports and chairman's addresses for 1926 and 1927; BMB/15, pp.13, 29, 30, 33, 36, 36, 58–9; 100/C/3; 110/A/1.
15. RTC BMB/15, pp.60, 66, 70, 73, 78, 103, 113, 121.
16. RTC annual reports and chairman's addresses, 1926 to 1929; 100/C/3; Checkland, p.202.
17. Information of Mr E. Bourne and others.

18. Checkland, pp.202–3; RTC BMB/15, p.378.
19. Checkland, p.203.
20. RTC 100/D/8; BMB/15, pp.149, 153, 158, 162, 165, 168–70.
21. RTC 100/D/8; BMB/15, pp.175, 180, 252, 258, 262, 268, 277, 326.
22. RTC BMB/15, p.168.
23. RTC annual reports and chairman's addresses for 1929 to 1931; BMB/15, pp.219, 237–41, 245, 249–50; 100/A/22 to 25.
24. RTC annual reports and chairman's addresses for the years 1929 to 1931; BMB/15, pp.217–8, 227, 228, 231–2, 236–7, 248–9, 256–7, 266.
25. RTC BMB/15, pp.279–80, 282–3, 285, 286, 288; annual report and chairman's address for 1931; H. Potts, memoirs.

## Chapter 17

## Tumult and civil strife

1. RTC BMB/15, pp.274–5; information of the late D. José Sánchez-Mora.
2. Information of Mr R. Wilton Brown.
3. H. Potts, memoirs.
4. Checkland, pp.216–7; information Mr L. U. Salkfield.
5. Information of Mr L. U. Salkield.
6. Information of Mrs E. Howell and others.
7. RTC annual reports and chairman's addresses for 1931–4; BMB/16, pp.10, 23.
8. H. Potts, memoirs.
9. RTC annual report and chairman's address for 1934; BMB/16, p.34; information of Mr L. U. Salkield, Mrs E. Howell, Mrs B. Gough, and the late D. José Sánchez-Mora.
10. Information of the late D. José Sánchez-Mora.
11. *ibid.*, and of Mr L. U. Salkield; RTC 100/B/25; annual report and chairman's address for 1934.
12. RTC 100/B/25; BMB/16, p.34; annual report and chairman's address for 1934.
13. RTC annual report and chairman's address for 1936.
14. RTC BMB/16, pp.60, 70, 71, 114, 117; annual report and chairman's address for 1935.
15. RTC BMB/16, pp.117, 125; annual report and chairman's address for 1936; information of Mr L. U. Salkield.
16. RTC annual report and chairman's address for 1936.
17. Information of Mr L. U. Salkield.
18. *ibid.*
19. L. C. Hill, letters in the form of a diary (in the possession of Mrs Nan Hill) hereafter referred to as 'L. C. Hill, diary'.
20. Information of Mr L. U. Salkield; L. C. Hill, diary; Luis Bolín *Spain: the vital years* (London) 1967, pp.177–9.
21. L. C. Hill, diary; information of Mr L. U. Salkield.
22. *ibid.*
23. *ibid.*
24. *ibid.*; information of the late D. José Sánchez-Mora; RTC BMB/16, p.129.
25. L. C. Hill, diary; Checkland, p.226.
26. L. C. Hill, diary.
27. *Evening Standard* 14 August 1936; *Daily Mail* 15 August 1936; L. C. Hill, diary.
28. L. C. Hill, diary.
29. *ibid.* and information of the late D. Juan Wilkins Rincón.
30. *Daily Telegraph* 26 August 1936.
31. Information of Mr L. U. Salkield and the late D. Juan Wilkins Rincón.
32. RTC BMB/16, p.131; annual report and chairman's address for 1936.
33. *ibid.* and information of the late D. Wilkins Rincón.

## Chapter 18

## The threadbare years

1. Information of the late D. Juan Wilkins Rincón and the late D. José Sánchez-Mora.
2. RTC annual report and chairman's address for 1936.
3. *ibid.* for 1937.
4. *ibid.* for 1936 and 1937.
5. RTC BMB/16, pp.131, 140; information of the late D. José Sánchez-Mora.
6. Information of the late D. José Sánchez-Mora.
7. RTC annual report and chairman's address for 1936
8. *ibid.* for 1938.
9. RTC BMB/16, pp.131, 140.
10. *ibid.* p.156.
11. RTC annual reports and chairman's addresses for 1936–9.
12. Public Record Office Cab. 24/268 and 23/87 (for the subject of requisitions of Rio Tinto produce and British government reaction); Bólin, pp 122–4 and appendix IV.
13. Hugh Thomas, *The Spanish Civil War* (London) 1965, p.351; RTC BMB/16, pp.130, 131, 133, 134; annual reports and chairman's addresses for 1936 and 1937.
14. Bolín, pp.222–3.
15. RTC BMB/16, p.143; annual reports and chairman's addresses for 1936–8.
16. RTC BMB/16, pp.134, 147.

17. Information of a British wife who wishes to remain anonymous.
18. RTC BMB/16, pp.156, 195, 196; annual report and chairman's address for 1937.
19. H. Potts, memoirs.
20. RTC 100/A/34.
21. 400 *House of Commons Debates*, 5s.
22. H. Potts, memoirs.
23. RTC minutes of group co-ordinating committee, April and September 1939.
24. H. Potts, memoirs; RTC 100/A/33.
25. *ibid.*
26. RTC minutes of group co-ordinating committee, October 1940.
27. RTC annual reports 1939 to 1945.
28. RTC 100/A/34.
29. RTC annual report and chairman's address for 1950.
30. RTC minutes of group co-ordinating committee, October 1940.
31. H. Potts, memoirs.
32. RTC 100/B/66.


## Chapter 19

## Climacteric


1. RTC annual reports and chairman's addresses for 1946–48; 100/A/34.
2. RTC annual report and chairman's address for 1946.
3. *ibid.* and for 1947.
4. RTC minutes of management committee, November 1947.
5. RTC 100/A/34; minutes of management committee January to March 1948.
6. *ibid.* February 1948.
7. RTC 100/A/34.
8. RTC annual reports and chairman's addresses for 1947 and 1948.
9. *ibid.*, 1950; 100/A/34.
10. RTC annual report and chairman's address for 1951.
11. *ibid.* for 1952.
12. Information of Sir Mark Turner and Sir Val Duncan.
13. Information of Lord Byers.
14. Information of Mr Roy Wright
15. Information of D. Antonio de Torres y Espinosa, Sir Val Duncan, Lord Byers; RTC 100/B/67 and 68, 100/F/1.
16. RTC 100/F/2.
17. Information of Sir Val Duncan and the late D. José Sánchez-Mora.
18. Information of D. Antonio de Torres y Espinosa and Mr L. U. Salkield; RTC 100/F/4, 5, 6 and 11.


## Chapter 20

## Epilogue


This chapter is based on the information of D. Antonio de Torres y Espinosa and Sir Val Duncan; RTC 100/F/3 and 4, 100/G/1 and 2 and statistics provided by officials of Unión Explosivos Río Tinto, SA.


## Appendix One


1. I Kings, X, 22.
2. Isaiah, XXIII, 1.
3. The identification was first made in the late sixteenth century by Juan de Pineda, professor of holy scripture at Seville, and is now commonly accepted.
4. E. Deligny, 'Apuntes históricos sobre las minas cobrizas de la sierra de Tharsis' in *Revista Minera*, vol. XIV (Madrid) 1863.
5. Pinedo Vara, pp.28–9.
6. Rua Figueroa, pp.44–6.
7. Pinedo Vara, p.29.
8. RTC 100/B/13. (Illustrated notes on the ancient history of Rio Tinto and the museum exhibits by Professor David Williams.)
9. *ibid.*; R. J. Forbes, *Studies in Ancient Technology*, vol. IX (Leiden) 1964, p.101.
10. J. C. Allan, 'A Mineração em Portugal na antiguidade' in *Boletin de Minas*, vol. 2, no. 3 (Lisbon) 1965.
11. L. U. Salkield, 'Ancient slags in the south-west of the Iberian peninsula', proceedings of the sixth International Mining Congress (published by Catedra de San Isidoro, Leon) 1970.
12. RTC 100/B/13; O. Davies, *Roman Mines in Europe*, (London) 1935, p.17.
13. E. M. Whishaw, *Atlantis in Andalusia* (London) n.d., p.122. (Mrs Whishaw's archaeological reports to the Rio Tinto Company – which assisted her financially – are contained in RTC 100/B/9 and 100/B/55.)
14. A. Schulten, *Tartessos: ein Beitrag zur altesten Geschichte des Westens* (Hamburg) 1950, p.171.
15. A. Blanco and J. M. Luzón, 'Pre-Roman silver miners at Río Tinto', *Antiquity*, vol. XLIII (1969), pp.124–31.
16. *ibid.* p.125.
17. Forbes, vol. VII, p.225.
18. J. M. Luzón, 'Contribución al estudio de la minería romana en la Bética occidental', unpublished university of Seville thesis (1966), pp.124–31.
19. F. Sanz, *Memoria antigua de Romanos en las minas de Río Tinto* (Seville) 1762.
20. Pliny III, 10; Ptolomy II, iv, 10; Rua Figueroa, p.45.
21. Luzón thesis, pp.11–14.

22. *ibid.* pp.105–38; R. E. Palmer, 'Notes on some ancient mining equipment and systems' in the *Transactions of the Institution of Mining and Metallurgy*, vol. XXXVI (1927) pp.299–336.
23. J. F. Allan, 'Pyrites deposits of the province of Huelva', in the *Transactions of the North-Eastern Institution of Mining and Mechanical Engineering*, vol. 37 (1887). (J. F. Allan was the father of the late J. C. Allan referred to above.)
24. Forbes, vol. VII, p.229.
25. E. Deligny, art. cit.
26. Many of the articles from the Roman necropolis and the area of the mines are preserved in the Rio Tinto museum. For the organization and laws of the Roman mining communities see T. A. Richards, 'The Mining of the Romans in Spain' in the *Journal of Roman Studies*, vol. XVIII (1928), and H. Lewis, 'Some aspects of Roman mining laws under the Roman empire' in the *Proceedings of the University of Durham Philosophical Society*, 1920–1.
27. Rua Figueroa, p.10.

## Appendix Two

1. RTC annual reports and chairman's addresses for 1927 and 1929; 140/A/2 and 3.
2. RTC annual reports and chairman's addresses for 1929–31; 20/D/6 to 20/D/6(4).
3. F. L. Coleman, *The Northern Rhodesia Copperbelt 1899–1962*, (London) 1971, pp.31, 33, 64–9; RTC 20/B/3; 20/A/1 to 20/A/1(11).
4. RTC annual report and chairman's address for 1929.
5. *ibid.*; RTC 20/A/1(8 to 10).
6. RTC annual report and chairman's address for 1929; *The Times*, 10 May 1929; RTC 10/F/31 to 36.
7. Coleman, pp.70, 72.
8. *ibid.* and RTC 20/A/1 and 20/A/1(2).
9. RTC annual report and chairman's address for 1930; 10/F/17 to 30.
10. RTC annual report and chairman's address for 1932; 20/D/6(3 to 6); 110/A/1–2; 140/A/2.
11. RTC annual reports and chairman's addresses for 1934 and 1935.
12. *ibid.* for 1936 and 1937
13. *ibid.* for 1937.

# ILLUSTRATIONS

Thanks are due to the following for permission to use the illustrations shown:

ABC, Madrid: 314, 320, 339, 346, 358, 362.

Fotografía Rojas, Rio Tinto: 213 (Jesús Alonso).

Mrs L. C. Hill: 199, 210, 212, 214, 220, 234, 235, 242, 256, 258.

Mrs E. Howell: 223.

J. Jowers, Esq: 296.

Mansell Collection: 27, 50, 101,

Museo Romantico, Madrid: 110.

Doña Maximina Prados: 324.

National Trust: 40.

A. and J. Pavia, Valetta: 361.

Sir Brandon Rhys Williams, bart., MP: 288.

Baron Guy de Rothschild: 157.

D. Carlos Sundheim Días Trechuelo: 149.

Unión Explosivos Río Tinto, SA: 206, 253.

Doctor John Shields took the photographs on pages 22, 213 (boar hunt), 245, 246, 248, 249, 250 and 252 in the years 1904 to 1910; Dmitri Kasterine the photographs on pages 17, 20, 24, 56, 90, 143, 149, 183, 191, 218, 241, 283 (war memorial), 298, 330, 367, and 387 in November 1972; the illustration on page 70 was drawn by Ronald Maddox.

# INDEX

**An Agreement** made on the eighteenth day of February One thousand eight hundred and seventy five **Between** Ludwig Gottfried Dyes the Deputy Chairman of and acting for and on behalf of The Deutsche National Bank, Bremen and Messieurs Matheson & Co of 3. Lombard Street Merchants of the first part Edwin Clark and William Henry Punchard of No. 5 Westminster Chambers Victoria Street Westminster Contractors for Public Works of the second part and Hugh Mackay Matheson the Chairman of an acting for and on behalf of The Rio Tinto Compan Limited (hereinafter called "the Company") of the third part.

**Whereas** by a Memorandum of Agreement made on the twenty fourth day of June One thousand eight hundred and seventy three and between or on behalf of the parties hereto it was agreed th the parties of the first and second parts (hereinafter called the Vendors) should sell to the Company the Mines of Rio Tinto and ot property and rights therein referred to at the price of One hundre and fifty thousand pounds sterling in cash with interest as the mentioned and Debentures of the Company of the description the mentioned for the nominal amount of Six hundred thousand pou sterling and it was also agreed that the Company should rep to the Vendors the sums paid or to be paid by the Vendors to the Spanish Government as therein mentioned and further that al moneys payable or repayable to the Vendors under the Agreement in recital should be paid to Messieurs Matheson & Co as and when demanded by them for account of the Vendors and the sa Debentures should be issued upon the request and direction of Messieurs Matheson & Co to such person or persons and in su